American Dynamite

AN
ILLUSTRATED
HISTORY

GERALD P. CESTKOWSKI

American Dynamite

AN
ILLUSTRATED
HISTORY

KEEP
LIGHTS
and
FIRES AWAY!

DANGEROUS

Name of contents of carload shipments must be inserted here or on commodity card.

HANDLE CAREFULLY

This car must not be next to a car containing
Explosives

Avoid contact with leaking acid or
corrosive liquid

Beware of fumes or
vapors!

WHEN LADING IS REMOVED
THIS PLACARD MUST
BE REMOVED OR
REVERSED.

SCHIFFER
PUBLISHING

4880 Lower Valley Road • Atglen, PA 19310

Edited by Ian Robertson
Designed by Beth Oberholtzer
Cover design by Jack Chappell

Type set in ITC Franklin Gothic/Times

ISBN: 978-0-7643-6847-9
Printed in India

Published by Schiffer Publishing, Ltd.
4880 Lower Valley Road
Atglen, PA 19310
Phone: (610) 593-1777; Fax: (610) 593-2002
Email: info@schifferbooks.com
Web: www.schifferbooks.com

For our complete selection of fine books on this and related subjects, please visit our website at www.schifferbooks.com. You may also write for a free catalog.

Schiffer Publishing's titles are available at special discounts for bulk purchases for sales promotions or premiums. Special editions, including personalized covers, corporate imprints, and excerpts, can be created in large quantities for special needs. For more information, contact the publisher.

FSC
www.fsc.org
MIX
Paper | Supporting
responsible forestry
FSC® C016779

CONTENTS

INTRODUCTION

For many of us, our familiarity with dynamite is limited to cartoon images. Most everyone has witnessed the demolition of an animated someone or something via the iconic, plunger-operated blasting machine. In cartoons, dynamite elicits excitement and suspense. Any second, a big boom could occur and dispatch a villain or temporarily waylay a hero. For the same reason, toy dynamite is found in product lines where you might expect it (G.I. Joe, and board games such as Dynamite and Dynamite Shack) and where you might not expect it (Hot Wheels, Playmobile, Lincoln Logs, LEGO, and Thomas & Friends). The internet domain name "dynamite.com" is owned by Dynamite Entertainment, a comic book publisher.

"Dynamite" is also used by advertisers as an adjective to indicate power and, by extension, high quality. A "dynamite dish detergent" must be good. If a motor oil promises "dynamite results," you can't go wrong buying it. In the current century, "dynamite" has largely become an abstraction for the majority of those who speak and hear the word.

Pinback buttons were patented in 1896 and were popular political campaign and advertising giveaways. This one measures 1¼" in diameter. *Author*

But dynamite was one of the most important tools of industry, enabling access to metallic ores and other minerals that would have otherwise been cost prohibitive to obtain. Without enough iron, copper, and other vital metals, the modernization of American society would have slowed to a stall. Vast areas of land were cleared with dynamite. Railways, highways, tunnels, and dams—all were accomplished with dynamite. Without it, the transformation of American society from agrarian to modern would have been prolonged by many decades. "Prosperity Follows Dynamite" went one DuPont slogan, and the explosive really did enable one man to do the physical labor of ten or more, as early advertisements proclaimed.

Most people might be surprised by the ½-pound heft of a standard stick of dynamite, and particularly by the fact that the body of the cylinder is not rigidly hard but feels like modeling clay wrapped in several layers of thick, waxed paper. Indeed, the wrapper, formally called a shell, *is* mere waxed paper, and there is no hard cardboard tube or glue involved. Many likely model their concept of a dynamite stick (technically referred to as a cartridge) around a firecracker, but a cartridge of dynamite does not come with a protruding fuse and would not detonate directly via a fuse anyway. Dynamite must be shocked into exploding; a lit fuse stuck into a cartridge of dynamite would simply fizzle out.

In the first half of the twentieth century, dynamite was an ordinary tool available at the local hardware

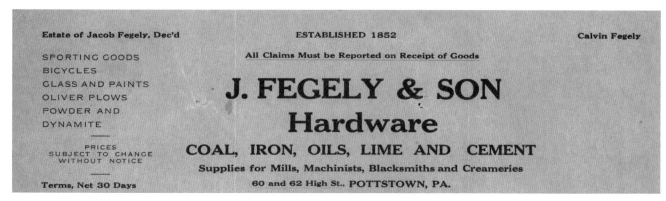

A century ago, a hardware store was the place to buy bikes for the kids and maybe pick up some dynamite for Dad. *Author*

store. One could also order by mail, no questions asked, direct from the manufacturer. Dynamite was regularly offered in advertisements in newspapers and magazines, and even in the legendary Sears, Roebuck catalog. Sixteen- and seventeen-year-old Frank and Joe Hardy are depicted detonating a stick, with their friends watching, in the 1935 Hardy Boys book *The Hidden Harbor Mystery*.

DuPont estimated in 1910 that 500,000 of the then-90,000,000 United States population used dynamite on a *daily* basis. In the 1930s and '40s, US consumption of dynamite averaged 1,000,000 pounds a day. Yearly dynamite use peaked at more than 800,000,000 pounds in 1955.

Those who purchased dynamite for "home use" were not blasting for minerals but rather were ordinary farmers and ranchers. These rugged men of the land used every means at their disposal to shape their properties for optimum efficiency and maximum productivity. Blowing up plowing obstacles such as tree stumps and boulders was much easier than manually removing them (the Hardy Boys used their dynamite to blast away a stump that was blocking access to a treasure chest). Dynamite producers realized early on that their product was being used this way, and many made a concerted effort to market to farmers.

Mining is what we usually think of regarding a real-life dynamite blast, and miners used the vast majority of dynamite for coal mining, hard-rock mining, and quarrying. There is an old expression, "If you can't grow it, you have to mine it." While water is not grown or mined, the most-important raw materials that make civilization run smoothly *must* be extracted via mining. And the mining of industrial metals and the quarrying of construction materials are wholly dependent on explosives.

Dynamite was developed by Swedish chemist Alfred Nobel, who obtained patents in 1867. The explosive was first made in the United States under his patent the following year. It is worth noting that Nobel's prior invention of the blasting cap, originally used to detonate liquid nitroglycerin, was crucial to the success of dynamite. This is because the concept of using a small explosion to initiate the larger detonation of a relatively more stable compound meant that powerful explosives could be formulated to resist shock and heat.

There were once scores of high-explosives producers in the United States; now only one, Dyno Nobel, still makes dynamite. The largest companies such as DuPont, Hercules Powder Company, and Atlas Powder Company sold not only explosives, but all manner of blasting gear, from blasting machines to connecting wire to blasting caps and fuse. All of these companies published magazines and instructional handbooks that are a remarkable historical record of the progress of the industry.

The first chapter of this book is an introduction to dynamite and explains what is in it and how it is labeled, packaged, stored, and transported. The second chapter outlines the basic implements and methods of the trade. The third chapter describes the many uses for dynamite. The fourth chapter is a historical overview of seventeen of the major companies that have produced dynamite in the United States. The fifth chapter is a quick look at some explosives that resemble dynamite. The sixth chapter gives brief mention of other dynamite producers—fifty-six in all.

The "Digging Deeper" section at the end of each chapter offers a chance to explore a little off the beaten path. When there was not room in the main narrative for something slightly off topic or a little too detailed, it went into the Digging Deeper file. These are not dry endnotes; there are loads of interesting stories and facts therein.

I gathered information from hundreds of sources, including the just-mentioned magazines and handbooks, plus newspaper articles, price lists, product brochures, trade manuals, and numerous books related to explosives and mining. Of particular note is the 1,132-page *The History of the Explosives Industry in America*, written in 1927 by Hugo Schlatter and Arthur Pine Van Gelder. This is still considered to be the most accurate and comprehensive work on the early years of US explosives, especially black powder. For many of the shorter histories of the (mostly) smaller entities in chapter 6, I started with Van Gelder and Schlatter. If there were discrepancies, I went with the preponderance of evidence.

Many of the illustrations you will see are harvested from instructional books and provide a feel both for the state of the art at the time and the methods of conveying best practices to users. Imagine looking at a diagram of how to prime dynamite while actually priming a stick of dynamite! Tens of thousands of those who used dynamite did just that.

Explosives are inherently dangerous, and an enormous amount of effort has been dedicated to increasing safety in their production and use. In dynamite's earliest days, accidents were common, and formulations were gradually altered to be less susceptible to detonation

This bumper sticker from the 1980s emphasizes the importance of mining. The Ensign-Bickford Company was a major producer of blasting supplies. *Author*

by minor shock. Dynamite also had an early association with terrorists, particularly in Europe, owing to the ease of use and the outsized amount of devastation wrought.

I have unapologetically chosen not to focus on the numerous explosions and loss of life that occurred during the production of dynamite. Mundane factory accidents killed many times the number of workers lost in headline-making disasters at explosive plants. By the 1920s an electrician or fishery worker was statistically more likely to perish in an accident than a dynamite plant worker. If you are interested, there are plenty of compilations of newspaper accounts of dynamite factory explosions online. In this work, such incidents are mentioned primarily as they relate to innovations in safety, which all but eliminated catastrophic explosions by the 1950s.

While dynamite per se has been largely displaced by other explosive compounds, the value of explosives has not diminished. Because of civilization's dependence on metals, no aspect of modern life would be possible without explosives. There would be no automobiles, no skyscrapers, and no computers. Mining prior to high explosives did not produce nearly enough raw materials to enable the workings of our complex society. One of the most important drivers of progress in the late nineteenth and early twentieth centuries was the power of dynamite—as important as steam power and electricity. According to the Institute of Makers of Explosives, "In the age between the closing of the Civil War and the end of World War Two no single engineering tool surpassed the achievement of dynamite."

This book explores the manufacture, marketing, and use of dynamite in the United States from its introduction to the current day. I tried to make the material accessible for those who know

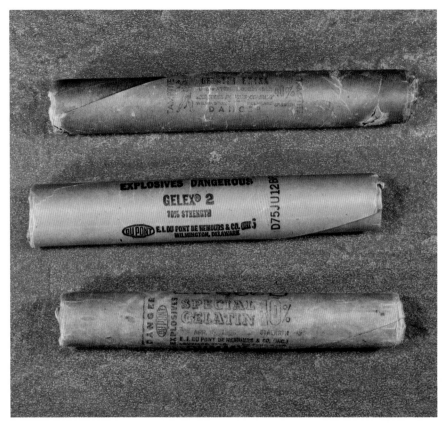

Almost all of the billions of wrappers that were used to contain dynamite were blown up decades ago. *Author*

nothing about dynamite, and useful for those who may think they already know everything. I hope you are entertained while you learn (or learn more) about one of the most consequential tools ever developed.

One major barrier to illustrating a history of dynamite is the fact that there are virtually no intact, vintage dynamite cartridges to photograph. A small number of inert sample sticks are known from rare salesman or convention display cases. The few empty shells that have made it out of the mine, quarry, or construction site are usually beat up and resemble the genuine ones shown here, which are mounted on wooden rods. Images of a handful of extant wrappers do not begin to convey a major theme of this book, which is the extraordinary variety of dynamite and its users. The solution was

to acquire images of dynamite sticks from all available sources and then create identical replicas to photograph. The downside to this strategy is the lack of images of pre-1900 cartridges after which to model replicas. Sadly, this means no very early sticks. Rest assured that every stick that is pictured in this book is indistinguishable from the cartridge shown in the image from which it was patterned. There are no "fantasy" designs. These historically accurate, facsimile sticks are intended to evoke a visitor's snapshots of exhibits in a hypothetical museum of American dynamite. (While most dynamite sticks bore dates or date codes, some of the images from advertisements and product catalogs do not have dates. Rather than try to guess at the placement, font style, and size of the date, I left it out for those images.)

Dynamite Fundamentals

This first chapter is an overview of dynamite and an introduction to its production, handling, transportation, and storage prior to use.

There are certainly more shades of paint, and more types of screws and bolts, but dynamite was a product with a wider variety of choices than almost any other tool of industry.

The first variable was the mixture of explosives and other ingredients in the cartridge. This affected the strength and speed of the explosion. Other variables included the diameter and length of the cartridge, and how the shell for the cartridge was prepared. As a result, dynamite was routinely sold by multiple producers in thousands of different varieties.

Paradoxically, the boxes used to ship dynamite came in only three standard net weights, and just one of those, the 50-pound case, was used almost exclusively. A surprise for many people is that dynamite was being shipped in cardboard boxes by the 1950s.

The wrappers containing the various dynamite mixtures were mostly drab, with utilitarian graphics. The shells do chronicle the evolution of practices regarding marking cartridges with warnings and other information.

Because explosives are dangerous, more care is exercised in making, transporting, and storing them. The standardization of prudent practices developed over time, but beyond safety considerations, dynamite was simply an efficient way to get things done.

Dynamite Cartridges

Why cylindrical sticks and not square bars or cakes? The short explanation is that drill holes are round. But why a standard size of 8" by 1¼"? Here the answer lies in the equipment originally employed to make those round holes. A 1" to 2" hole was the largest diameter within the practical capacity of a

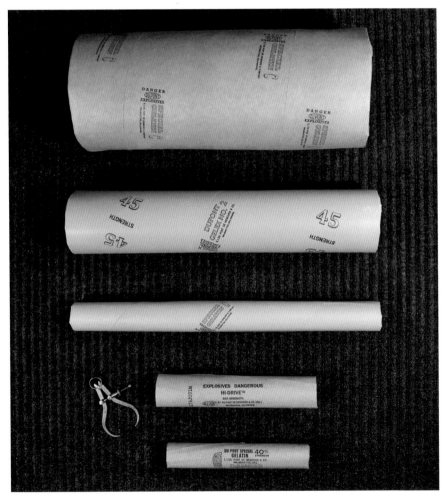

By far the most popular size of dynamite cartridge was 8" by 1¼" (bottom stick in the photo), which came to be known as the "standard size." Larger-diameter dynamite cartridges were used mainly for quarrying and strip mining. *Author*

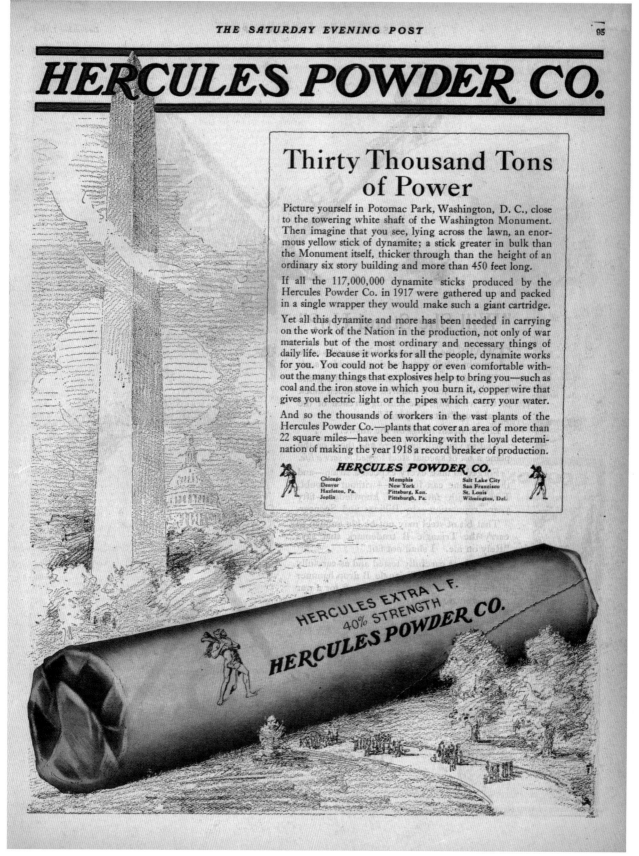

HERCULES POWDER CO.

Thirty Thousand Tons of Power

Picture yourself in Potomac Park, Washington, D. C., close to the towering white shaft of the Washington Monument. Then imagine that you see, lying across the lawn, an enormous yellow stick of dynamite; a stick greater in bulk than the Monument itself, thicker through than the height of an ordinary six story building and more than 450 feet long.

If all the 117,000,000 dynamite sticks produced by the Hercules Powder Co. in 1917 were gathered up and packed in a single wrapper they would make such a giant cartridge.

Yet all this dynamite and more has been needed in carrying on the work of the Nation in the production, not only of war materials but of the most ordinary and necessary things of daily life. Because it works for all the people, dynamite works for you. You could not be happy or even comfortable without the many things that explosives help to bring you—such as coal and the iron stove in which you burn it, copper wire that gives you electric light or the pipes which carry your water.

And so the thousands of workers in the vast plants of the Hercules Powder Co.—plants that cover an area of more than 22 square miles—have been working with the loyal determination of making the year 1918 a record breaker of production.

HERCULES POWDER CO.

Chicago	Memphis	Salt Lake City
Denver	New York	San Francisco
Hazleton, Pa.	Pittsburg, Kan.	St. Louis
Joplin	Pittsburgh, Pa.	Wilmington, Del.

HERCULES EXTRA L. F.
40% STRENGTH
HERCULES POWDER CO.

Explosives producers advertised in trade journals, newspapers, and general-interest magazines. This ad was intended to engender goodwill toward dynamite rather than marketing specific products. Indeed, Hercules reminds us that happiness itself is dependent on explosives. Hercules Powder, "Thirty Thousand Tons of Power," advertisement, *Saturday Evening Post*, December 7, 1918, 95.

hand drill for boring hard rock. When mechanical drilling became increasingly efficient in the 1920s and '30s, larger-diameter (up to 12") dynamite cartridges were used, mostly for quarrying and open-pit mining.

The 8" length was found to be convenient to handle and to simplify metering multiple and combined charges. The large variety of blasting situations required numerous choices regarding the strength and other characteristics of an explosive formulation. Small charges meant more options, especially for delicate work. The New York City subway was blasted with small charges to avoid disrupting city life.

Dynamite was and is manufactured in many other lengths. Dyno Nobel currently offers its Dynosplit line in 2' long cartridges that are designed to be coupled together and fired by a single blasting cap. Just three of those cartridges make for a 6' long stick. The largest single cartridges of dynamite resembled cardboard buckets and often had handles or attached bales for easier handling. Many had tapered ends to facilitate loading.

Larger cartridges tended to detonate more completely and were considered safer because fewer cartridges were handled overall. An article in the *Engineering and Mining Journal* in 1923 related: "The Institute of Makers of Explosives, in co-operation with engineers from the Bureau of Mines, in recent tests have demonstrated conclusively that both economy and safety in the use of explosives are lost by the use of dynamite in cartridges of ⅞" or 1" diameter, and therefore, that every effort should be made to drill holes of such a diameter as will accommodate cartridges of not less than 1¼" diameter."[1]

Another advantage of larger-diameter cartridges is that more explosives can be packed into a shell made of relatively less paper. A 2" diameter cartridge uses about half as much shell paper as two sticks of 1" diameter each. Although only about 5 percent of the

The ½", ¾", 1", and 1¼" diameter cartridges. Despite the advantages of larger-diameter cartridges, ⅞" diameter sticks were sold into the 1950s. Most were used for delicate underground blasting. *Author*

cost of a typical dynamite stick was for the shell, the figures added up for large quantities and could result in savings of tens of thousands of dollars a year for major consumers such as quarries.

The vast majority of dynamite cartridges were sold in diameters ranging from ⅞" to 6". According to the 1958 DuPont *Blasters' Handbook*, sizes in order of popularity at that time were 1¼", 1⅛", 1½", 1¾", and 2". That edition of the *Handbook* also noted that at the time, "there is a marked trend toward cartridges up to 24 inches"[2] in length. Since multiple cartridges are loaded into boreholes, longer cartridges make sense. Some modern water-gel-explosive "chubs" come in lengths of up to 15'.

The 1958 *Blasters' Handbook* also relates that at that time, of the larger diameters, 4" to 6" were the most popular in lengths of 16" to 30". Government regulations dictated a maximum length

of 36" and a maximum diameter of 12", except for dynamite of less than 10 percent strength, for which there were no technical length restrictions but obvious practical limits.

A 1936 price list from DuPont shows practically unlimited combinations of diameter, length, and formulation. For example, DuPont's Special Gelatin line was offered in standard sizes of ⅞" × 8", 1" × 4", 1" × 6", 1" × 7", 1" × 8", 1⅛" × 7", 1⅛" × 8", 1¼" × 4", 1¼" × 7", 1¼" × 8", 1¼" × 12", 1½" × 8", 1½" × 12", and 1¾" × 8". But wait, there's more! Custom lengths and diameters were also available for an additional charge. According to DuPont, "All sizes 2" and over in diameter and 8" in length" were offered in "gradations to be ½" for diameter and 2" for length."[3]

And that was just for one of what was then more than twenty brands of DuPont's extensive line of dynamites. Each brand was also offered in a

COLORADO—Counties of Dolores, Hinsdale, La Plata, Montezuma, Ouray, San Juan and San Miguel.

CARLOAD DELIVERED PRICES OF HIGH EXPLOSIVES

Strengths	Red Cross Extra	Du Pont Special Gelatin	Du Pont Gelatin Hi-Velocity Gelatin	Du Pont Nitroglycerin
20%	10.50		10.75	10.75
25%	10.75		11.15	11.25
30%	11.00	11.00	11.50	11.75
33%	11.15			
35%		11.40	11.90	12.25
40%	11.25	11.75	12.25	12.75
50%	12.00	12.50	13.00	14.00
60%	12.75	13.25	13.75	15.25
75%		15.75	17.00	
80%		17.00	18.25	

OTHER GRADES

Du Pont RRP (5%)	9.25	Du Pont Extras	12.50
R. C. Blasting No. 2 F. R.	10.50	Durox	12.50
R. C. Blasting No. 3 F. R.	11.00	Gelex 1, 2	13.00
R. C. Blasting No. 4 F. R.	11.25	Du Pont Ditching (50%)	14.00
R. C. Blasting No. 5 F. R.	12.00	Nitramon A, B, C, D	12.75
Agritol	12.50		

PERMISSIBLES

Duobels	12.50	Lump Coals	12.50
Monobels	12.50	Gelobels	13.00

See Page 6 for less than carload prices.

Above: Prices shown are dollars per 100 pounds. E. I. Dupont de Nemours, *High Explosives Price List No. 8* (Wilmington, DE: E. I. Dupont de Nemours, November 17, 1936).

Right: This 50-pound case of DuPont Gelex cost about $7.00 in 1936. Now, 50 pounds of dynamite will set you back $500.00. *Author*

variety of strengths. For our example of DuPont Special Gelatin, the strengths (or "grades") available in 1936 were 30 percent, 35 percent, 40 percent, 50 percent, 60 percent, 70 percent, 75 percent, 80 percent, and 90 percent. By 1952, DuPont boasted "approximately 220 grades and 125 sizes."[4] The possible combinations run into the tens of thousands.

Custom strengths could be had for an extra charge. If you wanted, say, 25 percent strength, you could get it, provided you met minimum order weight requirements.

If all that were not enough, a few brands of dynamite were even available in diameters of less than ⅞". Giant Powder was originally sold in diameters of ½" and ¾".

Finally, one could order cases with a specified number of sticks for an extra cost. The reason for the charge was that the sticks had to be specially made to weigh a total of either 10, 25, or 50 pounds, which were the standard case sizes offered. Either the diameter or length of the cartridge was adjusted. Ordering an exact number of sticks for a small project meant no leftovers to dispose of.

There are always sound business reasons for offering so many choices, and for dynamite the rationale was the diversity of its users. According to Giant Powder Company in 1927, "Each of the more than 100 grades of Giant Explosives has its own definite qualities, and is suited to some particular kind of work."[5] Farmers, loggers, builders, engineers, and miners all had different needs. Mining in particular required a broad range of dynamite for the wide spectrum of mined materials. Explosives manufacturers marketed their products accordingly.

The first dynamite sold in the United States in 1868 cost around $1.75 a pound. Within ten years, fierce price competition from blasting powder and increased efficiency of production had cut the price in half, and by 1883 dynamite averaged forty cents a pound. Sales of black powder in the United States crested in 1917 and decreased to 10 percent of peak levels by 1950, having been displaced by dynamite.

In the early years of the twentieth century, dynamite prices averaged between ten and twenty-five cents a pound, depending on the strength and type of formulation. Price generally increased with strength, and

nitroglycerin-based straight dynamites and gelatin formulations cost the most. Dynamites that were mostly other explosives such as ammonium nitrate were the least expensive.

Dynamite prices were affected by temporary conditions such as interruptions in chemical supplies and labor issues. Shipping restrictions during World War I led to a shortage of raw glycerin. This in turn led dynamite prices to rise from twenty cents a pound to thirty-three cents a pound for straight nitroglycerin dynamite during the last two years of the war.

In 1936, Hercules Powder Company offered its line of products in prices ranging from nine cents a pound for 15 percent strength straight dynamite, to sixteen cents a pound for 75 percent strength low-freezing gelatin, to twenty-two cents a pound for 100 percent strength blasting gelatin. Dynamite prices remained quite stable, averaging

between ten and thirty cents a pound until the late 1970s, when they began a climb culminating in today's price of $10.00 a pound.

Retail prices for small quantities (100–200 pounds) of dynamite were typically at least 30 percent higher than what was paid by large industrial consumers. Prices for just a few sticks from the hardware store could be twice the wholesale cost. For users such as farmers, the outlay was far offset by savings in labor and increases in crop yields.

A 1925 advertisement for Dumorite claimed a "⅓ more per dollar" cost savings over ordinary dynamite. However, the comparison was to high-strength, high-nitroglycerin-content formulas. Dumorite was mostly lighter-weight ammonium nitrate and black powder and was rated at 20 percent volume strength. Since all less dense formulations have higher cartridge-to-weight

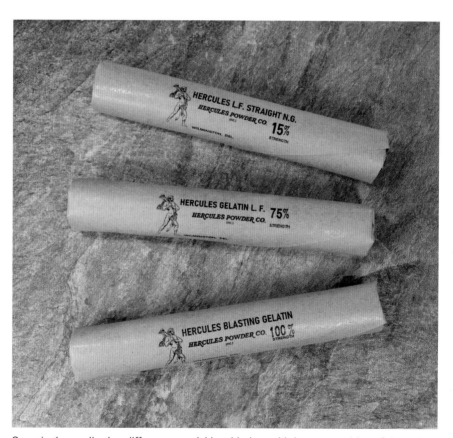

Seemingly small price differences quickly added up with large quantities of dynamite. Fortunately for mining operators, the most-expensive, higher-strength dynamites were used to blast the most-valuable metallic ores. High-value metals are found mostly in very hard rock. *Author*

NET PRICE LIST
AETNA DYNAMITE

No.	Per cent.		Per 100 lbs.
4,	20 Low Freezing		$14.40
4x,	25	"	$14.80
3,	27	"	$14.95
3x,	30	"	$15.20
2c,	33	"	$15.45
2b,	35	"	$15.60
2,	40	"	$16.00
2x,	45	"	$16.50
2xx,	50	"	$17.00
1,	60	"	$18.00

Nitro Glycerin and Gelatin Grades $1.00 per 100 lbs. more.

Contactor's Powder, . . $10.00

This 1911 Aetna Powder Company price list shows about a 100 percent difference in cost for the cheapest Contractor's Powder and the most expensive dynamite. Aetna Powder, *Aetna Dynamite and Blasting Supplies* (Chicago: Aetna Powder, 1911), 1.

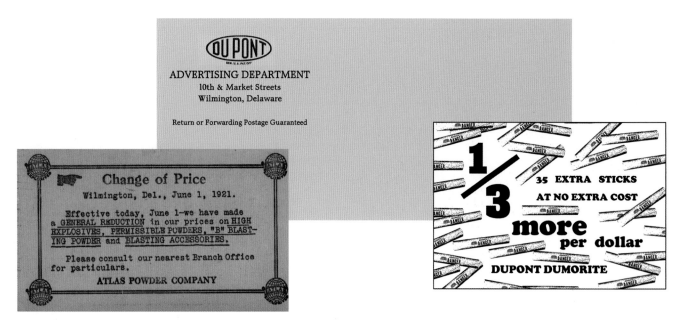

Price competition among dynamite producers evolved to include all manners of advertising. Envelope and postcard, *author*. Dumorite ad adapted from E. I. DuPont de Nemours, advertisement, *Rock Products*, February 25, 1922, 3.

ratios, this was marketing on DuPont's part. Farmers, who were the main users of Dumorite, found more cartridges to be preferable to stronger cartridges.

Prices varied by region, since safe delivery of product was included. Large companies issued regional price schedules that included higher rates for more-remote delivery areas. To make getting product to consumers quickly and cheaply, major producers operated multiple plants and warehouses in strategic locations throughout the country.

A weight of 50 pounds could comprise from eighty to 250 standard-size (8" × 1¼") sticks, depending on density. A chart by Atlas Powder Company in 1957 showed that more than three hundred ⅞" diameter cartridges containing a very low-density mixture were needed to make 50 pounds. Cases varied in width and height; taller or multiple cases were needed for the least dense mixtures. Sawdust and wood shavings filled empty space at the tops of cases. Large-diameter cartridges filled cases quickly; one of the wooden boxes pictured (p. 33) held only four 5" × 16" sticks!

Dynamite Formulations

Dynamite is officially defined as an explosive compound containing nitroglycerin. Trinitrotoluene, commonly known as TNT, does not qualify; neither does nitrostarch, about which we will learn more in the chapter on Trojan Powder Company. Dynamite formulations did contain many other ingredients in addition to nitroglycerin. (Nitroglycerin is still the most powerful industrial explosive ever used. Other, stronger chemical combinations exist, but none are suitable for blasting.)

The first dynamite compounds were called straight dynamites and consisted of nitroglycerin absorbed in an inert medium. Early straight dynamite formulations became known as granular, owing to their suspension in kieselguhr, which is a German word for diatomaceous earth. Diatoms are tiny sea creatures whose decomposing shells make for a chalky, crumbly soil that absorbs liquids readily. Sawdust, nut meal, and other fillers that soon replaced dirt were considered "fuels" since they burned and contributed marginal energy to an explosion.

The process of making a batch of straight dynamite through the 1910s was not complicated. Glycerin was nitrated in the nitrating building by being slowly mixed with nitric and sulfuric acids in large tubs. The acids were separated from the nitrated glycerin, which was then carefully transported to the mixing house, where "dope" in the form of wood pulp and a few other ingredients were added. The glycerin and dope mixture was loaded into shells, and the sticks possibly were sprayed or dipped in paraffin. Finally, the finished cartridges were packed into cases. (There is not room here to tackle the complex chemistry of explosives. If you are interested, a great place to start is Jacqueline Akhaven's reader-friendly *The Chemistry of Explosives*, now in its fourth edition.)

While the basic steps of manufacture remained essentially the same, dynamite production became increasingly mechanized in the 1920s, particularly for larger makers. By the 1950s, plants were completely modernized, with streamlined production lines resembling those of many other goods.

Initially, cartridges were marked with #1, #2, and #3, usually indicating

TABLE II

Atlas		Other Brands of Equivalent Strength							
Brand	Nitro-glycerine Per Cent.	Repauno Gelatine	Hercules Powder	Hercules Gelatine	Giant Powder	Giant Gelatine	Hecla Powder	Ætna Powder
A	75	A	No. 1 XX	No. 1 XX	No. 1+	No. 1+	No. 1 XX	
B+	60	B+	No. 1	No. 1	No. 1 A	No. 1	No. 1 XS	No. 1
B	50	B	No. 2 SS	No. 2 SS	New No. 1	No. 2	No. 1 X	No. 2 XX
C+	45	C+	No. 2 S	No. 2 S	No. 2 Extra			No. 2 X
C	40	C	No. 2	No. 2	No. 2	No. 3 C	No. 1	No. 2
D+	33		No. 2 C		No. 2 C		No. 2 X	No. 2 C
D	30		No. 3		No. 3		No. 2	No. 3 X
E+	27		No. 3 B		XXX		No. 3 X	No. 3
E	20		No. 4 B		XXXX		No. 3	No. 4

This 1906 chart shows the variety of early, company-specific grade markings on cartridges. Over the next ten years, these grades would be replaced with numerical strengths. International Library of Technology, *Rock Boring, Rock Drilling, Explosives and Blasting* (Scranton, PA: International Library of Technology, 1906), section 36, p. 18.

such as DuPont Dumorite and Atlas Ammite debuted in the early 1920s. These mixtures contained only about 5 percent nitroglycerin, meaning they were nonfreezing and did not cause headaches (nitroglycerin increases blood flow to the brain, which can cause harmless but severe pain; frozen nitroglycerin was unusable). Such mixtures had nowhere near the fracturing power and general strength of dynamites with more nitroglycerin. Lower-strength dynamites were popular with farmers, and with miners of softer materials such as coal and clay.

60 percent, 40 percent, and 30 percent nitroglycerin content, respectively. Soon other grades were added: #4 was 20 percent, 2X was 45 percent, and so on. Some companies also had lettered formulations: A, B+, C or 1A, 2C, etc.

The next mixtures added sodium nitrate and ammonium nitrate, which are explosive compounds in their own right. This kind of dynamite, first patented by Nobel in 1879 as "extra dynamite," was more powerful than other formulations containing the same ratios of nitroglycerin to inert or minor fuel ingredients. Ammonium nitrate is less expensive than nitroglycerin, so extra dynamites were economical. They also had a less violent action, making them ideal for softer rock and for blasting paths through hills.

Many ammonium-nitrate-based explosives were introduced in the 1890s. Unlike the earliest extra dynamites, "ammonia dynamites" were mostly ammonium nitrate. Ammonium nitrate by itself takes a much-larger shock to explode than nitroglycerin. Nitroglycerin was absorbed into ammonium nitrate to make an explosive that could be more readily detonated. Brands such as Ammonite, Robustite, and Securite were patented but did not gain wide use because they were difficult to detonate, particularly with the much-weaker blasting caps of the era. More-successful formulations

This Hercules Powder Company ad stresses the importance of strict attention to careful handling and proper mixing of explosive ingredients. Advertisement, *Century*, June 1920, 90.

Numerical strength markings made it easier to judge the power of a cartridge of dynamite but also introduced a new and expansive spectrum of strengths. *Author*

Dynamites were originally rated strictly by their nitroglycerin content, so a cartridge marked "40 percent strength" actually contained 40 percent nitroglycerin by weight. When ingredients with explosive properties were added to the mix, the strength of cartridges was still rated relative to the power of straight dynamite, regardless of total nitroglycerin content. A mixture containing ammonium nitrate might have only 20 percent nitroglycerin by weight but have the strength of a 40 percent straight dynamite. "Volume strength" became an equivalent term. Cartridges with the same volume strength were said to be equal in explosive power, irrespective of their nitroglycerin content.

According to the Giant Powder Company in 1924, 1 pound of 60 percent strength dynamite was equivalent to 1.10 pounds of 50 percent strength, 1.27 pounds of 40 percent strength, 1.73 pounds of 20 percent strength, and 2.09 pounds of 10 percent strength. Such comparisons are of an explosive's "weight strength." "Bulk strength" is a synonym for the above-described volume strength. However, the current meaning of bulk strength is an explosive's power relative to ammonium nitrate / fuel oil explosives (ANFO) rather than nitroglycerin.

Explosives are now assigned both weight and bulk strengths relative to ANFO, with ANFO being "1" on the scale. For example, Dyno Nobel's Unigel has a 1.09 relative weight strength and a 1.72 relative bulk strength. This means that pound for pound, Unigel is almost 10 percent stronger than ANFO. However, a given volume of Unigel is almost 75 percent stronger than the same volume of ANFO. Put another way, a 50-pound container of ANFO is 1.72 times as large as a 50-pound container of Unigel dynamite.

High explosives instantly explode or detonate, while low explosives such as black powder burn or deflagrate, albeit still quickly enough to explode. Scientifically speaking, detonation is chemical decomposition that occurs faster than the speed of sound, while deflagration takes place slower than the speed of sound. The United States Bureau of Alcohol, Tobacco, Firearms and Explosives currently defines a high explosive as one that can be detonated unconfined by a #8 blasting cap (blasting caps are numbered according to strength; #6 is the most common and #8 is the strongest commercial fuse cap produced). If an explosive needs a larger shock than that provided by a blasting cap, then it is not a high explosive. The "unconfined" part is crucial; black powder blows up only when confined in a small space. The classic example of this is a firecracker, the powder from which merely burns when removed from the cardboard tube encasing it. A blasting cap placed in a pile of blasting powder will almost always scatter the powder instead of detonating it.

Blasting powder explodes at about 3,000 feet per second, while dynamite's speed ranges from 5,000 to 23,000 feet per second (the speed being measured is that of the pressure wave front, which emanates from all explosions). Legally, dynamite was a class A explosive (the most dangerous) from 1931 to 1981; now it is classified as a division 1.1 explosive (still the most dangerous).

Another term for high-explosive compounds such as dynamite is "cap

High Explosives—Relative energy of different strengths; velocity; water resistance

One Cartridge	60%	50%	45%	40%	35%	30%	25%	20%	15%
60%.....	**1.00**	1.12	1.20	1.28	1.38	1.50	1.63	1.80	2.08
50%.....	0.89	**1.00**	1.07	1.14	1.23	1.34	1.45	1.60	1.85
45%.....	0.83	0.93	**1.00**	1.07	1.15	1.25	1.36	1.50	1.73
40%.....	0.78	0.87	0.94	**1.00**	1.08	1.17	1.27	1.40	1.59
35%.....	0.72	0.81	0.87	0.93	**1.00**	1.09	1.18	1.30	1.50
30%.....	0.67	0.75	0.80	0.85	0.92	**1.00**	1.09	1.20	1.38
25%.....	0.61	0.69	0.74	0.78	0.85	0.92	**1.00**	1.10	1.27
20%.....	0.55	0.62	0.67	0.71	0.77	0.83	0.90	**1.00**	1.15
15%.....	0.48	0.54	0.58	0.61	0.66	0.72	0.78	0.86	**1.00**

Table showing number of cartridges of any given strength required to equal one cartridge of any other strength.

This chart shows relative volume strength in terms of numbers of cartridges (of equal size) rather than weight. Arthur La Motte, *Blasters' Handbook* (Wilmington, DE: E. I. DuPont de Nemours, 1925), 11.

What is 40% Dynamite?

Many people use explosives without knowing exactly what is meant by the percentage markings on the cases and cartridges. They know that 40% dynamite is stronger than 30%, and that 50% is stronger than 40%, but they are in the dark as to what those figures actually represent.

When dynamite was first made the formula was simple. It contained nothing but nitroglycerin and an absorbent. In those days 40% meant that 40% of the contents was nitroglycerin. The straight nitroglycerin dynamites of today also contain the amount of nitroglycerin expressed by the percentage figures, and they contain other things that make them safer to handle than dynamites made on the old formula.

With the Ammonia or Extra dynamites the old definition of 40% is not literally true today. The percentage figures now refer more to the strength than to the ingredients, and it is the strength that is important to consumers. Any dynamite made by an honest manufacturer today is equal in strength to dynamite that actually contains an amount of nitroglycerin equal to the percentage figures.

Hercules 40% nitroglycerin dynamite contains 40% nitroglycerin and Hercules 40% Extra Dynamite has the same strength and will do the same amount of work. Hercules 40% Gelatin Dynamite also has the same strength pound for pound, but a cartridge of it will do more work because it is heavier.

With the exception of some special grades that are well known to those who use them, the strength of Hercules Dynamites is always clearly marked on the cases and cartridges, and the goods correspond exactly to the markings. In purchasing Hercules explosives you run no risk. Honest products, truthfully marked, have won universal confidence in Hercules goods.

If you are interested in goods of this character, we shall be glad to hear from you.

 HERCULES POWDER COMPANY *WILMINGTON, DELAWARE*

This early ad from Hercules Powder Company endeavors to explain the distinction between straight strength and relative strength. Hercules Powder, "What Is 40% Dynamite?," advertisement adapted from *United Mine Worker's Journal*, April 19, 1917, 2.

While DuPont used nitrocellulose for myriad nonexplosive products, this galvanized aluminum lid belonged to a drum of the chemical that was clearly destined for use for explosives. The lid measures 22" in diameter. *Author*

SEMI-GELATIN

RED CROSS   ATLAS   HERCULES

These recently developed explosives have already become invaluable in many kinds of work. Are so highly water resisting that they are giving satisfactory results in submarine blasting. Can be used in all kinds of work whether wet or dry. Red Cross Semi-Gelatin does not freeze in temperatures above 35° F.

E. I. DU PONT DE NEMOURS POWDER COMPANY ESTABLISHED 1802   WILMNGTON, DEL., U.S.A.

Upon their introduction in the 1890s, semigelatin formulas offered a less expensive alternative to full-gelatin dynamites. E. I. DuPont de Nemours, advertisement adapted from *Bulletin of the American Institute of Mining Engineers*, October 1910, advertising 7.

sensitive" (also called detonator sensitive), again meaning that the formulations can be exploded by a blasting cap of sufficient strength. Many modern slurries and emulsions, also known as blasting agents, are cap-insensitive, meaning they require a larger explosion to trigger them. Dynamite cartridges are still used as big blasting caps for explosives that cannot be detonated directly with normal blasting caps. Some water-gel and emulsion mixtures are also available in cartridged, cap-sensitive forms.

The combination of ammonium nitrate and fuel oil, known as ANFO, is markedly more stable and cheaper than dynamite. ANFO compounds and water-gel slurries are now much more widely used than dynamite because of their stability, and especially their ease of pouring into any size of hole. Yet another advantage of modern emulsions and water-gel slurries is their absolute water resistance. ANFO, on the other hand, cannot be used in wet holes.

Nobel devised gelatin dynamite in 1875, and early on it existed as

gelignite, blasting gelatin, and gelatine, depending on formulation, which consisted of varying ratios of nitroglycerin to nitrocellulose and a few other ingredients. Nitrocellulose is an explosive in itself but does not contain nitroglycerin, despite the name. Gelatin dynamites containing nitrocellulose were rubberlike in consistency and were valued for their water resistance and their plasticity for tamping in upward holes. Semigelatins consisted of ammonium nitrate sensitized with nitroglycerin with a little nitrocellulose added. Semigelatins were more economical but not quite as water resistant as full gelatins, but they had slightly lower freezing temperatures. The terms "gelatin extra," "special gelatin," and "extra gelatin" are synonyms for semigelatin.

Two 40 percent strength dynamites might behave quite differently, depending on their relative nitroglycerin content and a host of other factors influenced by differing formulations. Generally speaking, the less nitroglycerin by volume, the slower the explosion. Coal-mining dynamites tended to have lower nitroglycerin contents. Coal is dramatically softer than hard rock, and lower-strength dynamite mixtures were formulated to explode as slowly as possible. Dynamites specifically designed for blasting coal were an exception to the rule, in that they usually did not have strength markings, even into the 1970s. Instead they were rated as to speed of detonation and had lettered and numbered grades.

Most dynamite formulations sold in the United States were straight, extra, gelatin, semigelatin, and gelatin extra. Each had its niche. Straight dynamite was more sensitive to shock and was

Above: Nitroglycerin is the most dangerous ingredient in dynamite. Here, Hercules Powder Company conveys the message that safety is the company's top concern. Advertisement, *Sunset*, August 1920, 80.

Right: Hercules eventually offered ten different grades of its popular Gelamite brand gelatin-extra dynamite. *Author*

Straight (nitroglycerin) dynamite consists of nitroglycerin, sodium nitrate, an antacid, a carbonaceous fuel, and sometimes sulfur. The term "straight" means that a dynamite contains no ammonium nitrate. The weight strength indicates the approximate percentage of nitroglycerin or other explosive oil. The use of straight dynamite is limited because of its high cost and sensitivity to shock and friction. Fifty percent straight dynamite, by far the most common straight dynamite, is referred to as ditching dynamite and is used in propagation blasting.

High-density ammonia dynamite, also called extra dynamite, is the most widely used dynamite (in 1983). It is like straight dynamite, except that ammonium nitrate replaces part of the nitroglycerin and sodium nitrate. Ammonia dynamite is less sensitive to shock and friction than straight dynamite. It is most commonly used in small quarries, in underground mines, in construction, and as an agricultural explosive.

Low-density *ammonia dynamite* is manufactured in a weight strength of about 65 percent. The cartridge (bulk) strength ranges from 20 to 50 percent, depending on the bulk density of the ingredients. A high-velocity series and a low-velocity series are manufactured. Low-density ammonia dynamite is useful in very soft or prefractured rock or where coarse rock such as riprap is required.

Blasting gelatin is a tough, rubber-textured explosive made by adding nitrocellulose, also called guncotton, to nitroglycerin. An antacid is added to provide storage stability and wood meal is added to improve sensitivity. Blasting gelatin emits large volumes of noxious fumes upon detonation and is expensive. It is seldom used today (1983). Sometimes called oil well explosive, it has been used in deep wells where high heads of water are encountered. Blasting gelatin is the most powerful nitroglycerin-based explosive.

Straight gelatin is basically a blasting gelatin with sodium nitrate, carbonaceous fuel, and sometimes sulfur added. It is manufactured in grades ranging from 20 to 90 percent weight strength and is the gelatinous equivalent of straight dynamite. Straight gelatin has been used mainly in specialty areas such as seismic or deep well work, where a lack of confinement or a high head of water may affect its velocity. To overcome these conditions a high-velocity gelatin is available which is like straight gelatin except that it detonates near its rated velocity despite high heads of water.

Ammonia gelatin, also called special gelatin or extra gelatin, is a straight gelatin in which ammonium nitrate has replaced part of the nitroglycerin and sodium nitrate. Manufactured in weight strengths ranging from 40 to 80 percent. It is the gelatinous equivalent of ammonia dynamite. Ammonia gelatin is suitable for underground work, in wet conditions, and as a toe load, primarily in small diameter boreholes. The higher grades (70 percent or higher) are useful as primers for blasting agents.

The basic varieties of dynamite catered to specific needs. A dry salt mine didn't need water-resistant cartridges, a trenching crew wanted straight dynamite, and farmers preferred lower-strength, ammonium-nitrate-based formulas. *Author*

used for ditching, which relied on propagation of an explosion from one hole to the next. Gelatins, both straight and extra, were prized for their water resistance and shattering power. DuPont Gelex and Hercules Gelamite, both introduced in 1929, were two of the most popular gelatin-extra formulations. Semigelatins were less expensive than full gelatins and were ideal for medium-hard rock. Nongelatinous-extra formulations were economical and suited to blasting where a slower, less violent action was needed.

Formulations such as Atlas Amodyn were mostly ammonium nitrate, with nitroglycerin added as a sensitizer. A sensitizer is any cap-sensitive compound that is mixed with a cap-insensitive compound. Many nonnitroglycerin sensitizers have been developed for use in modern water-gel and slurry formulations.

From the 1870s to 1900, dozens of nitroglycerin-sensitized formulations

By 1983, much had changed regarding dynamite preferences. For example, blasting gelatin was "seldom used." Adapted from Richard A. Dick, Larry R. Fletcher, and Dennis V. D'Andrea, US Department of the Interior, *Explosives and Blasting Procedures Manual* (Washington, DC: USGPO, 1983), 7.

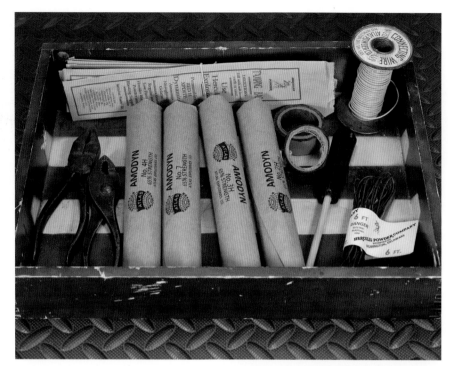

such as Borland's Carbo-Dynamite, Horsley's Powder, and Rutenberg's Explosive were patented. Running a successful explosives business was not easy; the vast majority of such concoctions lasted for only a few months or years, if they ever saw commercial production at all.

While some experts advised against loading cartridges of different strengths into a single borehole (called "combination loads"), Giant Powder Company maintained that "far better results frequently are obtained when the hole is loaded with a mixed charge."[6] The technique, which involved loading the stronger explosives at the bottom of the hole, was said to be useful for ore mining in very hard rock. Sometimes, particularly in large quarry explosions,

Atlas Powder Company introduced its Amodyn brand in 1938. The line was somewhat unusual because all cartridges had the same 65 percent weight strength. The various grades were dependent on velocity of detonation and differences in density. *Author*

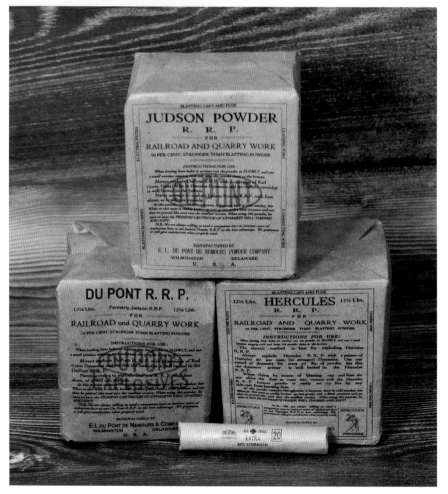

Blasting powder was first sold in wooden kegs, then metal containers such as this 25-pound can of DuPont "B" blasting powder dating from the 1930s. In 1936 a 25-pound can cost around $2.50. *Author*

Egbert Judson invented "Railroad Powder" in 1876. Midway in strength between black powder and dynamite, "R.R.P." was ideal for blasting away the hills that impeded the laying of smooth railway. Railroad powder was seldom used underground owing to the objectionable fumes produced upon detonation. *Author*

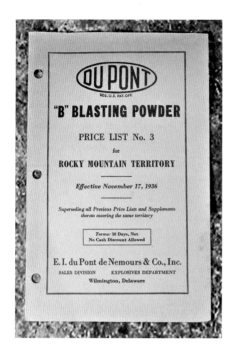

Blasting powder was sold by granule size instead of strength. E. I. DuPont de Nemours, *"B" Blasting Powder Price List No. 3*, November 17, 1936 (Wilmington, DE: E. I. DuPont de Nemours, November 17, 1936).

blasting powder was mixed with dynamite cartridges. The practice was frowned upon because oftentimes the blasting powder either scattered or caught fire without exploding.

One of blasting powder's advantages over cartridged dynamite is that powder is free running, meaning it can be poured into a borehole. This was not practical in upward holes; then the black powder was made into cartridges on-site, using blasting paper or any suitable scrap paper. The most commonly used "B" blasting powder is about half as strong as dynamite. "A" blasting powder is stronger than "B" blasting powder. Due to its much-higher cost, "A" blasting powder was seldom used for mining after the introduction of "B" powder in the mid-1850s. Free-running dynamites were developed as alternatives to blasting powder altogether.

There were two types of free-running dynamite formulations: nitroglycerin-coated black powder and nitro-glycerin-sensitized ammonium nitrate. Judson Powder, patented in 1876, was

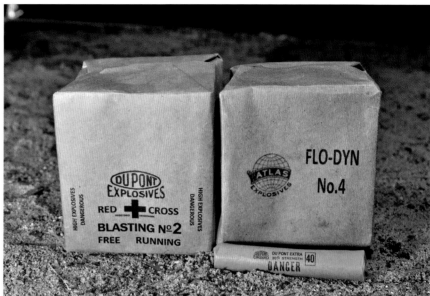

Free-running dynamites were easy to pour and tightly pack into the large boreholes common in open-pit mines and quarries. Funnels made of nonsparking metal or wood were used to facilitate loading. *Author*

the most famous example of the first. It came in 6.25-, 12.5-, and 25-pound bags as Railroad Powder (R.R.P.) and proved enormously popular. DuPont continued to carry the Judson brand after formally acquiring Judson Powder Company in 1903. By 1920, DuPont had phased out the Judson name, marking its bags of railroad powder "Formerly Judson Powder R.R.P."

Judson Powder and similar formulations were stronger than blasting powder. Judson Powder was also sold in cartridged form in strengths of 10 to 15 percent, rated on actual nitroglycerin content. Other companies offered their own takes. Aetna Powder Company labeled its version Contractor's Powder,

which became another synonym for this type of formulation in general. DuPont also sold a nonnitroglycerin black powder with mixed-size grains as Railroad "B" Blasting Powder.

One problem with nitroglycerin-coated black powder was its tendency to form clumps, especially in humid environments. For this reason, it could not be poured into damp boreholes. It was also difficult to detonate with just a blasting cap. In January 1921 the *Monthly Bulletin of the Canadian Institute of Mining and Metallurgy* pointed out the following:

The so-called dynamite R.R.P. ("railroad" powder) is a low-grade high explosive that may be said to

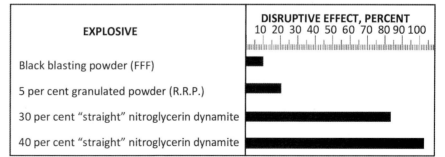

This chart shows the relative speed and disruptive properties of black powder, railroad powder, and dynamite. It was dynamite's vastly superior shattering power that transformed hard-rock mining into a mighty engine of progress. Adapted from *Engineering News*, September 11, 1913, 495.

rank between black blasting powder and dynamite. It is granular in form, the lower grades being coarse and free running, but the higher grades fine and, owing to their greater nitroglycerine content, inclined to pack or cake. This class of dynamite is particularly valuable for blasting in open work that is fairly dry and where black blasting powder would be too slow in action and not of sufficient strength but where dynamite would be too rapid in action and stronger than necessary. To obtain the best results when using this explosive a dynamite primer of 40 percent strength, or higher, is generally used. The fumes are rather objectionable and it is therefore an explosive intended for outside work such as railway construction, being particularly suitable in loose, shale rock.[7]

The second type of bagged, free-running dynamite debuted in the 1920s and consisted of granular ammonium nitrate sensitized with a small amount of nitroglycerin. DuPont's Red Cross brand was by far the most popular of this kind of formula and was ideal for filling the sprung holes used for quarrying, and for removing overburden in open-cut mines. Atlas Powder's Flo-dyn brand was also extremely successful and was sold into the 1960s. All free-running dynamites were best detonated with a regular cartridge of dynamite, which acted as a jumbo blasting cap. Free-flowing ANFO and other explosives were and still are detonated with dynamite primers, but cast (plastic-shelled), sensitized, PETN/TNT primers are now more commonly used.

DuPont, Aetna, and Hercules also sold cartridged nitroglycerin-coated blasting powder. The fineness of the powder grains varied from fine to very fine and was indicated on the wrapper with F, FF, FFF, and FFFF. This was in keeping with the classification scheme for black powder. Larger granules have less surface area overall and produce slower, heaving explosions. Finer particles have more total surface area and make for a faster, more powerful blast. Controlling the speed of an explosion is vital; too fast a blast can ruin certain materials, and too slow a blast is inefficient in hard rock.

Dynamite Shells

According to the United States Bureau of Mines in 1916,

A cartridge of dynamite consists of a covering of paper, sometimes waxed or parchmentized and often coated with paraffin, within which the dynamite is more or less tightly packed to give it the density desired. In the manufacture of dynamite the paper "shell" is first made, and is then packed with the explosive, after which the cartridge is sometimes "redipped," by which is meant that the cartridge is plunged into a bath of paraffin heated slightly above its melting point. The paraffin closes up any openings in the wrapper and tends to make the cartridge waterproof.[8]

Blasting powder has been rolled in paper cartridges since the late 1600s. Prior to that (and after; "putting up" powder in cartridges took time and was not needed for dry, vertical holes), black powder was often simply poured and packed into boreholes or natural crevices and ignited. One advantage of cartridges was that the more thoroughly compacted and confined powder produced a more complete explosion. Cartridges also protected the powder from moisture in wet holes. Perhaps most importantly, cartridged powder proved to be an ideal method of loading upward-slanting holes.

The wrappers of dynamite sticks were originally made of vegetable parchment and then manila and kraft papers. The latter papers roughly

To make the sticks sensitive enough for detonation via blasting cap, cartridged contractors' powders had up to 15 percent nitroglycerin content, making some formulas as powerful as low-strength dynamite. *Author*

By the 1930s, railroad and contractors' powders had been supplanted by free-running dynamite. Adapted from US Department of the Interior, *Civilian Conservation Corps Safety Regulations* (Washington, DC: USGPO, 1938), 121.

		Free-flowing bag powders			
Maker	Brand name	Grade/strength			
		20	30	40	60
Apache	Quarry Special	5	4	3	Bag
Atlas/Giant	Quarry Special	No. 25	No. 30	No. 40	No. 70
Austin	F.R.	No. 2	No. 3	No. 4	No. 5
Burton	Burflow	2	3	4	5
Columbia	Free Flowing	No. 2	No. 3	No. 4	No. 5
Du Pont	R.C. Blasting F.R.	No. 2	No. 3	No. 4	No. 5
Egyptian	Free Flowing	No. 2	No. 3	No. 4	No. 5
Equitable	Free Flowing	No. 2	No. 3	No. 4	No. 5
General	Free-flow	A	B	C	D
Hercules	Herculite	No. 2	No. 3	No. 4	
Hercules	Hercomite	No. 2	No. 3	No. 4	Bag
Illinois	Bag Powder	No. 2	No. 3	No. 4	No. 5

After filling, some cartridges were either dipped or sprayed with molten paraffin. Thousands of sticks that were packed vertically into large tubs or bins were treated and allowed to cool and dry. *Author*

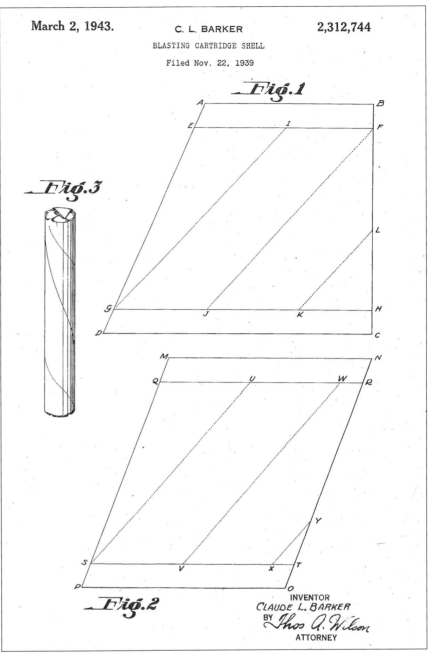

This is the patent for DuPont's perforated shell. Inventor Claude L. "Lindy" Barker was a 1931 graduate of the Colorado School of Mines who worked for Dupont for nearly four decades. Claude L. Barker, Blasting cartridge shell, US Patent 2,312,744, filed November 22, 1939, and issued March 2, 1943.

correspond to modern 60# to 110# manila text (not card stock) and kraft paper. "60#" is the basis weight and is calculated by weighing five hundred "parent" sheets of a certain size, depending on the paper. With 60# paper, a stack of five hundred 25" × 38" sheets weighs 60 pounds.

Vegetable parchment was invented in 1847 and is made by briefly dipping sheets of pressed plant fiber in sulfuric acid. The acid melts some of the pulp in the paper and converts it into a natural, waxy substance. Before wood pulp paper became dominant in the 1880s, most paper was made from old rags and straw.

While the color of a dynamite stick is often red in popular culture, with few exceptions most were some shade of brown, from creamy to dark and leathery. An article in the *Mining and Scientific Press* in 1915 noted that some large surface mines requested red wrappers for increased visibility and that "producers complied."[9] Apache Powder Company and Austin Powder Company used red wrappers for some cartridges, and shells made by IRECO in the 1980s were bright red.

Atlas Powder Company and Giant Powder Company, Consolidated, used white wrappers for several brands, but this was a rare exception, at least throughout most of the twentieth century. Early-twentieth-century Hercules Powder Company advertisements described the color of its cartridges as "yellow," but the sticks were a manila color and not a vibrant yellow (many modern cartridges, particularly those from other countries, are neon yellow or stark white). I do not think it is an overstatement to say that at least 95 percent of all dynamite cartridges ever produced in the United States were brown.

Most shells have a glossy look, owing either to paraffin-impregnated paper or a coating of wax applied by spraying. Sometimes the entire stick was dipped for further water resistance. Thus a shell could be unwaxed, prewaxed before filling, sprayed after filling, and dipped after filling or spraying (or both). In addition to water resistance, a prewaxed shell afforded protection from the internal contents leaching out. The trade-off was that the more paraffin in or on the shell, the more fumes and flames produced during the blast, which could be problematic in underground mines. One study by the Bureau of Mines in 1936 estimated that the average, standard-size waxed shell contained about 5 grams of paraffin. This meant that a 50-pound case of one hundred dynamite sticks could contain more than a pound of wax.

Manila paper is normally a cream color but darkens when dipped in paraffin. The same is true of kraft paper, and the change in shade is caused by the oil in the paraffin. Nitroglycerin is also oily. Before paraffined shell paper came into use, linseed oil was applied to rag paper for waterproofing.

There is no glue involved in the making of the standard-size dynamite cartridge. The paper forming the tube is rolled around a rod three times and then folded or crimped closed at one end. This forms a tube, the closed end of which is the bottom, into which the explosive mixture is loaded. The tube is then folded shut over the top (the other end). With rubbery gelatin mixtures, the shell is wrapped around an already-extruded cylinder of dynamite. The spiral or "convolute" winding of the paper means that there is only one corner to fold in at each end of the tube. This kind of self-closure is much more secure than one with a square sheet of paper and a single overlapping seam extending parallel the length of the tube. The shape of an empty, unrolled shell is a parallelogram or a rectangle with one angled edge, ensuring that the outer seam is diagonal rather than parallel with the length of the cartridge.

Thicker shells were employed for underwater blasting and for underwater seismic prospecting. Usually, a thicker cartridge was made with more wraps of thicker paper and additional moisture proofing in the form of paraffin. Cardboard tubes typically were used only for large-diameter (over 2") cartridges. A special, large shell designated the ICC 23-G shell was so rigid that it was approved in 1946 as its own shipping container as long as it conformed to the general maximum weight requirements, which allowed for a gross weight of 65 pounds.

Many different designs of the basic shell were patented, with most offering some minor improvements such as small slits or cutouts to facilitate twisting, folding, or crimping the cartridge ends closed. Improved papers were developed as well, both to increase moisture resistance and reduce fumes. Specialty shells were made of hard plastic for submarine work and seismic blasting beginning in the 1950s. Many modern cartridged explosives are plastic-wrapped in chubs.

Because dynamite cartridges (except for primers) were often slit prior to loading into a hole, companies began offering perforated or preslit cartridges in the 1940s. According to the 1958 DuPont *Blasters' Handbook*, the company's model of special shell was "provided with several rows of perforations, the number, location, and character of which have been carefully determined. The object of this design is to allow the cartridge to be broken easily in the borehole with the tamping stick but still be able to withstand normal handling before loading."[10] Slit or perforated cartridges were easier to expand via tamping. Filling the borehole wall to wall made for a much more efficient explosion than a borehole with air spaces between the cartridges and the wall of the hole. Another advantage of these shells was reduced skin contact with the internal explosives, since sticks did not have to be slit by hand.

While DuPont made perforated shells with visible surface perforations,

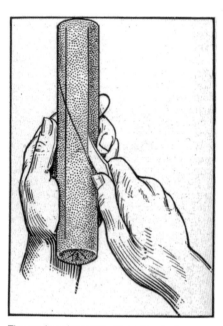

The main advantage of perforated or preslit cartridges is the time saved because cartridges no longer had to be slit before loading. Atlas Powder, *Better Farming with Atlas Powder* (Wilmington, DE: Atlas Powder, 1919), 48.

According to Claude Barker's patent application for this perforated shell, "The utilization of explosive cartridges in accordance with the invention makes it possible to load them into bore-holes and by comparatively mild tamping thrusts, to cause said cartridge to rupture along the lines of the perforations." *Author*

This metal die for printing shells was taken out of service because of wear and deformation. Note the misshapen "H" and "E." The die measures 3¾" × 1⅛". *Author*

Atlas and Hercules produced shells with slits on the inner wraps, sufficiently weakening the shell but offering more moisture resistance (later DuPont shells also had slits on the inner wraps). Atlas called its preslit cartridges Redi-Slit, while Hercules sold its cartridges under the name Tamptite. The Tamptite brand, introduced in 1942, was inherited by IRECO and Dyno Nobel and is still produced.

Unmarked cartridges were common through the 1870s. The first marked shells were hand-stamped with the company name and the lettered or numbered grade of dynamite prior to filling with explosive mixture. Letterpress and other direct lithography methods were soon used, and by the late 1910s the largest companies were employing web presses to imprint rolls of shell paper that were fed into cutting and forming machines.

US dynamite producers never applied lavish graphics or multicolor printing to their shells, with even two-color shells being exceedingly rare (European and Canadian shells were somewhat fancier). This was surely to keep costs low and to make the product less attractive to handling by the uninitiated. In 1913 DuPont was the first to mark its cartridges with the word "danger." The Institute for Makers of Explosives recommended that its members follow suit in 1914. Giant Powder Company never emblazoned its sticks with "danger," and Atlas Powder Company did so only on later shells.

In the 1890s, in response to export requirements, suppliers began stamping their cartridges with the date of production. Individual states started mandating dates on cartridges used within their confines, beginning with Arizona in 1902, followed by Colorado in 1903. The *Journals of the 21st Legislative Assembly of the Territory of Arizona* in 1901 decreed that

any person who shall, after January 1, 1902, knowingly sell or have in his possession any dynamite, nitroglycerin, or other highly explosive material, or any fuse, or who shall cause the same to be transported from point to point in this Territory, without having plainly marked in large letters, in a conspicuous place on the box or package containing such explosive material, the name and explosive characteristic thereof, and without having plainly marked upon the wrapper of each stick of dynamite or other explosive material, or package of fuse, the date of the manufacture thereof, is guilty of a misdemeanor.[11]

In 1970 the Institute of Makers of Explosives finally recommended that a formal "Date/Plant/Shift Code" be placed on each cartridge. While a standard format for the code was prescribed, in practice the arrangement of information varied from manufacturer to manufacturer. The recommendation was adopted into law and currently requires that the date, location, and shift number appear on all cases, bags, and cartridges of explosives of all kinds. No specific format is dictated. Applying dates and other specific identifying information to cartridges complicates

Word "Danger" On Every Cartridge

N line with our established policy to promoting safety in the use of explosives, we will, at an early date, print the word "DANGER" prominently on every cartridge of dynamite made by us.

In doing this our aim is to protect all as far as possible and particularly those not familiar with the appearance of a dynamite cartridge nor the precaution to be observed in handling it.

Such marking does not mean that our dynamites are any more dangerous than heretofore or more than competitive explosives not so marked.

The millions of pounds of our dynamite used without accident testify to our constant effort to increase their safety.

Such marking, we fell, will promote careful handling because it will indicate t o the general public as well as the regular user that there is an element of danger in handling cartridges of

The word "Danger" was directed mostly toward laymen rather than users of dynamite. Accidents involving dynamite and nonusers were rare because dynamite is difficult to detonate without the right tools. "Word 'Danger' on Every Cartridge," *DuPont Magazine*, February 1914, 4.

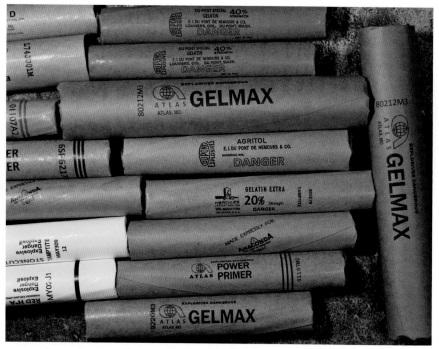

Dates on cartridges made it easier to rotate in fresh stock. Many jurisdictions prohibited the use of dynamite that was more than one year old. Both federal and local laws now also require strict control of explosives through detailed logs recording the dates of receipt and use of the contents of a company's magazines. *Author*

production because the manufacture of shells must be matched to a single day's output of dynamite.

The markings make it easier for explosives producers to track quality control and logistical issues. Another advantage is that explosives used in crimes are easier to trace. Toward that end, a proposal to put microscopic "taggants" in all explosives was considered in the 1970s but was deemed impractical.

Even before 1970, many cartridges were marked with the location of the plant where the dynamite was produced. Occasionally, custom-printed wrappers were provided for explosives destined for a specific large consumer, as with the Anaconda Copper Mine near Butte, Montana.

The dynamite shell is not as simple as it might seem. According to the 1980 *Blasters' Handbook*, "The overall weight-percentage and quantity and type

In the 1890s it was common for women, and even girls as young as twelve, to assemble dynamite cartridges. This was also true of blasting-cap factories. "Manufacture of Dynamite at Isleton," *Illustrated London News*, January 7, 1893, 13.

Making and packing dynamite in such close quarters is an invitation for disaster. The entrepreneurial spirit of the small producer was no match for the dangers of producing high explosives. Many a burgeoning business ended in a deadly explosion. "Manufacture of Dynamite at Isleton," *Illustrated London News*, January 7, 1893, 13.

of wrapper have an important influence on the dynamite's fume production, water resistance, and loadability."[12] Fumes are perhaps the main reason why most shells were their natural brown colors. The composition of the approximately 70 square inches of paper-wrapped surface of a standard cartridge was analyzed as part of the process for certifying a coal-mining explosive. At an average of 12.4 grams (including paraffin) per stick, the weight added up quickly with larger blasts. More chemicals involved in dyeing the paper meant more of a potential for undesirable gases produced during an explosion.

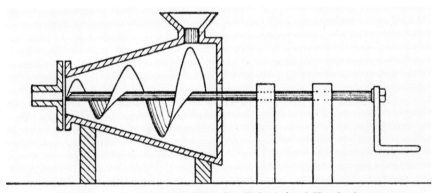

FIG. 19.—Sausage Machine for Gelatinized Explosives.

This Quinan shell-filling machine really does resemble a sausage maker and worked in a similar fashion. The explosive mixture was loaded into the top and pushed through a tube of the desired diameter. E. Barnett, *Explosives* (New York: D. Van Nostrand, 1919), 102.

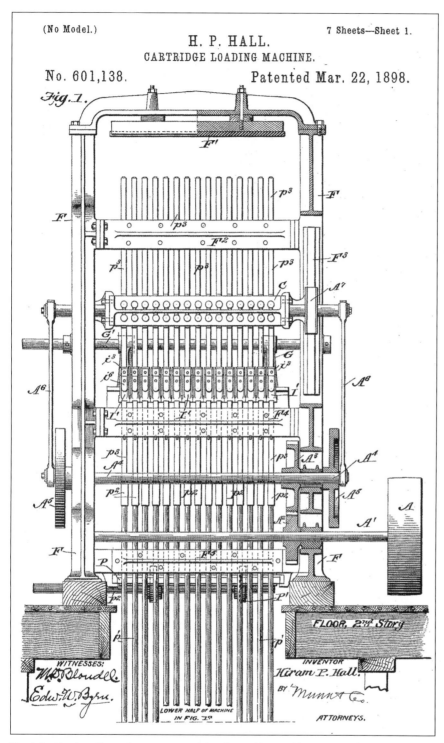

(No Model.)   7 Sheets—Sheet 1.

H. P. HALL.
CARTRIDGE LOADING MACHINE.

No. 601,138.   Patented Mar. 22, 1898.

Fig. 1.

WITNESSES: INVENTOR
M. P. Bloudll.   Hiram P. Hall.
Edw. W. Byrn.   BY Munn & Co.
                                                ATTORNEYS.

LOWER HALF OF MACHINE
IN FIG. 1.

The late nineteenth century saw a variety of cartridge-making contraptions, the most popular of which was the Hall Packer, which remained in use through the 1920s. H. P. Hall, Cartridge loading machine, US Patent 601,138, issued March 22, 1898.

Despite the introduction of shell-making and shell-filling devices, many small nineteenth-century dynamite operations remained at cottage level for their entire short existences. Chapter 6 relates the history of these minor but nonetheless fascinating players. DuPont engineered its own mechanized, assembly-line-style production at its plants beginning in 1906. Modern production of all types of explosives is highly mechanized and is controlled and monitored with the aid of advanced technology. This includes all manner of computer integration to govern ingredient proportions, product weight, and timely execution throughout the production and delivery processes.

Dynamite Boxes

In the early twentieth century, a combination of new government regulations and coordination among manufacturers led to the standardization of containers for all manners of goods. For example, cantaloupe boxes came in six different standard sizes, each intended to hold a specific number of standard-size melons. The goal was to increase efficiency and simplify the coordination of transportation and storage.

The lock-corner dynamite box was originally referred to in the trade as Style Number Five, and by the 1920s Style Number Six. While cases came in many sizes, dynamite producers were restricted by law to using smaller boxes suitable for approved loads, which limited a single container to 65 pounds gross weight. The standard net weight per box was 50 pounds, and 10- and 25-pound cases were used much less frequently. The number of sticks per box depended on the diameter and length of the cartridges, as well as the density of the explosive mixture.

Trade journals documented the development of every kind of product packaging imaginable and reveal that the Style Six boxes were to be constructed of ½-inch-thick single-board sides and ends, while the top and bottom

Commercial dynamite cartridges were assembled completely by hand until 1883, when the first cartridge-filling device appeared in the form of the sausage-maker-like Quinan machine. Before, granular mixtures were spooned into preformed cartridges, while rubbery gelatin-dynamite mixtures were extruded through a tube of the required diameter. These sticks were then cut and wrapped by hand with hand-stamped wrappers.

The lock-corner wooden box is an ancient invention that was made by hand until the 1870s, when the proper notching saws were perfected. *Author*

The standard box styles were devised by the National Association of Wooden Box Manufacturers in conjunction with the US Department of Agriculture's Forest Products Research Laboratory. US Department of Commerce, *Modern Export Packaging* (Washington, DC, USGPO, 1940), 16.

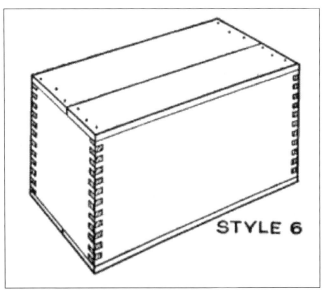

STYLE 6

Smaller containers such as this 25-pound box were intended more for farmers than miners. *Author*

could be composed of multiple boards. According to Eric Twitty, in his *Blown to Bits in the Mine*, most dynamite makers made the switch from nailed to lock-corner boxes between 1904 and 1907. Twitty reports that the thickness of box components was also increased from as little as a "downright flimsy"[13] ¼" to a new industry standard of ½". Illustrations from the 1890s reveal that some producers such as Aetna Powder Company had already changed to using lock-corner boxes.

Beginning in 1910, dynamite boxes were tested at the US Department of Agriculture's Forest Products Laboratory in Madison, Wisconsin. Cases with glued lock corners generally proved to be more rigid than those with nailed corners. Nailed cases were required either to be made partially of thicker wood or, starting in 1927, to have additional supports called cleats. Screws were used to close the earliest boxes. Nailing machines, first developed in the 1840s, were soon the tool of choice.

Early dynamite producers purchased boxes. Some small concerns obtained boxes from outside sources for their entire existence. Larger companies acquired boxes from local manufacturers until an on-site box plant was built. *Barrel and Box*, in November 1906, reported that "the Eastern Dynamite Co. has purchased the business and plant of the Sebago Box. Co. at

Sebago Lake, for a consideration of $30,000. The company will manufacture its own boxes for the shipment of dynamite, instead of purchasing them from outside parties, as had been the practice heretofore."[14] Sebago Lake is in southern Maine, northwest of Portland.

The Eastern Dynamite Company was actually a holding company operated by DuPont. By 1925, another article related that DuPont dynamite "is packed in boxes made at its own plants from logs cut from its own timberlands."[15] The woods most often used were pine,

This box-printing die measures 7" × 1¼". Made of brass, a die such as this was good for up to 500,000 impressions. *Author*

The largest dynamite makers such as DuPont and Hercules Powder made their own dies in-house. This one measures 6" × ⅞". *Author*

Brass dies of this type were mounted on rotary presses. Wooden box components are fed through just like a basic sheet-fed printing press. The main difference is that dies for printing on wood directly contact the wood. This die measures 7½" × 3⅜". *Author*

container's construction. The rules regarding wooden dynamite boxes were particularly detailed, specifying the type and dimensions of wood, maximum load capacity per box, packing, and even the type of nails to be used. Once fiberboard containers were approved, they were regulated as well, with particularly strict requirements for strength when stacked.

Government regulations required the following

> All boxes in which high explosives in cartridges, bags, sacks, or in bulk are packed must be lined with strong paraffined paper or other suitable material. The lining must be without joints or other openings at the bottom or on the sides of the box, and must be impervious to water and to any liquid ingredient of the explosive. In packing cartridges of nitroglycerin explosives at least one-quarter of an inch thickness of dry, fine wood pulp or saw-dust must be spread over the bottom of the lined box before inserting the cartridges, and all the vacant space in the top must be filled with this material.[16]

The wood dust at the bottom of cases served both as a cushion and an absorbent for leaks. Sawdust or excelsior was also used to fill empty space in cases. Many producers used a top covering sheet of kraft paper to provide instructions and advertisements. Aetna Powder Company and others packaged cartridges within two or four cardboard boxes enclosed by a standard wooden case.

Reusing empty dynamite boxes was extremely common but potentially dangerous. Nitroglycerin could leak from cartridges and be absorbed into the wood, creating the remote possibility of an explosion. In 1916, DuPont, in its *DuPont Magazine*, published an article titled "A Dynamite Birdhouse," which related how to make a birdhouse from an old dynamite box.[17] In the next issue of the magazine, DuPont issued what was essentially a retraction, admitting that the idea was "contrary to safety rules. Owing to the possibility of

spruce, hemlock, and cedar, with cheap pine far in the lead.

Two methods were used to mark wooden cases with producer name and content information. Initially, simple stencils were employed. In the 1880s a variety of machines became available for direct lithography from metal dies, producing an embossed image. (Plates for printing on paper leave an impression in ink on another cylinder,

which then transfers the image to the paper.) Box sides, ends, and tops were imprinted before assembly. Two-color images were produced either by running a box piece (called a shook) through a press twice or once through a press with multiple printing cylinders.

The Interstate Commerce Commission (ICC) was established in 1887, and by 1914 ICC regulations covered every aspect of a shipping

Lining boxes with waxed paper helped contain leaks. *Author*

nitroglycerin leakage due to improper storage, there is danger in reusing an empty dynamite box."[18] The US Bureau of Mines advised exercising extreme care in giving away empty dynamite cases and recommended that each one be thoroughly inspected for nitroglycerin stains by "a knowledgeable person."[19]

Officially, all dynamite boxes in the United States were made of wood until 1931, when the ICC adopted paragraph 23-F of its *Regulations for the Transportation of Explosives and Other Dangerous Articles by Freight and Express*. This stipulation allowed for the use of fiberboard (cardboard) dynamite containers that met strict requirements for strength and water resistance.

The first fiberboard (then spelled fibreboard since they debuted in England) boxes were produced in 1817, and corrugated cardboard was patented in 1856. The material was approved in the United States for shipping most goods in 1914, and cardboard gradually replaced wooden boxes and crates for many applications (the term "crate" is reserved for larger wooden containers and those with open sides).

Incredibly, some careless miners dropped dynamite cases corner-first to quickly burst the boxes open. Others used shovels or picks. The correct method was to use a wooden mallet to drive a wooden wedge under the lid. *Author*

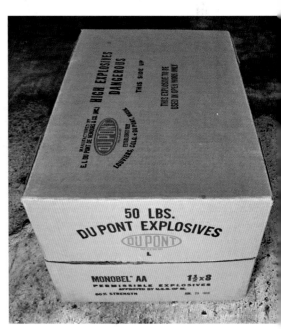

This half-telescope, fiberboard case is dated January 25, 1956. The corners are fastened with large brass staples. *Author*

According to the Institute of Makers of Explosives, the dynamite industry "began to investigate the use of fiberboard boxes"[20] in 1925. However, the material did not come into wide use for dynamite boxes until after World War II. Early in the conflict it was discovered that most fiberboard containers could not withstand the extremely moist conditions of the South Pacific. In response, improved, water-resistant, corrugated cardboards were developed. After the war, explosives producers began to gain confidence that they could ship their product to humid places.

The 1949 *DuPont Blasters' Handbook* reported that "marked improvement has been made in the quality of fiberboard, and an increasing amount of dynamite is being packed in these cases."[21] However, a 1949 article reported that Atlas Powder Company still used wooden boxes "as a precaution against moisture."[22] In 1958 the *Handbook* related that most boxes were now fiberboard, and that "the fiberboard case in most general use consists of only two pieces: a full cover and a bottom. This virtually constitutes a double container and eliminates the need for any additional support."[23] (This is the full-telescope style box.) By the 1966 edition of the *Handbook*, wooden boxes were no longer in use.

This full-telescope case has the corners reinforced with large brass staples. The box was made before Hercules Powder Company updated its logo in 1963. The box held 50 pounds of 2" × 16" sticks of 60 percent strength, gelatin-extra dynamite. *Author*

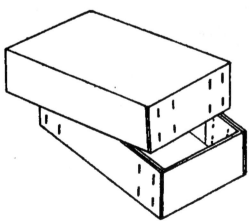

The "full-telescope" style of box was the most popular style of fiberboard dynamite case. C. A. Plaskett, *Principles of Box and Crate Construction* (Washington, DC: USGPO, 1930), 29.

DuPont gave the full-telescope design high marks for its ease of opening and reclosing, high compressive strength, and the fact that the top can be used as "an auxiliary box."[24] This type of case is now officially designated a "Full Telescopic Design Style Container," with standard codes of 0301, 0302, 0303, 0304, 0305, 0307, 0308, and 0309. The codes are developed by the International Corrugated Case Association, an industry trade group, and vary depending on construction elements of the box.

Fiberboard boxes were tested for compressive strength and moisture resistance. According to the ICC,

Fiberboard containers must be able to pass tests as follows:

Three samples must be subjected to and pass all tests. Containers for tests must be tightly packed with dummy contents similar in shape and weight to the explosives to be shipped therein. Tests must be conducted on samples immediately after exposure to at least two weeks to humidity of at least 90 percent at 75° F. Loaded containers must withstand at least 200 drops in the standard 7' revolving test drum with pointed hazard in place, before spilling of contents of containers occurs. Loaded containers must withstand an end[-]to[-]end pressure of at least 1,600 pounds without deflection of more than 1½". Empty containers must withstand top[-]to[-] bottom pressure of at least 500 pounds without deflection of more than ½".[25]

The latter rule meant that the bottom box of a stack of ten 50-pound cases suffers almost no deformation.

Fiberboard boxes for dynamite were lined either with waxed paper or, beginning in 1953, polyethylene sheeting. The introduction of this new type of interior lining contributed to the adoption of fiberboard cases because leaks from cartridges were contained by the waterproof plastic. Modern packing is in a large plastic bag within the cardboard box.

Some designs of cardboard cases needed double-walled bottoms to pass strength tests, but the full-telescope design met the requirements because the nesting pieces doubled the thickness

This austere, 1980s Atlas Powder Company box has double-thick sides and glued corners. The box held 50 pounds of 1¼" × 8" Atlas Farmex Ditching dynamite. *Author*

ICC SHIPPING CONTAINER
SPECIFICATION NO. 14.
WOODEN BOXES.

1. These boxes must comply with the following specifications:

CONSTRUCTION.

2. Must be made of good, sound white pine or any wood of equal or superior strength, dry and well seasoned, and with no loose knots or knots liable to get loose in any part.

3. When sides, ends, tops, or bottoms are made of more than one piece the pieces must be tongued and grooved and glued, and the joints in making up the boxes must be staggered.

4. All lock and dovetail corner joints must be glued.

5. Nails driven through sides, tops, and bottoms into ends must be at not greater than 3-inch centers for boxes not more than 12 inches in width and at not greater than 4-inch centers for boxes of width greater than 12 inches.

6. Nails driven through tops and bottoms into sides must be at not greater than 6-inch centers for boxes not more than 24 inches in length and at not greater than 8-inch centers for boxes of length more than 24 inches.

7. Gauge of nails used shall be not less than the following sizes, depending upon the thickness of lumber into which they are to be driven.
2-penny into 5/16-inch lumber.
2-penny into 3/8-inch lumber.
4-penny into 7/16 to 1/2 inch lumber.
5-penny into 9/16 to 5/8 inch lumber.
6-penny into 11/16 to 13/16 inch lumber.
7-penny into 7/8 inch or thicker lumber.
For example, nails driven through a 1/2-inch side into a 3/4-inch end must be 6-penny. Screws of equal efficiency may be used in place of nails.

8. When boxes are set up the bottom and lids must fit evenly on the frame.

9. Each box must be plainly marked with a symbol consisting of a rectangle, as follows:

ICC-14

The letters and figures in this symbol must be at least ½ inch high. This symbol shall be understood to certify that the package complies with all the requirements of this specification. When offered for shipment the package must also bear the wording prescribed by the regulations for the particular article contained therein.

10. Thickness of lumber in the finished box must not be less than the following;

BOX AND CONTENTS NOT OVER
75 POUNDS GROSS WEIGHT.

	ENDS	SIDES	TOP AND BOTTOM
For nailed boxes	7/8 inch	1/2 inch	1/2 inch
For lock or dovetail boxes	1/2 inch	1/2 inch	1/2 inch

11. When explosive material of equal weight is substituted (fine and dry sand for granular explosives, dummy cartridges for high explosive cartridges) for the contents of the package, and the outside package is dropped two times on its end to solid brick or concrete from a height of 4 feet, the outside package must not open or rupture or must any portion of the contents escape therefrom.

Regulations also required all dynamite cases to be marked "This Side Up" on the top. Shipping cartridges horizontally helped prevent leaks from the ends of cartridges. Interstate Commerce Commission, *ICC Regulations for the Transportation of Explosives and Other Dangerous Articles by Freight and Express* (Washington, DC: USGPO, 1914), 246.

of the sides. The quality of corrugated cardboard, lighter and cheaper than wood, continued to improve through the 1950s. An empty 50-pound-capacity wooden dynamite box weighs about 9 pounds, versus 2 to 3 pounds for a fiberboard box of the same capacity.

In addition to the full-telescoping box, a few other designs of cardboard cases were (and still are) used for shipping explosives. Boxes with "glue-sealed top flaps" were best opened "by sliding a wooden wedge under the flaps," according to the 1952 *Blasters' Handbook*.[26] These flaps are as large as the box, as opposed to the half flaps used for most other commodities. The cases have double walls or reinforced corners. Another design, the half telescope, is similar to the cases now used for boxing reams of copy paper. Both half and full telescope-style boxes were "half-sealed with tape. Such boxes can be opened very quickly because the tape may be stripped off with the fingers"[27] per the *Handbook*.

Regulations in 1909 stipulated that all wooden containers used for shipping dynamite be marked with the words "High Explosives – Dangerous" and, beginning in 1914, "ICC 14." "ICC 14" refers to the commission's "Shipping Container Specification No. 14," contained in its *Regulations for the Transportation of Explosives and Other Dangerous Articles by Freight and Express*.

All manufacturers recommended burning empty boxes, both wooden and fiberboard. Because small amounts of nitroglycerin were occasionally absorbed into a box, tiny explosions sometimes occurred. According to DuPont, such detonations were usually small and harmless but "can be dangerous however, at close range, such as when men might be warming themselves at the fire."[28] Later fiberboard boxes were marked "Do Not Reuse This Box."

It is a minor myth that lock-corner wooden cases were used to avoid the use of sharp, spark-producing nails. (The ICC regulations mention the term "dovetail," but dovetail corners have wedge-shaped "teeth," are trickier to assemble, and were not used by dynamite producers. The lock corner that

Note the difference in the sizes of these cases (although one is shown facing forward and one is lying sideways). The boxes for less dense explosives such as railroad powder are much taller. E. I. DuPont de Nemours, *High Explosives, Second Section—Kinds, Grades and Brands* (Wilmington, DE: E. I. DuPont de Nemours, 1915), 2.

was used is also called a box joint or finger joint.) The intermeshed corners were simply sturdier. An article in 1913 noted that the lock-corner box "is now used extensively, principally on account of its superior strength and neat appearance."[29] Fewer nails did mean that there were fewer chances for a cartridge to be damaged and its contents leaked.

It does not appear that the ICC explicitly mandated lock-corner cases. The regulations specified that lock-corner joints be glued (unglued joints failed in strength tests) and that the ends of nailed boxes be made of thicker wood. The regulations even stipulated what size of nails to use and their spacing, stating, "Nails driven through sides, tops and bottoms into ends must be at not greater than three-inch centers for boxes not more than twelve inches in width."[30] This statement directly relates to the lock-corner mandate issue. If nailed cases were banned,

then driving a nail through a box side into a box end would not be addressed in the 1914 regulations.

In 1927, boxes with cleated ends were allowed. The extra weight of the nails and additional bulk, weight, and cost of the required thicker boards or cleats made them less preferable. Consequently, only a few producers used cleated-end boxes. Many boxes from Illinois Powder Company are cleated, while only the occasional box from Atlas Powder Company and Hercules Powder Company came with cleats. DuPont very rarely used cleated boxes.

The edicts were constantly updated until the ICC was abolished in 1996. The Bureau of Alcohol, Tobacco, Firearms and Explosives now regulates the explosives industry. Current law requires that explosives producers mark cases and cartridges with the maker's name and the location, date, and shift when produced.

Dynamite Magazines

Regulations regarding the storage of dynamite were the purview of local jurisdictions until the passage of the Explosives Control Act of 1941. Oversight then became the job of the US Bureau of Mines. The Bureau of Alcohol, Tobacco, and Firearms (BATF) took over supervision and enforcement in 1972, with authorization provided by a new Explosives Control Act, part of the Organized Crime Control Act of 1970. The BATF, an arm of the US Treasury, was renamed the Bureau of Alcohol, Tobacco, Firearms and Explosives in 2002.

Current requirements for a safe and secure magazine fill several pages. Generally, an explosives storage building should be as remote as possible, be bulletproof, have proper drainage, and have a strong, locked door. Ample ventilation, access only by authorized persons, and regular inspection are other requirements. Magazines must be made of concrete, wood (with sand between double walls), or nonsparking metal.

Underground magazines usually consisted of a carved-out chamber or dead-end tunnel. Larger operations built wooden or tin structures to fit strategically dug rooms to provide locked, controlled entry. Predictably, many miners took no precautions whatsoever regarding storage. Explosives were kept in mining offices, miners' homes, parked wagons, and working tunnels. Cases of dynamite were often simply stacked near the main entrance of the mine.

Before modern federal oversight began in the 1940s, the US Bureau of Mines and the private Institute of Makers of Explosives (IME) provided guidance for the construction and proper operation of explosives storage facilities. The larger producers also offered instructions, blueprints, and hands-on training.

Many recommendations on the IME's list were pure common sense, including, "Do not throw dynamite

MAGAZINE RULES FOR DYNAMITE AND POWDER MAGAZINE.

Recommended by Institute of Makers of Explosives.

1. Store only dynamite and powder in this magazine. Do not store blasting caps or electric blasting caps, inflammables, metal tools or other implements in this magazine.

2. Explosives should be handled carefully.

3. Store dynamite boxes flat, top side up. Store powder kegs on ends (bungs down) or on sides (seams down). Corresponding grades and brands should be stored together in such manner that brand and grade marks will show. All stocks should be stored so as to be easily counted and checked and so the oldest stocks can be shipped, delivered or used first.

4. Always ship, deliver or use oldest stocks first.

5. Do not throw packages of explosives violently down or slide them along the floor or over each other or handle them roughly in any manner.

6. Do not open packages of explosives or pack or repack explosives in a magazine or within 50 feet of a magazine.

7. Use a wooden sledge and mallet in opening or closing packages of explosives. Open powder kegs by removing the slide or unscrewing the (top) bung,

8. Do not use metal bale hooks in handling, or metal tools to open packages of explosives.

9. Do not have loose dynamite or powder in this magazine.

10. If artificial light is needed use only an electric flash light or electric lantern. Do not use oil-burning or chemical lamps, lanterns or candles in or around this magazine.

11. Do not carry or allow others to carry matches or smoke in or near this magazine.

12. Do not allow shooting or allow anyone to have firearms or cartridges in or near this magazine.

13. Keep this magazine clean.

14. If leak develops in magazine roof or walls, repair it at once.

15. Keep ground around magazine clear of leaves, grass, trash, stumps or debris to prevent fire reaching it.

16. Do not allow unauthorized persons in or near magazine.

17. Keep constant watch for broken, leaky or defective packages.

18. Do not use emptied dynamite cases or powder kegs.

19. If any packages of dynamite or powder are received in leaky or damaged condition, put packages to one side in magazine and make full report in detail to manufacturer, giving probable cause of damage.

20. Powder kegs should be thoroughly shaken by hand sufficiently often to prevent caking. Don't knock against floor or against each other.

21. Keep door of this magazine securely locked when not engaged in it.

Note. (It is suggested copy of above rules be posted in magazine used for storing dynamite and powder for guidance of persons in charge of magazine.)

While these suggestions from 1920 are inadequate by today's standards, the core safety procedures prescribed certainly helped save lives. Adapted from Institute of Makers of Explosives, *Standard Storage Magazines Recommended by the Institute of Makers of Explosives* (New York: IME, 1920), 14.

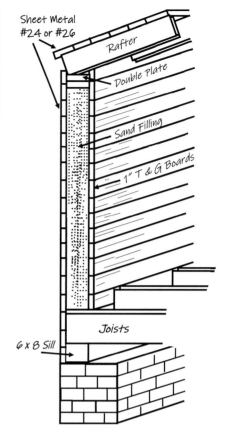

Large, sheet-metal magazines required elevation off the ground and sand-filled walls. Adapted from John Cosgrove, *Rock Excavation and Blasting* (Pittsburgh, PA: National Fireproofing, 1913), 176.

boxes violently down or slide them along the floor or over each other or handle them roughly in any manner."[31]

Large companies sold ready-to-assemble steel magazines. In 1957, Atlas Powder offered five different portable box magazines, three portable field storage magazines, and seventeen sizes of semipermanent magazines. Capacities ranged from a single 50-pound case to 1,650 cases, and even "larger houses" were "available upon special order."[32]

Stacking cases of dynamite more than ten high was not recommended. Even so, this meant that the bottom case of the stack was subject to almost 500 pounds of pressure.

The 1960s-era magazine at the Hoover Dam typifies the permanent, bunker-style storage facilities of larger

This small, portable, sheet-metal magazine is correctly elevated by a brick foundation. The precaution is an obvious one and avoids damage to the explosives from rainwater. Institute of Makers of Explosives, *Standard Storage Magazines Recommended by the Institute of Makers of Explosives* (New York: IME, 1920), 14.

A magazine at the Hoover Dam. Construction of the dam consumed more than 8,500,000 pounds of dynamite. *Library of Congress Prints and Photographs Online Catalog*, HAER NV-27-V-2.

operations. Blasting contractors had compelling incentives to build safe and secure explosives magazines. The contract for explosive work at the dam stated that the contractor was "liable for all injuries to or deaths of persons and damage to property caused by blasts or explosives."[33]

Embedding a magazine into a hillside was a marvelous option if available, because three sides of the structure were absolutely protected. Concrete or steel magazines were sometimes partially buried. Cut-and-cover magazines were emplaced in flat ground with access sloping down from ground level to the structure. Another common practice was to build dirt hills between multiple magazines.

Many early mining catastrophes were caused by unsafely storing large quantities of explosives underground. Fires ignited by miners' lamps, candles, and cigarettes could create the concentrated heat needed to explode dynamite. Thawing dynamite underground was

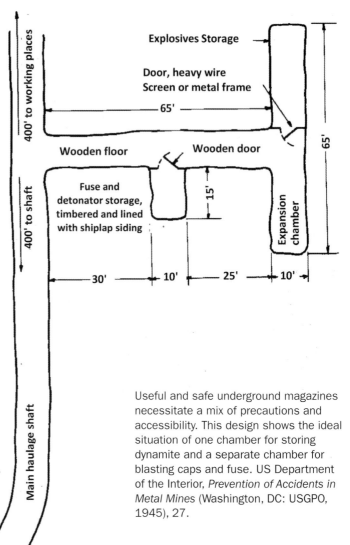

Useful and safe underground magazines necessitate a mix of precautions and accessibility. This design shows the ideal situation of one chamber for storing dynamite and a separate chamber for blasting caps and fuse. US Department of the Interior, *Prevention of Accidents in Metal Mines* (Washington, DC: USGPO, 1945), 27.

particularly hazardous but occurred on a regular basis.

Most early underground magazines were not well thought out, and convenience and cost concerns often superseded safety measures. Current government regulations applying to underground storage of explosives govern the distance between magazines for blasting caps and explosives, and also the distance to other potential hazards such as electrical wires and ore trolley ways.

A standard table of distances was introduced by the explosives industry in 1909. The table prescribes safe distances between explosives storage buildings and highways, railways, inhabited structures, and other magazines. For example, two freestanding magazines containing 1,000 pounds of explosives each should be separated by 78 feet. If natural or artificial barriers are between the magazines, the distance is cut in half. Larger operations build their own barriers by dumping fill dirt around magazines. Sometimes the buildings are entirely surrounded by purpose-made berms. The standard table of distances has been updated over the years and is still in use today.

Dynamite Transportation

Early black-powder mills relied on waterways both for power and to transport explosives out from the mill. When railway systems were completed, transportation by rail became essential, and nearly every dynamite plant laid a spur of tracks from the plant to the main line. Coal-fired locomotives were not allowed on the spur near the plant, so horse- or mule-drawn railcars and man-powered pushcarts were initially used to move the dynamite a safe distance away. In the 1890s, fireless locomotives were developed that used compressed steam or hot air. These were charged at a station a prudent distance from the plant and had a range of about 5 miles.

This photograph was originally captioned "A large magazine for explosives with barricade to prevent fragments of magazine from damaging adjacent structures on the occurrence of an explosion." E. I. DuPont de Nemours, *High Explosives, First Section* (Wilmington, DE: E. I. DuPont de Nemours, 1925), 66.

Placards like this were produced privately by railroad companies but were required to have specific, government-mandated verbiage. One such placard was to be on all four sides of every railcar laden with explosives. This one measures 14" × 16". *Author*

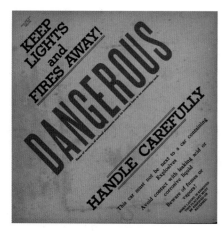

This placard was posted on a railcar carrying hazardous materials other than explosives, including flammable liquids and items such as fireworks and matches. This one measures 14" × 14", bears a date of February 1953, and was produced and distributed by the Interstate Commerce Commission. *Author*

Because of the instability of early straight dynamites, dozens of disasters occurred during the loading and transport of dynamite by rail, and many railroad companies refused to carry high explosives into the 1890s. In 1907 the American Railway Association, an independent trade group, formed its own Bureau of Explosives and issued rules for the safe transportation of dynamite by rail. These were incorporated in the Interstate Commerce Commission's first list of rules covering the transportation of explosives, published in 1908.

As with dynamite boxes, the regulations were extensive. Dynamite could not be transported on a railcar adjacent to a passenger car or next to the engine or caboose. A single car could not carry more than 70,000 pounds of explosives. The interior of the car was to be of nonsparking material. Any protruding bolts on the walls and floors were to be covered.

Sparks from the wheels entering a railcar full of explosives could start a catastrophic fire, so doors were to be tight and have stripping. In addition, the roof of the car could not be prone to ignition from sparks. Cars were required to be placarded with specific warnings and were never to be left unattended while stationary.

The rules worked, and railway accidents involving dynamite were virtually unknown from the 1920s on.

Many producers had their own fleets of railcars complete with engines specially designed for the safe movement of high explosives. In the 1960s, DuPont had almost five thousand cars and Hercules about two thousand. While most of these particular cars were for products other than dynamite, in 1939 one Bureau of Transportation report estimated that there were five thousand railcars containing explosives on the tracks at any given time in the United States and Canada. Another bureau report that same year related that 500,000,000 pounds of explosives of all types were safely transported by rail in the United States in 1938.

When possible, mines of any size built their own rail spurs to move material to the main line, and to receive shipments of explosives and other supplies. Mules, wagons, and then trucks were used to reach remote mines and to stock resellers such as hardware stores. Until the ICC issued its first rules in 1908, state and municipal governments regulated the transportation, storage, and use of dynamite within their respective bailiwicks. These laws were sometimes very detailed, sometimes vague, and sometimes nonexistent.

DuPont reported on the development of fireless gasoline trucks in 1913. Here "fireless" means fireproof. The trucks had conventional engines but featured a special compartment for carrying explosives. Trucks of any kind were not routinely employed to carry dynamite until the 1920s. Combining gas-burning vehicles and explosives was still generally deemed too dangerous, and horse-drawn wagons and electric vehicles were used.

Motorized trucks were heavily regulated by the ICC, beginning with its Transportation of Explosives Act of 1921. Most of the edicts were born of common sense. The interior of a truck must be made of nonsparking material. Blasting caps cannot be transported with dynamite, and explosives cannot be transported with any other cargo. The dynamite must be kept in its original containers. Trucks must be plainly marked, cannot be left unattended, and cannot be serviced or repaired while loaded.

The Pipeline and Hazardous Materials Safety Administration, a division of the US Department of Transportation, currently oversees the transportation of explosives within the United States.

The largest producers not only had their own railcars but operated their own terminals. This was especially true in company towns. *Author*

Note the recommendation that a "stuffed mattress" be used as a safety measure. US Interstate Commerce Commission, *ICC Regulations for the Transportation of Explosives and Other Dangerous Materials* (Washington, DC: USGPO, 1914), 18, 19, 26.

When not in transit, explosives are regulated by the Bureau of Alcohol, Tobacco, Firearms and Explosives.

Current regulations prohibit transport of dynamite by air without obtaining a US Department of Transportation Explosives Special Permit. Reasons for obtaining such a permit include avalanche-control flights and ferrying explosives to remote places in Alaska. A few small air carriers such as Phoenix Air Group Inc. specialize in the transport of dangerous goods, including explosives.

Digging Deeper

Dynamite Cartridges

At least one explosives expert thought that the large number of choices regarding dynamite was excessive, opining this in 1922:

The one thing we think that would help the efficiency of the explosives industry in the United States today is the standardization of the product. There is nothing new or novel about this statement, and several attempts have been started in the past to bring about a reduction in the almost numberless grades, sizes, styles, etc., in which explosives are placed on the market. The fact is, however, that today there is a greater variety of explosives offered to the trade than is needed or necessary. We very much question if any practical user can tell any difference between a 17 percent dynamite and a 20 percent dynamite, or a 27 percent dynamite and a 30 percent dynamite in actual use. Furthermore, we find straight dynamite, ammonia dynamites, high-freezing and low-freezing dynamites, permissible dynamites, both high and low freezing, straight gelatin dynamites, high and low freezing, blasting gelatin, etc.[34]

Remarkably, the above was written by G. M. Norman, then technical director of Hercules Powder Company, who went on to lament, "Who is responsible for this condition—the salesman or the customer—and how it can be remedied we are unable to say."[35] It is true that there isn't much difference between a 17 percent strength straight dynamite and a 20 percent strength straight dynamite. However, permissibles saved countless lives, and essentially equating ammonia dynamite with blasting gelatin is a very strange position to take, especially for an expert. As DuPont put it, "Coal is softer than granite. An open exposed quarry face presents an entirely different situation than the heading of a tunnel a thousand or more feet underground."[36] The fact is, producers produced and consumers consumed, indicating that offering an expansive spectrum of dynamite options was sound business.

This is a thoroughly insulated railcar designed for moving explosives from the main rail line to the mine. Metal parts such as the coupler and bumpers were covered with rubber to prevent static electricity from entering the car. US Bureau of Mines, *Explosive Accidents in Bituminous Coal Mines* (Washington, DC: USGPO, 1955), 62.

By the 1970s, all remaining major explosives producers operated their own trucking divisions.

This image was originally captioned "A Dupont Powder Wagon in Colorado" and dates from 1918. E. I. DuPont de Nemours, *High Explosives, Volume One* (Wilmington, DE: E. I. DuPont de Nemours, 1918), 19.

While some of the variety could be characterized as superfluous, most was the result of extensive research. DuPont developed a special dynamite for salt mining that would not stain the salt, extremely slow-acting dynamites formulated specifically as an alternative to black powder, and special pressure-resistant gelatins for seismic prospecting underwater. Low-freezing dynamites and permissibles in particular were certainly not advertising gimmicks.

The earliest references I could find explicitly listing 8" as the "standard" length for cartridges are from the mid-1870s. My own theory for the 8" length is that for the earliest crumbly dynamite mixtures, anything much longer might tend to bend when held in the hand.

By 1961, large-diameter cartridges had become so prevalent that some brands of dynamite were not even sold in small diameters. An ad that year by Hercules boasted, "The Hercol series of low-cost, cap-sensitive, high-ammonia-content explosives are now available in cartridges of small diameter and standard lengths, as well as in the large diameter and special packages."[37] The rationale was making Hercol also "available to the underground operator."[38]

Here are some historical spot prices per pound for 40 percent strength dynamite. The figures come from many sources and are for (rail)carload quantities of 20,000 pounds or more. In most cases the type of dynamite (straight, gelatin, etc.) was not indicated, but we can assume straight dynamite for the earliest figures. In some later cases the figure is an average of all types sold. Prices are per pound. (Not enough data are available for a nice, smooth chart. Figures for later years are particularly elusive.):

1868, $1.50; 1873, 90¢; 1877, 60¢; 1882, 40¢; 1883, 30¢; 1890, 14¢; 1892, 13¢; 1893, 12¢; 1895 20¢; 1897 16¢; 1898, 15¢; 1905, 16¢; 1913, 11¢; 1914, 14¢; 1915, 22¢; 1917, 19¢; 1918, 25¢; 1919, 33¢; 1920, 15¢; 1921, 17¢; 1922, 25¢; 1924, 23¢; 1936, 12¢; 1937, 15¢ 1940, 15¢; 1945, 12¢; 1950, 24¢; 1954, 13¢;

1959, 22¢; 1965, 30¢; 1968, 20¢; 1970, 21¢, 1971, 25¢; 1978, 40¢; 1980, 49¢; 1981, 48¢, 1991, $1.10; 2000, $3.00; 2021, $10.00

There was actually always a wider range of prices owing to all the varieties available. For example, in 1921, Hercules 40 percent strength gelatin sold for 17.5 cents a pound, while Hercules Red H permissibles were bargain-priced at only five cents a pound.

Dynamite Mixtures

An article in 1890 gave this description of operating a gelatin dynamite machine:

> The girl who works the machine lifts a lump of gelatine and pushes it into the opening with her left hand, at the same time turning the handle with her right hand. The threads of the screw push the gelatine rapidly forward to the nozzle, and in a few revolutions the whole of the gelatine is converted into a long rope. When the girl has expelled a yard or two of this rope from the machine, she seizes it with her left hand, and with a wooden knife in her right quickly cuts it into cartridge lengths, which are taken up by other girls and rolled in cartridge paper.[39]

A 1949 patent application revealed that cartridges with diameters of 8" and larger were still filled by hand at that time.

Before more-modern temperature-measuring methods, monitoring the temperature of a vat of nitroglycerin was an important but tedious job, basically consisting of staring at a thermometer. To prevent workers in this capacity from dozing off, one-legged stools were the only manner of seating allowed!

A 2016 article related that at the Apache Powder Company, "Onsite movement of dynamite involved the Angel Buggy, which could carry up to 400 pounds of nitroglycerin. The vehicle looks like a barbecue grill on wheels and was aptly named. That's because any bump in the road causing sudden movement might set off an explosion, and the driver would become an angel."[40] Despite maximum caution, accidents still occurred on occasion. In 1954, a newspaper reported that two men were killed pushing a "rubber-tired buggy"[41] filled with nitroglycerin at the Austin Powder dynamite plant in McArthur, Ohio.

Absorbents and fillers throughout the years included wood pulp, sugar, nut meal (ground whole nuts), cornmeal, sponge, plaster of paris, charcoal, ground nutshells, and rice hulls.

Dynamite has a usable shelf life of two to four years if properly stored (many states prohibit the sale or use of dynamite more than one year old). Beyond that time it will begin to discolor and change in texture and hardness. White crystals often form on the wrappers of very old dynamite. Contrary to popular belief, these are not crystals of nitroglycerin but are other ingredients that have leached through the wrapper and crystallized in the air. While the crystals themselves are not dangerous, they are a sign that the dynamite has begun to decompose and could potentially be unstable.

Old dynamite is usually burned for disposal. The sticks are first laid out in a single layer and doused with kerosene or diesel fuel. This desensitizes the nitroglycerin. The dynamite is then ignited, and the bonfire is monitored from a very safe distance. Because dynamite becomes markedly more sensitive when heated, many dynamite explosions were actually caused by falling fragments of flaming sticks from stacks of burning dynamite. Even a short fall was enough shock to cause detonation.

Dynamite Shells

The *Dictionary of Paper*, published by the American Paper Institute, defines "Dynamite-Shell Paper" as "a well-formed, highly finished sulfite or kraft sheet, usually in basis weights of sixty and ninety pounds to be converted into tubes for packaging of dynamite, powder, etc. Tensile and tearing strength and moisture resistance, as well as finish and uniformity, are important characteristics."[42] Sulfite and kraft papers are named after the processes used to produce the wood pulp from which the papers are made.

According to *Black Powder and Hand Steel: Miners and Machines on the Old West Frontier*, written by Otis Young and Robert Lenon and published by the University of Oklahoma in 1973, putting up blasting powder (then potassium-nitrate gunpowder) in cartridges was devised by German bookbinder Hans Luft in 1687. The book also relates that the concept of stemming had come from German mine supervisor Carl Zumbe two years earlier. Paper at that time was made out of old rags and other scrap textiles, and of pressed plant matter such as straw.

In 1952 the Crown Zellerbach Company related that its plant in Lebanon, New York, employing two hundred men, devoted 40 percent of its production capacity to making paper for dynamite shells. Other notable makers of dynamite cartridge paper included Hollingsworth & Whitney Company of Alabama, Knowlton Brothers of New York, and Minosink Paper Company and Anolomink Paper Company, both of Pennsylvania.

Numerous accounts have appeared in newspapers detailing the hazards of dynamite wrappers to livestock. When wrappers were removed for shaping charges into odd spaces, the wrappers were supposed to be burned. When these were carelessly discarded in a field, cattle were attracted to the sweet taste of nitroglycerin, often with fatal results.

While some countries such as Poland at one time required stripes on permissibles and even reserved certain colored shells for certain dynamites, the United States never had any laws regarding shell colors or stripes.

In 2014, Dyno Nobel conducted a cost-benefit analysis of printing black-and-red stripes on its permissible dynamites and other explosives. The vagaries of printing, cutting, and rolling convolute shells meant that the stripes had to be printed on the paper separately. The company calculated the number of shells wasted due to misprinting of the stripes, and added in the shells consumed during routine reinking of the rollers, not to mention finished product that was rejected at inspection due to smeared or crooked stripes. Dyno Nobel concluded that there were no benefits to printing stripes on shells, and reportedly decided to discontinue the practice for a savings of about $1,000 a week. However, current product data sheets have images with striped shells. This means either a reversal of the decision, a new way of applying the stripes to the shells, or outdated images on the data sheets. The price of shells was pegged at fifteen cents each for the analysis, with an average output of 16,000 shells a day.

Government contracts for dynamite destined for use in the Panama Canal project stipulated that all cartridges "were to be wrapped in a distinctive color of paper and marked with I.C.C. in common red letters."[43]

Dynamite Boxes

Another reason for the preference of lock-corner over nailed boxes is that the original ICC requirement for nailed boxes listed two minimum thicknesses of wood: ⅞" for ends and ½" for sides, tops, and bottoms.

In 1927 the US Department of Commerce advised that "certain thicknesses of rough lumber are standard in every section of the country; for box-making purposes, these are known as 4/4, 5/4, 6/4, and 8/4, being 1", 1¼", 1½", and 2", respectively. The thinner box boards most in use are ⅝", ½", ⅜", 5/16", and ¼" lumber, of which all but the last can be economically resawn from American Standard thicknesses."[44] A ⅞" thick board is in effect a 1" thick board because it would be a waste of wood and time to cut it down to the tune of ⅛" (tolerances for shrinking or

"seasoning" and waste from cutting are built into the figures). An extra 1" length per case adds up quickly in railcar quantities. In 1939 the ICC added a provision allowing for ⅜" nailed sides and end pieces, which coincidentally can be cut from boards of standard thickness.

In 1931, wooden-box producers lobbied Congress for the continued use of wooden cases for shipping dynamite. Naturally the Wooden Box Institute and the National Wooden Box Association maintained that their product was sturdier and safer than fiberboard. Regardless, the transition from wood to cardboard was gradual. A story in the August 1977 *Engineering and Mining Journal* pegs the transition as happening "in the 1950s."[45] The latest date I have personally seen on a wooden case is 1959.

The same *Engineering and Mining Journal* article relates that empty wooden cases were used by miners and their families for bookshelves, toy wagons, bedposts, and sleds, and "In Bisbee, Arizona, their use for kitchen cupboards was almost standard practice."[46] Despite the supposed dangers of reusing dynamite boxes, I could find no accounts of a repurposed case exploding.

Fiberboard boxes were required to be marked "High Explosives – Dangerous" and the tops marked "This Side Up." Also required was "ICC 12-H 65" on the bottom of the box. The statutes were amended in 1978 to change the symbol from "ICC 12-H 65" to "DOT 12-H 65." "DOT" is the US Department of Transportation, and "12-H" is the paragraph of the regulation referenced ("ICC 14" also refers to a paragraph and not the year 1914).

The regulations covering fiberboard boxes include almost an entire page devoted to the tape used to seal the boxes. According to the rules, "Tape used for closing must be pressure sensitive, filament reinforced. Tape backing shall have a minimum longitudinal tensile strength of 160 pounds per square inch

and a minimum elongation of twelve percent at break. The tape shall have sufficient transverse strength to prevent raveling or separation of the filaments. Tapes shall have an adhesion of eighteen ounces per inch minimum."[47]

In 1945 the Bureau of Mines advised, "The safest implements for opening wooden cases are a hardwood wedge and a mallet. One approved method of opening the case is to drive the wedge against the edge of the lid to loosen it and then drive it under the lid. Another method is to split the sides of the box with the wedge on a corner at the third dovetail from the top. A metallic slitter may be used safely for opening fiberboard cases, provided it does not come into contact with the metallic fasteners of the case, but it would be preferable to use one made of plastic or some other nonmetallic substance."[48] ("Dovetail" is a misnomer, as we have learned.)

In addition to making dynamite shell paper, Crown Zellerbach made fiberboard dynamite boxes. Other makers included Weyerhaeuser Company and Canadian firm MacMillan Bloedel. Both shell- and box-printing dies were attached to rotating metal cylinders mounted on printing presses. This allowed for the change-out of dies and for additional dies to be mounted to print dates and other variable information.

Dynamite Magazines

The headline in the August 29, 1915, *New York Tribune* read, "Manhattan Indifference to Danger Shown by the Throngs Which Daily Pass Boxes on Sidewalks Filled with Enough Explosive to Wreck Five Skyscrapers." The story was accompanied by a remarkable photograph showing pedestrians walking among 3" square, locked wooden trunks emblazoned with the words "MAGAZINE BOX." According to the piece, "In the 'magazine' box, and there was another one just like it twenty feet away, marked 'Danger,' was 200 pounds of dynamite containing enough explosive force to reduce the wall of a modern fort. Unobservant Gothamites and transient out-of-towners walk by these explosive centres every day, never pausing to think what would happen if a fulminate cap should make connections with the candles of paste."[49] The magazines were used to conveniently supply workers building the New York subway with dynamite and were limited by law to 62 pounds per magazine. Chicago had a similar law, restricting the amount of dynamite in any one magazine to 50 pounds.

Apparently no one in New York City had learned from an accident in January 1902, when a larger wooden dynamite magazine used for subway blasting exploded, killing six and injuring 125. The blast severely damaged the eight-story Murray Hill Hotel and several other buildings. While the cause was never determined, the amount stored (585 pounds) well exceeded any safe limit for storage in such a densely populated area.

Lightning strikes have detonated dynamite magazines, including in Kingswood, New York, in 1895, killing six; Lowellville, Ohio, in 1898, killing eight; an 1886 explosion at a Laflin & Rand magazine near Chicago that killed five; and a 1903 explosion that killed six Ohio miners and their carpenter. Nonetheless, explosives experts did not recommend placing lightning rods near magazines, because the rods needed maintenance and, if not installed properly, actually posed a greater danger by providing a high point for lightning to strike.

The Institute of Makers of Explosives advised operators of magazines to "always post clearly visible 'EXPLOSIVES—KEEP OFF' signs outside of the magazine."[50] However, be sure and "locate signs so that a bullet passing directly through them cannot hit the magazine."[51] Shooting signs is not just born of malice or mischief. According to an article in 1957, hunters were known to use road signs to "sight in" or "target in" their rifles, a process involving firing a small grouping of rounds into a target at a known distance (rifle sights are adjusted to compensate for gravity's effect on a bullet).

Current federal regulations regarding dynamite magazines define three types of magazines: type 1: Permanent; type 2: Mobile or Portable; and type 3: Attended Storage. Types 1 and 2 must be bullet resistant, defined as "resistant to penetration by a bullet of 150-grain M2 ball ammunition having a nominal muzzle velocity of 2,700 feet per second fired from a .30[-]caliber rifle from a distance of 100 feet perpendicular to the wall or door."[52] This means a construction requirement of at least ¼" thick steel lined with at least 2" of wood.

Federal law now also requires stringent record keeping through a "Daily Summary of Magazine Transactions," which must include the explosive manufacturer's name, the date when the explosives were originally received, the total quantity received in and removed from the magazine during the day, each and every day, and the total

remaining at the end of each day. Yearly inventory must be taken and records retained for five years. Many similar rules have been enacted in years past by state governments or implemented by mining companies.

Dynamite Transportation

An article detailing the history of Apache Powder Company gave this information about the company's rail spur: "The El Paso & Southern Railroad serviced the company's transportation needs with a 6,986-foot spur from Curtiss Station to the Apache Powder Works."[53] Originally, mule-drawn carts were used, and "later, a 'fireless' locomotive, manufactured by the H.K. Porter Co. of Pittsburgh, provided greater horsepower. It relied on an insulated steam reservoir rather than a conventional boiler that emitted sparks, which could prove deadly when handling dynamite."[54] The spur also had its own two-story manned depot at the juncture to the main track.

ICC regulations called for railcars to be coupled "with no more force than is necessary to complete the coupling."[55] This was to prevent a load of explosives from shifting due to a sudden jarring motion.

In 1933 an article reported that from 1927 to 1933, 2.5 billion pounds of explosives had been transported on the railroads of the United States and Canada with "no loss of life, with no accidents from explosives, and with a total property loss of less than $100.00."[56]

CHAPTER 2

Dynamite Tools and Techniques

The implements used for blasting were simple at first. In the 1880s, most excavation with explosives was accomplished with a lighted fuse. As electric firing was perfected over the next three decades, blasting evolved into an enormously complex process. Dozens of variables were evaluated, including explosive strength; hole depth, diameter, and spacing; the material being harvested; the location of the deposit; the logistics of transport; and all the costs associated with every step along the way.

Some tools and techniques such as hand drilling seem downright primitive today. But drilling by hand, or "hand jacking," was utterly indispensable to early mining, and the tools were as durable and efficient as they could be. There was a considerable amount of skill involved, and fast hand drillers were in high demand. Contests offering substantial prize money were held to see who could make a hole of a certain diameter and depth in the least amount of time.

Machine drilling revolutionized mining because larger, deeper boreholes could be drilled in a fraction of the time for hand drilling.

Numerous other tools were used for blasting, including tamping bags, rods, and plugs; blasting caps and crimpers; blasting machines; and various means of transporting small amounts of dynamite.

Basic Tools

The use of dynamite was scalable from a one-man stumping job to a multimillion-dollar quarrying operation. Farmers, miners, loggers, and engineers employed the tool in countless different ways, but the very basic principles of preparing and exploding a dynamite cartridge were essentially the same in all cases.

Fig. 14.—Double roll of safety fuse with paper wrapping removed.

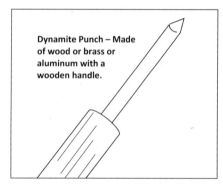

Dynamite Punch – Made of wood or brass or aluminum with a wooden handle.

Fig. 12.—No. 6 Blasting Cap— exact size.

To correctly place and detonate a simple charge, several steps were followed. The first, in most cases, was to drill a hole. The borehole was then cleaned of debris. Next, a dynamite cartridge was primed, which meant that a hole was punched into it, and a wire or fuse tipped with a blasting cap was inserted. The fuse was tied to the cartridge with string, or the wires

Fig. 15.—The cap crimpers are necessary for attaching blasting caps to fuse. They are also useful for making holes in cartridges of high explosives when making primers. The No. 2 type has a fuse cutter.

The US Mine Safety and Health Administration currently states "at least a 36" length of 40-second-per-foot safety fuse or at least a 48" length of 30-second-per-foot safety fuse" must be used "to allow sufficient time to evacuate the area."[1] Arthur La Motte, *Blasters' Handbook* (Wilmington, DE: E. I. DuPont de Nemours, 1925), 24–26. Punch. *Author*

NEVER USE ANY

THING WEAKER

THAN HERCULES

No. 6 BLASTING

CAPS OR

ELECTRIC BLAST-

ING CAPS.

THE UNITED

STATES BUREAU

OF MINES SAYS

IN BULLETIN 59

THAT WEAK

DETONATORS

SACRIFICE OVER

10% OF THE

STRENGTH OF

EXPLOSIVES.

GOODS USED

MUST BE

PAID FOR

IN FULL.

RULES AND INSTRUCTIONS
TO BE FOLLOWED IN
TRANSPORTING, STORING, HANDLING AND USING HIGH EXPLOSIVES

Transportation—Our High Explosives may be hauled in wagons, mine cars, etc., or carried by persons on the surface or in the mines.

Storage—High Explosives should be stored in dry, well ventilated, bullet-proof buildings, isolated from other buildings and public or private thoroughfares. Cases should be kept right side up so that cartridges will lie on their sides. Do not store explosives in the same building nor transport them in the same vehicle with blasting caps or electric blasting caps.

Freezing—Most High Explosives freeze at temperatures between 45° F and 50° F and must be thawed before using.

Hercules Low Freezing (E.L.F.) Grades do not freeze until *after* water freezes. They freeze more slowly and when frozen are thawed more quickly than other grades.

Hercules Nitro-Glycerin and Extra (Ammonia) Grades are standard for general work.

Hercules Gelatin is especially adapted for tunneling, submarine and hard rock work. When properly detonated it develops less objectionable fumes than any other High Explosive.

Hercules Permissibles—Among the many grades and kinds will be found one especially adapted to each class of work.

Following is a List of Some of Our Principal High Explosives:

STRAIGHT NITRO-GLYCERIN	STRAIGHT NITRO-GLYCERIN (Low Freezing)
EXTRA (AMMONIA)	EXTRA (AMMONIA) (Low Freezing)
GELATIN	GELATIN (Low Freezing)

PERMISSIBLES:

Red H   Bental   Xpdite   Guardian

In addition to the above we manufacture a number of special kinds of high explosives to meet specific requirements.

For Full Information About Any or All of Our Explosives Write the Company's Nearest Office.

Making Primers—There are two approved methods of priming dynamite cartridges with blasting cap and fuse or with electric blasting caps.

1. Unfold the paper shell at one end of the cartridge and punch a hole directly downward into the dynamite. Insert an electric blasting cap (figure 1) or a blasting cap crimped on a freshly cut end of fuse (figure 2) and tie the open end of the paper tightly with a string around the wires or fuse as shown.

2. Punch a slanting hole through the shell into the side of the cartridge near the end. Insert a blasting cap and fuse or an electric blasting cap and tie the fuse or wires directly to the cartridges as shown in (figure 3).

If the work is wet, soap or tallow should be placed over the joint between the blasting cap and fuse.

Fig. 1

Thawing— If cartridges are frozen thaw slowly and carefully by removing them from the cases and spreading them out in a warm room. If this is not possible cartridges may be thawed in a thawing kettle consisting of a water-tight vessel surrounded by water of a temperature not to exceed 150°F. A booklet on proper methods of thawing high explosives will be sent on request.

Do Not thaw by immersing cartridges in water, exposing them to steam or placing them on hot surfaces, such as pipes and stoves. Never expose them to the direct heat from a fire.

Loading the Bore Hole—Insert the charge and primer carefully in the bore hole and tamp with damp clay or other loose material which cannot burn, (never use coal dust or "bug" dust). Press the tamping material down gently for the first eight inches, after which greater force may be used. Be careful not to disturb the exploder in the primer when tamping. The bore hole should be filled completely to the mouth and should be packed so closely that no air spaces remain.

In a coal mine do not use more than one kind of exposives in a bore hole.

Fig. 2   **Tamping**—Always use a wooden rod with no metal parts.   **Fig. 3**

Misfires---Never draw a charge. A missed hole should be fired by the detonation of a charge in a new hole drilled close to but at a safe distance from it.

The great importance of strong detonators leads us to urge our customers to use nothing weaker than Hercules No. 6 blasting caps or electric blasting caps.

The explosives in this package were manufactured and packed under the most careful supervision and are guaranteed to have been in perfect condition when shipped from the factory.

If you have trouble with our goods set them aside and notify our nearest office. No complaint can be investigated or allowed unless samples of unsatisfactory goods are available for test and analysis.

HERCULES POWDER COMPANY

Instruction sheets such as this were included in cases of dynamite, both to offer recommended procedures and to further promote the company. This one dates from the early 1920s, because it still lists low-freezing formulas that don't freeze "until after water freezes." *Author*

were looped around the cartridge. The cartridge was loaded into the hole, followed by other nonprimed cartridges. The column of cartridges was tamped after each new cartridge to expand the cartridges to the diameter of the borehole. Then the hole was filled with dirt or clay, which materials are called stemming. To make this process easier in upward-slanting holes, sand-filled tamping bags and wooden plugs were loaded after the live cartridges. The fuse was lit, or an electrical impulse was sent via wires connected to a blasting machine.

The procedures grew more complex as more holes and charges were added. Multiple fuses or wires were initiated in sequence to achieve maximum efficiency. Modern explosive-charge sequences are planned, programmed, and initiated via computer.

But in the early decades of dynamite, a few simple and reliable tools were used. Cartridges were routinely slit or cut into pieces with brass-bladed knives. A sharp knife for cutting fuses was also mandatory. A pair of cap crimpers firmly fastened cap to fuse, and most crimpers had a built-in dynamite punch or fuse cutter. Separate dynamite punches for making priming holes in cartridges were made of wood, brass, or aluminum to avoid sparks, although the likelihood of a mere spark by itself initiating the explosion of a dynamite cartridge was remote. Blasting caps were another matter; an open box could be detonated by sparks, static electricity, or cigarette embers.

Safely transporting explosives into narrow mine shafts or up steep hills requires special gear. Even in mines with underground railways, getting the cartridges of dynamite to the final

In addition to the items listed, the farmer needed a soil auger, stump drill, and tamping stick. Adapted from Atlas Powder, *Better Farming with Atlas Powder* (Wilmington, DE: Atlas Powder, 1919), 10.

Farming dynamite ready for a day of stumping. A 1920 Aetna Explosives Company ad proclaimed, "A few sticks of Aetna Dynamite here and there do the work of a dozen farm hands, and do it quicker, better and cheaper."[2] *Author*

The basic accoutrements of farm blasting (all producers and safety engineers frowned on carrying matches or lighters with blasting caps and fuse). The 1915 edition of DuPont's *Farmers' Handbook* contains more than four hundred testimonials extolling the benefits of dynamite for increasing crop yields. *Author*

Collecting Materials and Tools. The first thing is to collect in a small box, easily carried, or in a basket, the following:

Blasting caps (keep in original box).
Roll of fuse.
Cap crimper.
Knife – must be sharp.
Twine or string.
Tallow or soap (only when charges go in damp or wet ground).
Small wooden punch.
Gloves for handling the powder.
These materials should be kept together at all times. If in a box, the box should have a bail or handle so it may be picked up and carried easily. If a basket is used, it should have a piece of blanket or canvas bag in the bottom, or better still, oil cloth or rubber sheet to keep out dampness from the ground. Another piece of waterproof material should be carried along to cover the basket from sun and rain and dew. Carry nothing but matches in pockets.

In another basket or box carry along what powder may be required. It is best to keep the caps and powder in different boxes. When a large number of sticks are needed in the job, carry the powder in its original box and do not disturb the sticks until the actual priming and loading is done.

location where they were needed was always a challenge. Miners stuffed cartridges into waistbands, boots, shirts, and jacket pockets. Local and state statutes gradually prohibited such foolhardy practices.

The Bureau of Mines advised in 1945 that "where explosives or detonators are not transported in the original containers they should be brought into the mine or taken to the working places in separate insulated containers. Canvas bags are used widely for carrying loose explosives to working places, and where the conditions do not favor the use of rigid containers, specially constructed strong bags with straps or handles may be suitable for this purpose."[3]

DuPont, Bemis Brothers, Mine Safety Appliances Company, and others manufactured explosives-carrying bags, backpacks, wooden boxes, and hard-shell containers (Bemis Brothers also made tamping bags). Mining companies also contracted to have custom bags produced locally. Early bags were usually made of thick canvas or finely woven brattice or jute cloth, which is similar to fine-weave burlap. Hard containers were made of heavy-duty plastic or wood. Wooden carriers were often homemade or made by a mine's carpenter.

The Mine Safety Appliances Company was founded in Pennsylvania in 1914 and still operates today. The

Bag for Dynamite and Fuse

Wallace McKeehan, safety inspector of the Copper Queen Consolidated Mining Co. and a leading member of the American Mine Safety Association, with the approval of the central safety committee of the company, recently designed the bag here illustrated for the carrying of

BAG FOR EXPLOSIVES

powder and fuse. The sack is made of heavy jute and is sewn with flax thread, so that it is exceptionally durable. From 50 to 70 sticks of dynamite can be carried in the bag proper, the fuse being placed in the outside pocket. Large brass eyelets are fastened into the top of the sack for convenience in handling and carrying. It was noted by the management of the Copper Queen mines that all dynamite when handled in quantities of less than a box was carried in any way that would suit the convenience of the miner, it generally being tied together with a piece of fuse. Capped fuse was carried in the miner's hand or wrapped around his hat or thrust loosely in his shirt front. All these methods were exceedingly dangerous, and since the bag has been introduced these bad practices have ceased. The sack is made by the Bemis Bros. Bag Co., St. Louis, Mo., and sells for 35c. in 100-bag lots. The size of the bags is 15x22 in. outside measurement.

Founded in St. Louis in 1858 by Judson Moss Bemis, Bemis Brothers is now multibillion-dollar Bemis Inc. Adapted from "Bag for Dynamite and Fuse," *Engineering & Mining Journal*, October 15, 1915, 679.

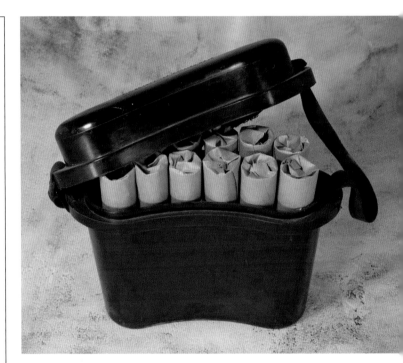

This carrier is made of Bakelite, which was invented in 1907 by Belgian chemist Leo Baekeland, who had moved to New York in 1899. Bakelite, known chemically as polyoxybenzylmethylenglycolanhydride, is considered the first all-synthetic plastic. *Author*

company has produced an enormous array of equipment and supplies, including first-aid kits, helmets, and "explosimeter" gas-detection instruments. The company's specialty has always been respirators, and their current catalog features hundreds of different breathing devices for every imaginable use.

The pictured carrier was introduced in 1938 and has a capacity of twelve 8" × 1¼" cartridges. In write-ups of the invention upon its introduction, the design was touted for numerous features, including being moistureproof and nonconductive. According to one report, the carrier "provides a safe and convenient means of carrying explosives" and is "kidney-shaped so it may be carried comfortably."[4] The adjustable shoulder strap had specially designed eyelets that would accommodate a small lock, "if such a precaution is necessary."[5] The carrier won an award from *Modern Plastics* magazine in 1939 for the innovative design.

DuPont began offering explosives-carrying bags in 1927 and at one point even produced a small catalog featuring them. The bags were made of another DuPont product called Ventube, a thick, extremely flexible yet tough, plasticized cloth that resisted moisture. True to its name, Ventube, introduced in 1921, was used mainly to make flexible ventilation ducts for mines and industrial applications. Bags were offered in backpack and shoulder strap satchel styles with capacities from fifty to 125 standard-size cartridges. In addition to the two bags shown, the 1949 *Blasters' Handbook* lists a larger "No. 3" backpack measuring 12" × 20" × 6", which held 125 sticks. DuPont touted its bags as being nonconductive and "highly resistant to acid water, fungus, rot and powder fumes."[6]

In 1939 the American Brattice Corporation of Indiana began selling a similar product made of the company's Minevent tubing material. Its version featured a locking zipper closure.

The bags and backpacks were labeled as "powder bags," even though they were designed for carrying sticks of dynamite rather than black powder. This was in keeping with the tradition of referring to dynamite and black powder interchangeably as powder. While modern explosives bags and backpacks are made of bright-yellow nylon, many are still marked "powder," although they are now used mainly for carrying cartridged emulsion explosives. (DuPont also made two special bags for carrying cartridged pellet powder. The #4 Pellet Powder Bag held twenty-five sticks, while the

This is DuPont's backpack-style "No. 2 Powder Bag." It measures 12" × 20" × 4½" and was designed to hold one hundred standard 8" × 1¼" cartridges. *Author*

DuPont's "No. 1 Powder Bag" was an over-the-shoulder satchel measuring 6" × 10" × 12" with a capacity of fifty standard-size sticks. *Author*

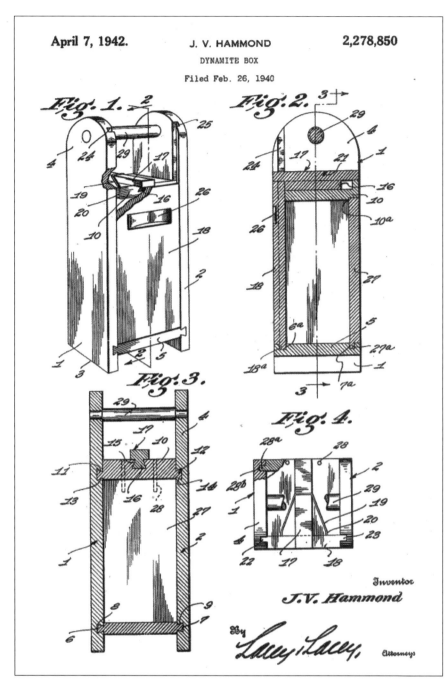

April 7, 1942.   J. V. HAMMOND   2,278,850

DYNAMITE BOX

Filed Feb. 26, 1940

Inventor
J.V. Hammond

By Lacy & Lacy, *Attorneys*

John V. Hammond also obtained patents for a wooden "scooter sled" and an "endless conveyor." John V. Hammond, Dynamite box, US Patent 2,278,850, filed February 26, 1940, and issued April 7, 1942.

A competent man at the powder house shall place such dynamite in such container. Carrying by box should be stopped immediately."[7] In 1920 the *Coal Industry* journal added,

Transportation of explosives from the magazine into the mine should receive careful consideration. It is not advisable to transport large quantities of explosives into the mine, or from one part of the mine to another, in cars propelled by electric power, unless the cars are thoroughly insulated in every way, and the hauling of explosives is done at a time when very few if any other men are in the mine. The most general, and probably the safest practice, is to have each individual miner carry the explosives he is to use each day into the mine, in a case which is a non-conductor of electricity, as well as being water-proof and fireproof.[8]

A popular wooden dynamite carrier was produced by the J. V. Hammond Company of Spangler, Pennsylvania, which was launched as the Hammond Lumber Company. Lawrence Hammond patented the original version of his box in 1923. A modified version with a clever, rubber-band-powered "locking" mechanism was patented by his son John in 1942. The boxes were touted as having all-wood, nonsparking components and were sold in several sizes. Because of the simplicity of construction, many containers of this exact design were privately made.

The pictured box holds fifty 8" × 1¼" cartridges and cost $6.40 in 1949. In 1951, J. V. Hammond advertised eight different sizes of dynamite boxes designed to hold from nine to seventy-two standard-size cartridges. The Hammond Company also offered seven sizes of wooden tamping poles from 1" to 2" in diameter, sold by the foot. In addition, Hammond made mallets and wedges for opening dynamite cases, as well as other mining tools such as sounding sticks, trolley poles, and brake chocks. ("Sounding the roof" of a mine was accomplished by tapping

#5 bag held fifteen sticks. Pelletized black powder was used mainly for small, underground blasts, so fewer sticks were needed.)

Tripping while carrying a partial box of dynamite caused numerous explosions. In 1919, the US Department of the Interior observed, "In the carrying of dynamite by men, we discover a very dangerous practice in that the high explosive is put into boxes and so carried by said men. We recommend that dynamite shall be deposited in canvas bags, reinforced by leather, with two catches to fasten cover, a hook or ring to hold the miner's ticket, and a long strap to place over the shoulder for convenience in carrying.

with a brass-tipped sounding stick to check the stability of the rock structure above. If the roof sounded hollow, then the room or tunnel was reinforced with timber beams and trusses.)

Some miners were employees of a mining operation, while others were akin to contractors. The freelancers had to provide their own equipment and materials, including blasting caps, fuse, and dynamite.

Some mining operations paid their employees and contractors in tokens and notes called scrip, redeemable for merchandise at the company store. Other tokens were called explosive-control tokens and were "GOOD FOR FIVE STICKS POWDER" or "GOOD FOR 25 CENTS IN MONOBEL." The tokens, made of aluminum or brass, were purchased by contract miners and were supplied free of charge to company miners. In either case, the tokens provided a way to track the use of explosives.

Above left: The larger versions of the Hammond-style box were used for transporting dynamite, while smaller ones were for blasting caps, fuse, and the squibs that were used to ignite blasting powder. Author

Left: When holding dynamite, the box was laid on its back before opening. The smaller boxes for carrying blasting caps and fuse were kept upright. Author

Control tokens discouraged waste and theft, kept money in the mining company, and were convenient for miners. Tokens could be advanced against pay so freelance workers could get needed supplies on day one of employment.

How much dynamite some tokens would purchase depended on the "cartridge count" and the price of dynamite at the time. Cartridge count is the industry term for the number of sticks per 50-pound case. Per DuPont, the cartridge count for Monobel No. 4 L.F. (low freezing) was 135, or 2.7 cartridges per pound. If this grade of Monobel was selling for twenty cents a pound, a 50-pound case containing 135 sticks cost $10.00, or about 7.5 cents a stick. Thus, the pictured sixty cents of Monobel tokens would have been worth eight sticks.

Booklets of paper coupons were another way to keep track of explosives. Some booklets were either issued or purchased ahead of time and redeemed as needed. Others had blank lines to be filled in with the amount and cost of supplies.

Large explosives producers offering free and paid training used a variety of tools, including inert cartridges like the ones shown. Inert blasting caps, fuse, wire, blasting machines, and other nonfunctioning training aids were

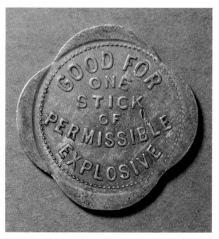

This token measures 24 mm in diameter and was issued in the 1920s. The Mt. Pleasant Supply Company was affiliated with a coal mine near Beatty, Pennsylvania. *Author*

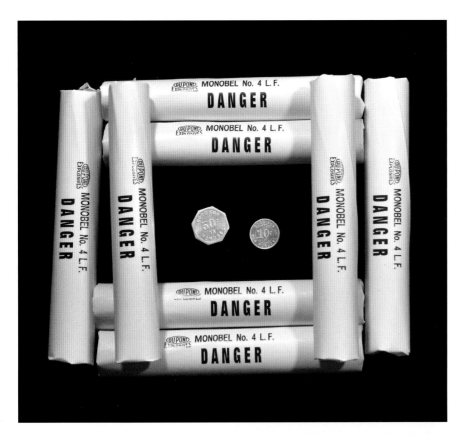

Because of fluctuations in the price of explosives and other blasting necessities, most explosive-control tokens were denominated in an amount of supplies rather than dollars and cents. Tokens marked "Good for five sticks powder" were easier to deal with than tokens marked "Good for ten cents in Monobel." *Author*

also employed to teach proper priming, loading, and tamping techniques. DuPont even used a replica blast hole encased in a hinged box. Students would prime, load, and tamp dummy cartridges, then the box was opened to look for issues such as crimped or dislodged fuse or damaged wire.

Nowadays, dummy or inert cartridges are made by suppliers of bomb-detection-and-disposal training aids. Some of these are designed to be identical to the real thing when x-rayed or broken open. Oddly, on the outside a few of these are replicas of brands of dynamite that have not been sold in fifty years!

The most important tool of all was knowledge. Wisdom was handed down from miner to miner, particularly regarding blasting with cap and fuse. Miners who worked during the initial transition from black powder to dynamite passed along their techniques, and most of the drilling processes were the same both for high and low explosives. When it came to complex electric blasting, an entire new set of

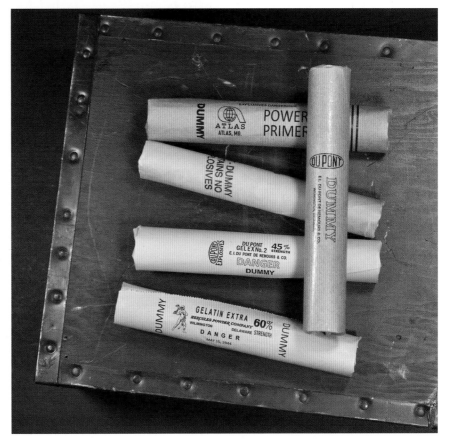

Dummy sticks made for training often contain an inert substitute for dynamite that feels like the real thing. Their use was once so common that the Interstate Commerce Commission directly addressed the transportation of dummy cartridges, albeit pronouncing them "exempt from regulations."[9] *Author*

procedures was needed (electric squibs were used to detonate black powder but were limited to small numbers of crudely timed shots). DuPont, Atlas, and Hercules went to great lengths to properly educate their customers. DuPont produced its vaunted *Blasters' Handbook* from 1918 to 1980, and an updated version is still available from the International Society of Explosives Engineers. More than 30,000 copies of the fourteenth edition were printed. Atlas likewise issued a variety of instructional material, including the comprehensive *Explosives and Rock*

Blasting in 1987. DuPont and Hercules published magazines containing tips, tricks, testimonials, and the latest recommended practices.

On-site training was an important way to gain customers, with Giant proclaiming, "The Giant Powder Company's Service Division will send a man to answer your questions."[10] Atlas Powder Company implored readers of advertisements to "Let the Atlas Service Man help you determine what grades will effect the GREATEST economy for YOUR work."[11] Even the smallest early dynamite firms

gave free exhibitions of stumping and boulder removal.

Both DuPont and Hercules Powder Company operated an "Agricultural Extension Section" to aid farmers in the selection and use of dynamite. According to Hercules, its section was "organized to give an efficient service to all users of explosives for agricultural purposes" and had "at its command the services of a corps of practical agricultural engineers."[12] These included both DuPont employees and outside consultants. In the 1930s, L. F. Livingston, president of

The physics behind an explosion is so complicated that scientists still haven't settled on a comprehensive explanation. Atlas Powder Company's 1987 *Explosives and Rock Blasting* devotes fifty pages to competing theories. Fortunately, users of explosives didn't need nearly that much detail. *Author*

the American Society of Agricultural Engineers, also headed DuPont's Agricultural Extension Section.

Other commercial entities do their part to educate the users of high explosives. The Institute of Makers of Explosives (IME) is a trade group founded in 1913. Its mission is "to promote safety and security for the commercial explosives industry."[13] *The History of the Explosives Industry in America*, cited many times in the current work, was produced under the auspices of the IME. The International Society of Explosives Engineers (ISEE) was founded in 1974 and shares similar goals, often working with the IME. The ISEE boasts over four thousand members in more than ninety countries.

The United States Bureau of Mines issued hundreds of bulletins and other educational publications from 1910 to 1973. The Department of Labor's Mine Safety and Health Administration has taken up the helm since, informing explosives users on the latest approved procedures and safety guidelines.

Finally, innumerable experts imparted their know-how in textbooks, guides, handbooks, and especially articles in scores of trade journals. These included publications you might expect: *Coal Age, Good Roads, Rock Products, Highway Engineer and Contractor, Cement Mill & Quarry*, the *Engineering and Mining Journal*, and *Brick and Clay Record*, among dozens of others. And, because dynamite was used to clear land for farming, articles also ran in such magazines as the *Farm Journal, Better Fruit*, and *Poultry Success*.

Drilling

Drilling by hand was the only way to make boreholes until the 1870s and was decreasingly used through the 1920s. The process, called hand jacking, was restricted to small-diameter holes, particularly in hard rock.

Hand-drilling contests were common in early mining camps and towns. Because hand jackers were literally the engine of some mines, their skill was held in high regard. A contest in Butte, Montana, in 1907 drew 25,000 onlookers. The winning two-man team split a $1,250.00 prize, equivalent to more than $30,000 in 2022.

The shallower holes for blasting powder tended to be as large as possible, owing to the need for more of the weaker explosive per hole. In softer rock these holes could be up to 3" in diameter, but in hard rock, 1½" was about the largest feasible bit that could be employed, and that for a comparatively shallow hole.

The bits (called "steels") for hand jacking were chisel-shaped and were made in sets of graduated length. The shorter steels were used until not safe to hold, then the next-longer bits were used until the hole was finished. The diameter of the steels decreased slightly with each increase in length, to avoid jamming in the hole. The deepest hole that could be made by one man (a single jacker) was 4 feet, and for a team of double jackers, about 8 feet.

Some holes were not even an inch in diameter. In the 1870s, dynamite cartridges with diameters as small as ½" were made by companies such as the original Giant Powder Company and the Judson Powder and Dynamite Company. Mechanized drilling quickly became more refined, leading to larger-diameter holes.

The short-handled hammers employed by single jackers averaged 4 pounds in weight. When there was enough room and manpower, triple jacking was employed, with two men alternately swinging sledgehammers while one man held the bit. Double and triple jackers could swing with both hands, so their hammers had longer handles and were twice as heavy. Because of underground space constraints, triple jacking was most often employed in aboveground operations such as quarries.

Despite the apparent risks, injuries from hand jacking were relatively rare. One study of mining mishaps

Here, the miner is striking the steel with the hammer inverted. The most difficult technique was for drilling a low ceiling. Then the hammer was held with index finger near the end of the handle. Try holding even an ordinary household hammer that way! *Library of Congress Prints and Photographs Online Catalog*, LC-USZ62-29850.

recorded no injuries to hands or arms, but several to feet and eyes. A flying piece of rock could take out an eye, and safety goggles did not catch on in mines until the 1940s. Another danger from hand jacking that made it into the statistics was detonation of a nearby misfire via the concussion of the hammer blows. In one case this occurred with just an unexploded portion of a dynamite cartridge.

Handheld machine drills, while quicker and more economical, were still constrained by the tyranny of hard rock. Holes of 1" to 2½" were the standard with early steam and pneumatic hand drills. The best one-man jackhammer rigs could achieve hole depths of about 9 feet. By the 1920s, truck-mounted well drills made holes exceeding 150 feet in depth.

the mechanisms for delivering the blow to the bit were multitudinous.

Churn drills were also percussion drills, but of a totally different design. Developed in the late 1600s, this type of drill raised a heavy bit in its entirety and released it to fall and chip away at the bottom of the hole. The first churn drills, also called drop drills, were operated purely by muscle power, with two or more men repeatedly lifting and dropping a weighted bit.

An important early innovation was the introduction of hollow shafts and perforated bits, which allowed for water to be jetted into the hole while drilling to keep it free of debris. Note the holes in the detachable drill bits pictured (see p. 60). Another general improvement was to increase the distance between

Triple jacking could be employed only when there was enough room to swing long-handled sledgehammers. The sanguine expression of the pipe-smoking miner holding the drill steel is likely not true to life. International Library of Technology, *Rock Boring, Rock Drilling, Explosives and Blasting* (Scranton, PA: International Library of Technology, 1906), section 35, p. 9.

It took the development of tricone bits in the 1930s and special hardened-alloy bits in the 1950s to advance technology to its present state of quick, economical rotary drilling. Rotating bits had been used for softer material such as clay and coal but were unsuitable for hard rock until more-durable cutting surfaces were engineered. Large pneumatic drills were replaced by hydraulic drills beginning in the 1970s.

Drills fall into two basic categories: percussion and rotary. Hand jacking is the most elementary type of percussion boring. The chisel-shaped "steels" were rotated a quarter turn after every blow with the hammer. Mechanical jackhammer drills, invented in 1844, worked essentially the same way, but

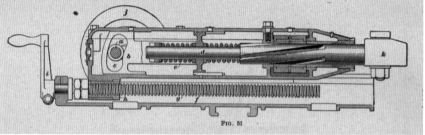

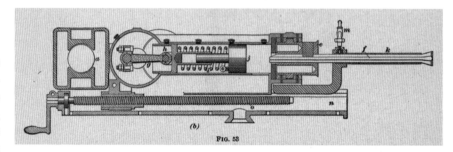

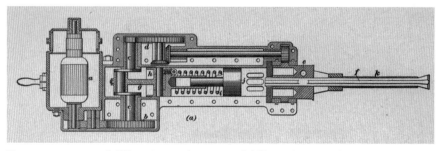

There were scores of different manufacturers of drills at the turn of the twentieth century. Every aspect of drilling was addressed, from economy to portability to safety to speed to durability. International Library of Technology, *Rock Boring, Rock Drilling, Explosives and Blasting* (Scranton, PA: International Library of Technology, 1906), section 35, pp. 16, 48, 50.

the drill and its power source, whether pneumatic, hydraulic, or electric. This allowed for the use of drills in deeper shafts and for multiple drills to be run off a single power supply.

Without the introduction of mechanical drills, the use of dynamite and the progress associated with high explosives would have been stymied. According to Eustace Weston, in his 1910 *Rock Drills: Design, Construction and Use*,

It is not too much to say that without the rock drill[,] society would have long before this have starved for its industrial and precious metals. The rock drill has added enormously to the wealth of the world. Mines, some

Above: Hand-cranked, rotary coal drills sometimes employed braces to steady the drill. Here we see the use of a "grip bar," which was emplaced in a 6-inch-deep hole below the main borehole. International Library of Technology, *Rock Boring, Rock Drilling, Explosives and Blasting* (Scranton, PA: International Textbook, 1907), section 35, p. 21.

Left: Angled boreholes were used for many blasting patterns. The column works like a car jack to wedge itself tightly between two surfaces. E. I. DuPont de Nemours, *High Explosives, First Section* (Wilmington, DE: E. I. DuPont de Nemours, 1925), 66.

of them the richest, and quarries in hard rock, where labor was scarce and dear, have been rendered workable only by its use. The world's greed for gold, silver, lead, copper, and tin could never have been satisfied without it. The wonderful mining undertakings of the past, the great drainage adits of Saxony and Cornwall, the deep shafts and drives into hard ground were accomplished only with an expenditure of human effort and time that we cannot afford in these days of large demand and large outputs. The rock drill with its hundreds of blows per minute, doing the work of ten or twenty men, came to help the miner extract bodies of low-grade ore and to enable the engineer to tackle problems undreamed of before.[14]

Despite the increasing efficiency of mechanical drills, very basic tools were still employed in certain situations. In tight spots where machine drilling was impossible, hand jacking was the fallback. When practicable, pumps, pressurized air, and water were used to clean holes of debris, but in underground mines dynamite spoons were still used. Dynamite spoons, also called chambering spoons, were made of brass or steel and were either forged by the mine's blacksmith or purchased. Some had one pointed end for smoothing the walls of boreholes. An ad in 1925 listed spoons made of "Norway iron" in lengths from 5' to 7'. For soft deposits, coal miners used hand-cranked coal augers that resembled huge hand drills for wood. Farmers used iron soil punches that were driven into the ground and winnowed free. Holes made with a soil punch were made deeper with coal augers or long-shafted wood-beam drills. Another type of soil punch had a small, posthole-digger-type tip. Loggers used long-shafted versions of barn-beam drills and barrel tappers.

Above right: The basic tools for single jacking: hammer, graduated drill steels, and a spoon for cleaning the hole. A container of water to make mud of drill hole debris made for easier removal. Wrench-like devices to free bits that were stuck in the hole were commercially available or made by a mine's blacksmith. *Author*

Right: The crank of a coal auger consisted of a wooden cylinder called a muff that fitted around the handle of the drill to enable a miner to smoothly rotate the tool. International Library of Technology, *Rock Boring, Rock Drilling, Explosives and Blasting* (Scranton, PA: International Textbook, 1907), section 35, p. 14.

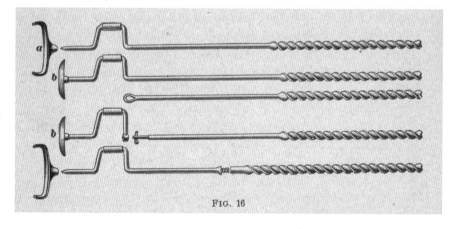

FIG. 16

Fig. 73.—Soil punch or drill for making bore holes in clay, hardpan and rotten shale.

A practical soil punch could not be much longer than 4 feet, so it could be pounded into the ground with a sledgehammer. Arthur La Motte, *Blasters' Handbook* (Wilmington, DE: E. I. DuPont de Nemours, 1925), 66.

Spacing of holes loaded with dynamite varied with the situation. In quarries, small-diameter holes were generally spaced 6 to 15 feet apart, while holes greater than 3" in diameter were distanced as much as 30 feet apart. Fragmentation between holes can be incomplete if holes are spaced too widely, and undesirable pulverization can occur with holes spaced too closely.

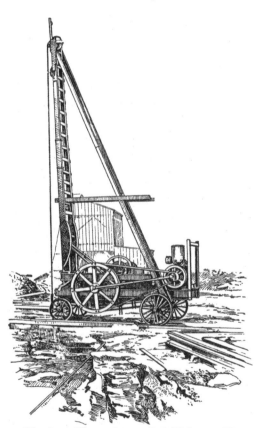

Fig. 79.—View of a well drill for making deep blast holes.

Early well drills were moved on temporary tracks. A well drill was ideal for rapidly boring rows of large-diameter holes in quarries. Arthur La Motte, *Blasters' Handbook* (Wilmington, DE: E. I. DuPont de Nemours, 1925), 69.

Hole patterns evolved for blasting in quarries and for tunneling. Angled boreholes were necessary for many situations; the front row of charges in a quarry bench was often angled to direct material outward. For underground work, angled boreholes were used to blast out wedges of rock without affecting the stability of the surrounding structure. While early blasters used their eyes to estimate angles, laser devices are now available for laying out patterns of precisely slanted holes.

The 1910s saw the first truck- and trailer-mounted drills. Vehicles with tracks were perfected in the 1910s, with the famous Caterpillar brand debuting in 1925. Large modern rock drills are mounted on bulldozers or similar heavy rigs. The advent of big, mechanized drills transformed mining by enabling blasts with deeper, larger-diameter holes and enormous numbers of timed charges. Open-pit mining became possible on a never-before-imagined scale. Holes grew wider and deeper with every improvement in technology, and today's boreholes can reach more than 200' in depth and exceed 2' in diameter (such holes are not used for loading dynamite but, rather, are filled with modern slurries and prills). Drills in general can produce even-deeper holes of much-larger diameter, but such holes are not practical for blasting. Currently, most surface-mining boreholes average 30' to 60' deep and range from 6" to 22" in diameter. Underground, small-diameter holes are still a necessity.

As important as advances in the mechanization of drilling were

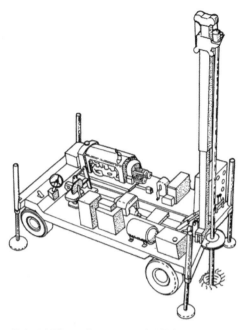

This 1970s trailer-mounted drill has its own power, air, and water supplies. The trailer is stabilized by four hydraulic cleats. US Department of the Army, *Multiservice Procedures for Well-Drilling Operations FM 5-484 / NAVFAC* (Washington, DC: USGPO, 1995), A-5.

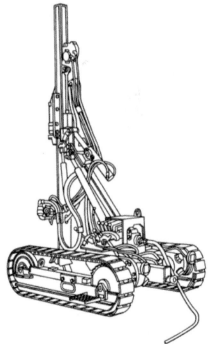

Tracked drilling platforms are the most popular and versatile method of drilling for strip mines and quarries. This one is connected to its water supply via a hose. US Department of the Army, *Equipment Data Sheets for TACOM Special Purpose Equipment* (Washington, DC: USGPO 1985), 7–34.

The two most common configurations of drill bits for mining are the cross and button designs. Examples of both are shown. The knobs on the button bits are actually tungsten-carbide inserts. Carbide button bits are reserved for use in very hard rock. *Author*

Eccentric billionaire Howard Hughes patented the roller bit, first with two cones in 1909, then with three cones in 1933. These bits became widely available in the 1950s and are still the industry standard for boring large holes. *Author*

improved bits. Early bits for hammer and churn drills consisted of long shafts, the tips of which were sharpened into various cutting shapes. Detachable bits were a major improvement because they were easier and safer to carry and less cumbersome to change out and sharpen. Note the apertures for air and water to jet out and clean the borehole. The shafts for these bits are hollow and are hooked up to pressurized water and air. Detachable coal-mining bits (see p. 108) appeared in the 1890s, and commercially successful, detachable hard-rock bits debuted in the late 1910s. Single-use bits began to see use in the 1940s.

Thawing

Nitroglycerin freezes between 45° and 50° Fahrenheit. Frozen dynamite cannot be reliably detonated; in colder regions and seasons, it had to be thawed before use. This was the situation from the 1860s until the 1920s, when truly low-freezing dynamite formulations were finally offered.

Thawing dynamite was accomplished by innumerable risky means and a few safe ones. These ranged from placing a case of dynamite near a heat source such as a campfire or blacksmith's forge to warming in a double boiler. According to Thomas Foster's 1916 *Coal Miner's Pocketbook*, "The common practice of thawing dynamite cartridges by passing them over a lighted candle is very dangerous."[15] The commercial dynamite-thawing kettles and sheds that were available were safe if properly used.

Some thawing-house designs incorporated manure in the walls and ceilings. However, according to one early textbook, "The method of thawing dynamite by placing it in a tight box surrounded by manure is a good one if the manure is fresh so that it is giving off heat, but it is useless to place the dynamite in an old manure heap from which no heat is being given off."[16]

According to the 1922 *Blasters' Handbook*, "The fundamental rule of thawing dynamite is to thaw slowly with the cartridges lying on their sides

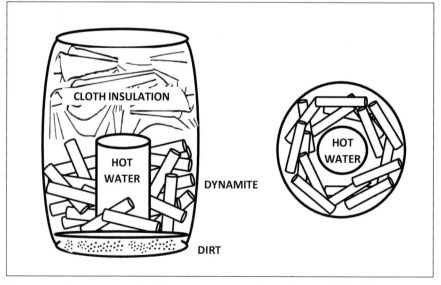

This suggested makeshift thawing setup, while totally safe, was likely seldom used. Despite the simple design, there were too many quicker ways to thaw dynamite. Adapted from *Brick and Clay Record*, February 1, 1916, 247.

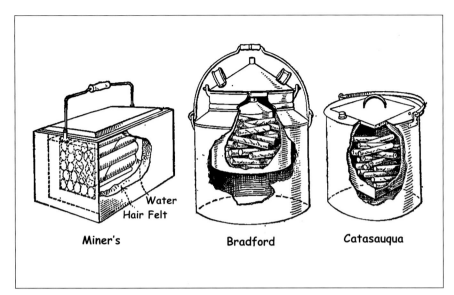

Miner's   Bradford   Catasauqua

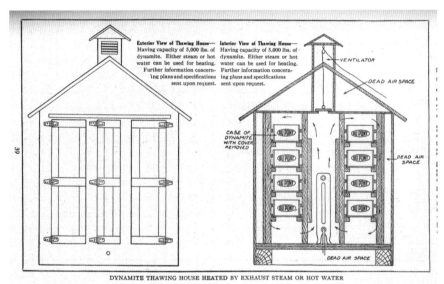

DYNAMITE THAWING HOUSE HEATED BY EXHAUST STEAM OR HOT WATER

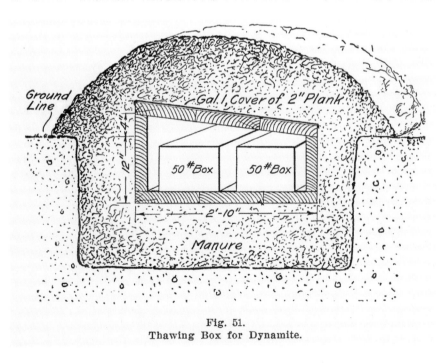

Fig. 51.
Thawing Box for Dynamite.

Top left: The Catasauqua kettle is named for a town in coal-rich Lehigh County, Pennsylvania. A 60-pound-capacity Catasauqua thawer cost $7.25 in 1916. International Library of Technology, *Rock Boring, Rock Drilling, Explosives and Blasting* (Scranton, PA: International Textbook, 1907), section 36, p. 20.

Center left: Because of their expense, thoughtfully designed commercial thawing sheds were not widely used. The majority of mines, especially start-ups, lacked the capital for a fancy dynamite-thawing house. Arthur La Motte, *Blasters' Handbook* (Wilmington, DE: E. I. DuPont de Nemours, 1925), 39.

Bottom left: One of the most popular low-tech, safe ways to thaw dynamite was to lay it on or cover it with manure! Decomposition produces enough heat to thaw cartridges relatively quickly without the danger of detonation. John Cosgrove, *Rock Excavation and Blasting* (Pittsburgh, PA: National Fireproofing, 1913), 130.

and not to place the dynamite over an active source of heat. DuPont thawing kettles are all made with a water-tight compartment for the explosives, which is surrounded by the receptacle for the hot water used to furnish the heat for thawing."[17] DuPont admonished, "Under no circumstances must the water be heated in the kettle itself."[18] However, thawing kettles were still misused and were often heated by placing over an open fire. In 1925, DuPont declared that "the use of thawing kettles has been almost entirely done away with by the development of low-freezing explosives."[19] This statement would prove somewhat premature.

The first patent for low-freezing dynamite was procured in 1895 by the Nitro Powder Company, but the Repauno Chemical Company advertised a low-freezing formula as early as 1893. The first low-freezing formula to enjoy commercial success was DuPont's Red Cross line. The advertisement (see p. 99) ran in August 1906. In addition to nitroglycerin, the Red Cross formulation included TNT, which is a crystalline

Despite the introduction of no-freezing dynamite formulations in the 1920s, some cartridges continued to be marked "Low Freezing" through the 1930s. *Author*

This kind of thawer used hot water that had been heated elsewhere to safely warm the tubes into which dynamite cartridges were inserted. An earlier, similar design relied on a single miner's candle placed underneath the apparatus to provide heat. Aetna Powder, *Handling Explosives* (Chicago: Aetna Powder, 1913), 40.

ANNOUNCEMENT
TO USERS OF EXPLOSIVES MANUFACTURED BY
ATLAS POWDER COMPANY

To simplify some of the markings previously used by us, the following changes in those markings will be made for the Explosives below specified manufactured after January 1st, 1940:

As all our High Explosives are very low freezing and will continue to be so manufactured, the letters "L. F." and "V . L. F." have become unnecessary, and will hereafter be omitted on case and cartridge markings.

The following changes are also being made:

POWDERS NOW MARKED	WILL BE MARKED
Giant Gelatin V .L. F.	Atlas Gelatin
Giant Special Gelatin V. L. F.	Giant Gelatin
Gicodyn	Apcodyn

You may be assured that these new names will be more convenient for easy identification of the respective brands. We are not changing the composition and guarantee that the explosives packed under such new names will be found fully up to the high standard we maintain for our products.

ATLAS POWDER COMPANY.

Giant Powder Company, Consolidated, was the last major producer to mark its cartridges with "Low Freezing" and "Very Low Freezing," or the initials "L.F." and "V.L.F." *Adapted from author's copy*

solid unaffected by cold temperatures. Nonetheless, "low-freezing" in this context meant lowering the freezing point to about 35°F. While this was a marked improvement, dynamite still had to be thawed in colder weather. Very low-freezing formulations froze at only slightly lower temperatures.

Finally, DuPont developed a practically no-freezing formula in 1921. In February 1924, *Municipal and County Engineering* reported the following:

Tests were recently conducted at Hibbing and Virginia in Northern Minnesota which prove that ordinary straight 40 percent dynamite now being made by the du Pont Company

on a low freezing formula was practically unaffected at a temperature of thirty-five degrees below zero. A continuous series of tests which had been made during the comparatively mild winter weather at those points when the temperature was at zero or a few degrees below, proved the low freezing dynamite would detonate without difficulty. In the test made at thirty-five degrees below zero, it was shown that one cartridge would detonate another at one foot.[20]

Formulating nonfreezing dynamites involves building antifreeze into the structure of nitroglycerin to make chemicals such as nitroglycol, diglycerol, and ethylene glycol dinitrate (ethylene glycol is the primary ingredient in automotive antifreeze). Around the same time, Atlas debuted its Ammite brand, which was billed as nonfreezing and non-headache-producing. Ammite contained ammonium nitrate, nonnitroglycerin sensitizers, and a tiny amount (about 2 percent) of nitroglycerin. Ammite was also sold under the Giant name by Atlas Powder Company. Trojan Powder Company had introduced its nonnitroglycerin, no-freezing, cartridged nitrostarch powder in 1905.

Atlas Powder Company took over the Miner's Friend brand from the original Giant Powder Company in 1915. Because they often contained less nitroglycerin, many permissible dynamites were somewhat lower in freezing temperature to begin with. *Author*

many regions of the country did not need low-freezing explosives. Apache Powder Company even made a "summer formula" series of dynamites for very hot environments.

Priming

Priming is the process of rendering a dynamite stick ready to detonate. Because dynamite needs to be shocked into exploding, a blasting cap is inserted into at least one cartridge for each hole drilled. The explosion of the primed stick detonates all other sticks in the hole. Blasting caps are also called exploders, detonators, or initiators. Early on, caps were called primers and electric caps were called fuzes, pronounced "fuzees." Currently, fuse cap and fuse detonator are the correct designations for ordinary, conventional blasting caps.

In 1939 the Institute of Makers of Explosives removed all references to thawing dynamite, since by that time all dynamites were nonfreezing. An unintended consequence of discontinuing the markings was that less knowledgeable miners continued to thaw dynamite because the sticks were *not* marked "low-freezing."

In 1938 the US Federal Power Commission reported that "enormous quantities" of ethylene glycol were used in the production of nonfreezing dynamites. Ethylene glycol was discovered by accident in 1856, and its use in dynamite predates its use in automotive antifreeze. Alcohol was the main ingredient of early automotive antifreezes. Ethylene-glycol automotive antifreeze debuted in 1926, but it was not until the 1940s that the new formula outsold alcohol-based antifreeze.

Low-freezing dynamites revolutionized blasting in cold climes, but

Two primed cartridges (themselves called primers) were sometimes inserted into deep holes. The goal was to avoid a costly misfire, which entailed time and hazardous labor during finding and retrieving the unexploded cartridges. An extra primer was insurance against one wire or fuse becoming dislodged or damaged during loading and tamping.

Retrieving a dynamite charge that did not fire is called "drawing the shot" and is so hazardous that it was prohibited by many mines and jurisdictions. A miner had to dig through stemming and drag the unexploded cartridge out with a chambering spoon. All too frequently a blasting cap was crushed and detonated. Many misfires could be remedied by carefully removing some of the stemming material and refiring the hole with another charge. Another tack is to drill and charge another hole about a foot away from the hole containing the misfire. The concussion from the explosion in the new hole will usually detonate the unretrievable charge in the old hole. When available, pressurized jets of water can be used to flush everything out of the hole.

Ads for Atlas Ammite proclaimed, "It will not freeze under any condition."[21] As an added advantage, due to the low-nitroglycerin content, Ammite "will not cause headaches from handling."[22] *Author*

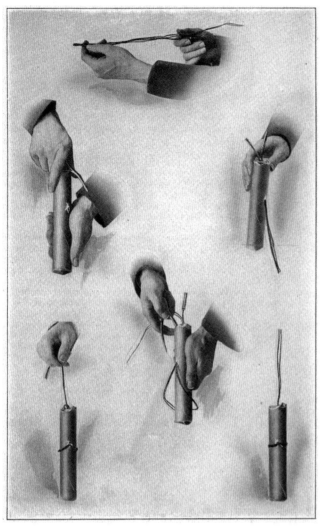

Here a hole is punched from the top of the cartridge diagonally through the side so that the wires can be looped around the cartridge. E. I. DuPont de Nemours, *High Explosives, First Section* (Wilmington, DE: E. I. DuPont de Nemours, 1920), 48.

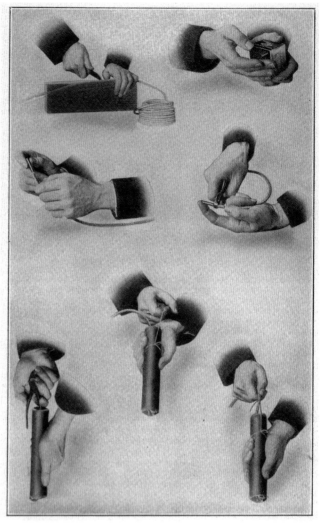

With this method of end priming, the string was tied around the cartridge first, then around the fuse. E. I. DuPont de Nemours, *High Explosives, First Section* (Wilmington, DE: E. I. DuPont de Nemours, 1920), 46.

Waiting several hours before even approaching a misfire was universally recommended. This is because, in rare instances, a misfire was actually an even more dangerous "hangfire." A hangfire is a charge that goes off after it is supposed to. Damaged, smoldering fuses and partially detonated, burning cartridges were the most-common causes of hangfires.

Basic end priming involves punching a hole in the end of the cartridge with a wooden, brass, or aluminum punch about the diameter of a pencil. The capped fuse or electric blasting cap is inserted into the hole. The wire or fuse is then either looped around the cartridge or secured with string or tape.

Side priming entails inserting the cap into a hole punched diagonally into the side of the cartridge.

Another technique involves punching a second hole through the cartridge perpendicularly and threading the wire or fuse through that hole. The goal is to provide a secure emplacement of the blasting cap, a secure connection to the fuse or wire, and, for vertical holes, a way to lower cartridges into the hole without mishap. Improper side priming with electric caps could cause the wire end of the cap to rub along the side of the borehole and either detach or prematurely detonate the cap. Primed cartridges were sometimes sheathed with a couple of

layers of blasting paper to protect the cartridge and fuse or wire assembly.

Priming with cap and fuse requires additional preliminary steps. A sharp knife is used to trim the fuse end so that it will abut the ignition charge in the cap. If the fuse is incorrectly cut, or if debris is caught between the end of the fuse and the ignition charge of the cap, a misfire may occur. Debris in the cap can also cause an explosion during insertion of the fuse via friction with the explosive charge in the cap.

The cap is firmly crimped to the fuse with purpose-made crimpers, although miners were known to use pliers, knives, and even their teeth. Large mines used bench-mounted

This miner is preparing primers at a dedicated station in an underground mine. If the mine followed proper procedures, this area was well removed from other boxes of dynamite. *Library of Congress Prints and Photographs Online Catalog*, LC-USF34-036934-D.

Note the dynamite punch in the foreground. Purpose-made punches were used instead of the punch provided on blasting-cap crimpers for one main reason. If, as here, the operator is not wearing gloves, using the separate punch meant less contact with nitroglycerin. *Library of Congress Prints and Photographs Online Catalog*, LC-USF34-036933-D.

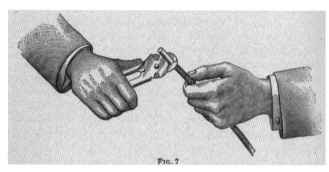

Left: The simple act of crimping a blasting cap onto the end of a length of safety fuse could be dangerous. International Library of Technology, *Rock Boring, Rock Drilling, Explosives and Blasting* (Scranton, PA: International Textbook, 1907), section 36, p. 42.

Below: A representative sampling of blasting-cap crimpers (*left to right*): Prince, Hercules Powder Company, Crescent Tool Company, Aetna Powder Company, Ensign-Bickford, and DuPont. Rounding out the group is a modern crimper sold by Blasters Tool & Supply Company of Lawrenceburg, Kentucky.

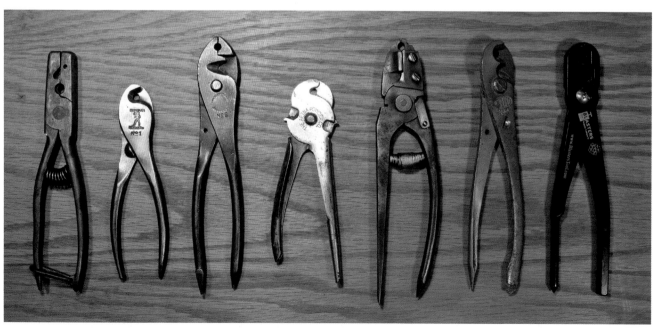

A cap affixed to safety fuse. While everything looks fine on the outside, a host of problems may lie hidden within. *Author*

cap crimpers, awls, and fuse cutters to punch holes in sticks and affix caps to fuse. In this way an entire day's worth of primer cartridges could be rigged at one location. Some mines prohibited carrying primed explosives, and government regulations outlawed the carrying of primed cartridges into coal mines in 1941. A primed stick of dynamite was more dangerous to transport because of the possibility of initiation of the blasting cap via shock.

Cap crimpers were sold by all major dynamite producers. Makers of other blasting gear such as Ensign-Bickford and a few toolmakers such as the vaunted Crescent Tool Company also sold crimpers. Rounding out the available selection were devices such as the Prince cap crimper, which was independently designed and produced via contract with larger toolmakers.

Some cap crimpers had fuse cutters on the sides or elsewhere. Most had a punch as one of the handles. The spring-handled Ensign-Bickford model was akin to the Cadillac of blasting-cap crimpers. It sported a crimper, fuse cutter, fuse-end splitter, wire cutter, wire stripper, punch, and screwdriver / pry bar. The tool even had a special mechanism to securely close the handles together, and a hole in one handle for hanging on a belt.

The two types of crimps that could be put on caps are known as cutthroat and sleeve crimps. Sleeve crimpers, also called broad-jaw crimpers, emboss a flattened indentation around the circumference of

the cap. The most common kind of cap crimpers is the cutthroat variety, which are used to make one or two thin, sharply indented crimps into the cap. Cutthroat crimpers are said to make a more watertight crimp.

The interiors of a fuse cap and an instantaneous electric cap are similar. In a basic blasting cap, the fuse or wires ignite an ignition charge, which ignites a priming charge, which ignites the base charge, which explodes the cap, which detonates the dynamite. With delay electric blasting caps, there are more elements that vary in length depending on the time of the delay. Early delay caps had short pieces of fuse serving as delay elements.

Dynamite producers sold their own branded blasting caps for two main reasons. The first was to provide convenience for those ordering explosives, providing "one-stop shopping." The second, more important reason was to ensure the proper detonation of the company's main product. Imported caps were considered unreliable, particularly as dynamite mixtures became less sensitive for safety reasons. The largest dynamite makers produced their own blasting caps. Smaller firms resold caps, usually in tins marked with their own company logos.

Nonelectric initiation came full circle with the invention of Nonel signal or shock tubing in 1967 by Swedish engineer Anders Persson. Nonel is short for nonelectric. The interior of the plastic tube contains a small amount of explosive powder. When ignited, a tiny explosive wave travels along the tube at a rate of 6,500 feet per second without destroying the tube. A blasting cap comes affixed to the end of the tube in the same manner as with electric blasting caps and leg wires. The tubing is flexible, and the impulse easily travels through kinks and even knots. The mechanics of

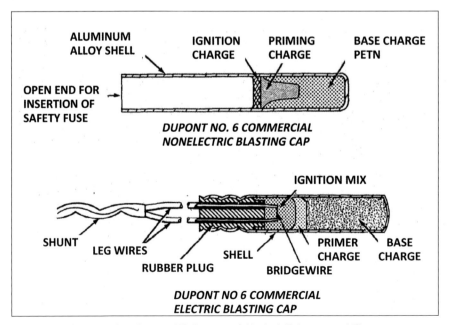

Basic fuse cap and electric cap. US Bureau of Alcohol, Tobacco and Firearms, *Modular Explosives Training Program* (Washington, DC: USGPO, 1976), 36.

Fuse caps were sold in metal cans until the 1940s. Caps in cardboard containers were said to be easier to remove from the box and be less likely to be fouled with debris. *Author*

Electric blasting caps come with leg wires varying from 6' to 50' in length. The wires resemble very long, very thin legs. *Author*

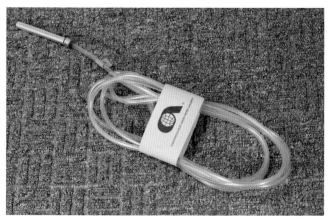

Above: Nonel signal tubing became commercially available in 1972. *Author*

Right: Here are two basic ways to prime with an electric blasting cap: "a" and "b" show basic end priming; "c," "d," and "e" depict side priming. Aetna Powder, *Handling Explosives* (Chicago: Aetna Powder, 1913), 56.

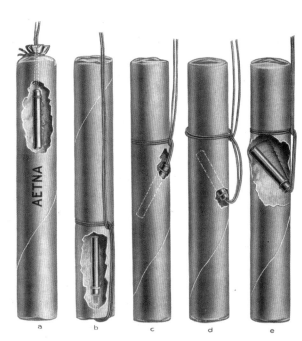

priming with shock tube is essentially the same as with electric blasting caps, with tubing instead of wires. A major advantage of the tubing is that extraneous electricity cannot initiate it. Machines for blasting with signal tubing range from small, handheld mechanical devices to sophisticated, computerized control systems.

There were more bad ways than good to prime a cartridge. Incorrectly primed sticks caused numerous problems. These include potentially damaging the fuse or wires, and the possibility of pulling the blasting cap out of the cartridge during loading and tamping. In addition, caps placed insufficiently deep into the cartridge could become damaged by contact with the wall of a borehole.

One final step for wet holes was to apply some cartridge soap (Peet's Cartridge Soap was a popular brand) to the openings in the cartridge. Sometimes the entire cartridge was smeared, rendering it relatively moisture resistant for a short time. Commercially available blasting sealants included Kapseal, Celakap (both sold by Hercules), P and B Paint, and DuPont Cap Sealing Compound. The tip of an end-primed cartridge was dipped in the compound, while the apertures in side-primed

**Blasting Accessories—For Firing High Explosives—Cap crimpers;
du Pont Cap Sealing Compound**

that blasting caps crimped tightly on the fuse are much more effective than those fastened loosely or not crimped at all. Crimping can be accomplished successfully only by the use of an instrument made especially for the purpose. Cap crimpers are essential for safety and for efficiency. Du Pont Cap Crimpers No. 1 and No. 2 make a flat sleeve crimp leaving a small air vent so that for use in wet work the cap and fuse must be dipped in some waterproofing substance to protect the cap charge from water. The No. 3 du Pont Cap Crimper makes a water-tight crimp on a smooth-covered fuse, so that no waterproofing is necessary. To assure detonation with this type of crimp, the freshly cut end of the fuse must be in contact with the explosive in the cap.

Du Pont Cap Crimpers are so made that they cannot squeeze the blasting cap far enough into the fuse to interrupt the burning

Fig. 21.—Cap crimped to fuse with flat sleeve crimper.

Fig. 22.—Cap crimped to fuse with air-tight crimper.

of the powder train and cause misfires. They are well made of good material, and if used only for the purposes intended will last for a long time.

Du Pont Cap Sealing Compound.—This is a material for sealing water-tight the space between the shell of a blasting cap and the fuse which is inserted into the blasting cap.

However well the cap may be fastened to the fuse by the crimper, it is almost impossible to make a joint that will prevent water from leaking in and spoiling the cap.

After the blasting cap is crimped on the fuse, the cap with two or three inches of the fuse is dipped for a second into the Cap Sealing Compound and hung up to dry. It is not desirable to soak the cap in the compound. By the time the compound has dried for about thirty minutes, a water-tight joint is formed which will resist almost any amount of water commonly encountered in blasting with cap and fuse. The cap should be used soon after it is dry as the Cap Sealing Compound becomes brittle after a few days and is likely to crack and admit water.

Fig. 23.—Du Pont Cap Sealing Compound.

Du Pont Cap Sealing Compound is put up in half-pint, pint, quart and gallon cans.

**Blasting Accessories—For Firing High Explosives—Multiple shots;
rotation shots; electric blasting caps**

Fig. 24.—Cap and fuse waterproofed with Cap Sealing Compound.

Multiple Shots.—When the charges in a number of bore holes are fired simultaneously, one of the following detonating agents must be used:

(a) Electric blasting caps for instantaneous firing under normal conditions.

(b) Waterproof electric blasting caps for instantaneous firing under water, or for a long series connection where the material to be blasted contains even minute quantities of mineral salts in solution.

(c) Cordeau for detonating charges instantaneously throughout their whole length.

Rotation Shots.—Where the separate charges are to follow each other in sequence, the following are applicable:

(a) Delay electric blasting caps for firing blasts in quick succession with a single application of the electric current.

(b) Delay electric igniters for firing blasts in quick succession with a single application of the current.

(c) Safety fuse and blasting caps as previously described, the fuse being cut in different lengths to give the desired rotation.

Electric Blasting Caps.—Electric blasting caps, or detonators, as the name implies, are special detonators fired or exploded by an electric current. Figures 25 and 26, show electric blasting caps complete and in section.

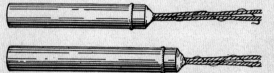

Fig. 25.—No. 6 and No. 8 du Pont Electric Blasting Caps. The wires are of different lengths to suit different depths of bore holes.

PRESSED CHARGE   ASPHALT & SULPHUR   ASPHALT   SULPHUR   LOOSE CHARGE

Fig. 26.—Du Pont Electric Blasting Cap (Section).

Du Pont Electric Blasting Caps are manufactured in the same strengths as ordinary blasting caps, designated as No. 6 and No. 8, the latter being twice as strong as the former.

DuPont Cap Sealing Compound was a type of asphaltum paint. Arthur La Motte, *Blasters' Handbook* (Wilmington, DE: E. I. DuPont de Nemours, 1925), 31. Arthur La Motte, *Blasters' Handbook* (Wilmington, DE: E. I. DuPont de Nemours, 1930, 39, 40.

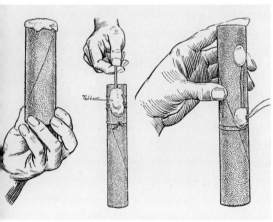

Here, cartridges are sealed with tallow, a waxy animal fat that is used for many purposes, including making candles. Institute of Makers of Explosives, *The Use of Explosives for Agriculture and Other Purposes* (New York: IME, 1917), 144.

cartridges were painted with a brush, or the corner of a rag. Thick lubricants such as axle grease were also used to waterproof cartridges. Some miners dripped hot wax from candles on cartridges to seal them.

Loading and Tamping

Tamping refers both to the act of compacting charges and the filling of the remainder of a hole with nonexplosive material to confine the charge. Any nonexplosive material loaded into the hole is called stemming. Vertical holes were usually easy to fill with dirt from the site and debris from drilling.

Wet clay was also used, particularly in farming applications. As long as paraffined cartridges or gelatin mixtures were fired within a reasonable amount of time, water did not affect the efficacy of the blast. Heavy dynamites with maximum "sinkability" were needed for water-filled holes.

With deep upward holes the primer cartridge was tied to a pole with light string so that it could be guided to the back of the hole. The pole was then jerked free and additional cartridges were added, again with care so as not to disturb the fuse or wires. One-third to one-half of the borehole was filled with dynamite, the rest with stemming.

Above: Miners using a tamping stick with a wooden block attached to tamp large-diameter cartridges. *Library of Congress Prints and Photographs Online Catalog*, LC-USE6-D-007360.

Above right: These miners are tamping a deep diagonal hole in a quarry. *Library of Congress Prints and Photographs Online Catalog*, LC-USF34-063979-D.

Right: Despite the apparent jumble of fuses, they were cut as precisely as possible to effectuate roughly timed detonation sequences. Electric blasting caps enable precision blasting. E. I. DuPont de Nemours, *High Explosives, First Section* (Wilmington, DE: E. I. DuPont de Nemours, 1920), 72.

More explosives than that "usually represent wasted energy,"[23] according to DuPont. Some miners loaded boreholes nearly to the top (called the "collar") of the hole for blasting very hard overburden.

Long, thin cartridges that could be coupled together made loading easier but made tamping more difficult. Tamping could reduce the height of a column of cartridges by up to 50 percent.

Placing the primer and other cartridges in the hole was always done with an eye on the fuse or wires, lest they be dislodged, frayed, or severed. Some tamping poles came with a groove that was intended to accommodate fuse or wires. The physical act of tamping required a firm but light touch. Miners who used too much force, especially with metal tamping poles, were sometimes seriously injured or killed when a blasting cap exploded due to the force of tamping. While foolhardy on their face, steel tamping bars were common because they were easy to make in the blacksmith's shop. Nonsparking copper tamping rods were also used, but almost all producers advised that only wood be used for tamping sticks. It was recommended that miners mark tamping poles to ensure that all holes in a given blast were loaded to equal height. Either marks were placed at measured intervals, or a rubber band or string was positioned at the appropriate height.

Whether to place the primer at the top, bottom, or center of the column of charges depended on the situation and also reflected practices recommended by various explosives producers. With cap and fuse, placing the primer at the bottom of a column could ignite the wrappers of the charges loaded atop the primer. For electric blasting, California Cap Company recommended an equal amount of explosives above and below the primer, while DuPont advised placing the primer at the top for basic blasting. Atlas recommended placing the primer at the bottom of the column, the point of "maximum confinement"[24] of the charge.

Tamping bags were crucial to properly loading upward holes and provided convenience for loading vertical holes. Tamping bags could be purchased empty or prefilled with sand and consisted of tube-shaped bags of the correct dimension for the hole. For example, 1¼" diameter bags were used for 1½" diameter holes. Some companies sold sand-filled cartridges as "tamping shells." Tamping bags and plugs were the only practical way to add stemming to up-angled holes.

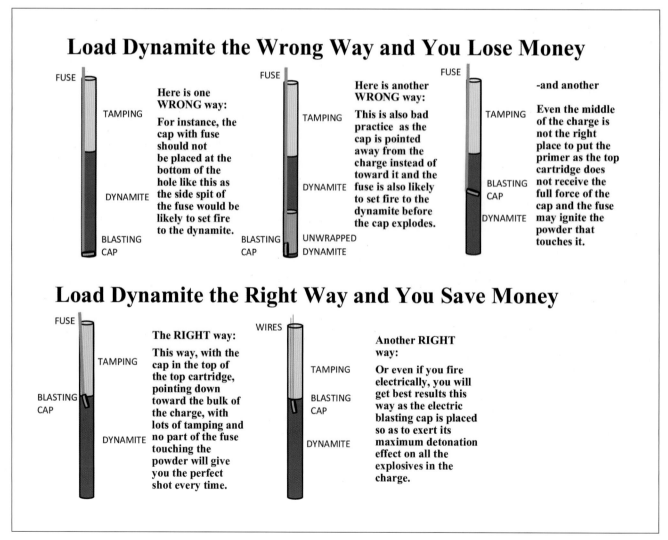

Load Dynamite the Wrong Way and You Lose Money

FUSE
TAMPING
DYNAMITE
BLASTING CAP

Here is one WRONG way:

For instance, the cap with fuse should not be placed at the bottom of the hole like this as the side spit of the fuse would be likely to set fire to the dynamite.

FUSE
TAMPING
DYNAMITE
BLASTING CAP
UNWRAPPED DYNAMITE

Here is another WRONG way:

This is also bad practice as the cap is pointed away from the charge instead of toward it and the fuse is also likely to set fire to the dynamite before the cap explodes.

FUSE
TAMPING
BLASTING CAP
DYNAMITE

-and another

Even the middle of the charge is not the right place to put the primer as the top cartridge does not receive the full force of the cap and the fuse may ignite the powder that touches it.

Load Dynamite the Right Way and You Save Money

FUSE
TAMPING
BLASTING CAP
DYNAMITE

The RIGHT way:

This way, with the cap in the top of the top cartridge, pointing down toward the bulk of the charge, with lots of tamping and no part of the fuse touching the powder will give you the perfect shot every time.

WIRES
TAMPING
BLASTING CAP
DYNAMITE

Another RIGHT way:

Or even if you fire electrically, you will get best results this way as the electric blasting cap is placed so as to exert its maximum detonation effect on all the explosives in the charge.

There was some disagreement among engineers and producers about the best practices for loading a column of dynamite. These are DuPont's recommended methods as of 1922. Adapted from E. I. DuPont de Nemours, "Dynamite Efficiency Depends on Proper Priming," advertisement, *Engineering and Contracting*, September 27, 1922, advertising section, 42.

The Advantages of Tamping

TAMPING PLUG

Exhaustive tests by the Bureau of Mines as to the efficacy of stemming on the effect of explosives have proved conclusively that confinement of the charge in the hole by the use of some inert substance on top increases its efficiency to a marked degree. The Bureau chooses to use the word "stemming" in place of the term "tamping," so commonly used, to differentiate the latter from the name of the operation of placing the stemming. In other words the Bureau favors the adoption of tamping the stemming in the hole. There is always a certain amount of opposition among miners to tamping a hole and some very prominent mining engineers question the advisability of so doing. It is probable however, that the data on which they formed these conclusions were not carefully and exhaustively drawn.

Tamping a hole has the same effect as putting a wad in a shot-gun shell or inserting the bullet in a rifle cartridge and is easily determined. For instance, take two ordinary rifle cartridges and crowd the bullet 1/16 in. nearer the base in one than in the other. Then fire the two. The "kick" against the shoulder will readily demonstrate the appreciable difference between the loose and the tight tamping of the powder charge. The same thing happens on a much larger scale when confining the explosive in a drill hole, and the kick of the explosive against the rock to be broken is much increased by close tamping in the throat of the hole.

It is almost always pointless to load explosives all the way to the collar of a hole, because the energy from the uppermost cartridges is dissipated into the air. Confinement with stemming materials directs energy toward the material that is being blasted. Adapted from *Engineering & Mining Journal*, July 17, 1915, 103.

Ordinary tamping bags came in a range of sizes, from 0.875" × 8" to 2" × 18". DuPont also sold jumbo bags measuring 3.5" × 16" and 4.5" × 16" for the larger holes bored by strip miners. There even were companies specializing in tamping bags, most notably the Tamping Bag Company of Illinois, which offered moisture-resistant bags in "sixty-five stock sizes"[25] in 1951.

Commercial tamping bags were available beginning in the 1890s. The early bags were usually made in sizes from 12" to 30" in length. Longer bags were harder to fill and could more easily get caught too high in a hole during loading. Most later-model tamping bags were less than 16" in length.

In the 1950s, companies began offering bags that were designed to be filled with water and closed by melting the open end, or with a self-sealing feature. Water-filled tamping bags were made of plastic and, along with extruded clay, were the only type officially approved for use in underground coal mines. Other stemming materials beyond dirt from the site were prohibited in subterranean coal operations due to the potential for fumes, dust explosions, and larger flames.

The Tamping Bag Company of Mt. Vernon, Illinois, was founded in 1935 by Alfred E. Pickard. Pickard also operated the Central Mining Supply Company in Mt. Vernon. By the 1960s, both companies were divisions of Pickard Industries, which had expanded to supply mining machinery of every description. *Author*

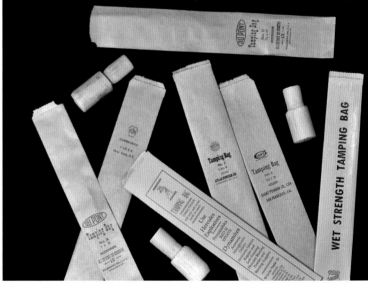

Major producers sold tamping bags emblazoned with their logo. Hercules Powder Company, one of the most prolific advertisers in magazines, trade journals, and newspapers, took advantage of the space on tamping bags to provide information about its other products. *Author*

As with all facets of mining, small operations could not afford luxuries such as commercial tamping plugs and prefilled tamping shells. Tamping bags and blasting paper were also used for loading railroad powder into upward holes. *Author*

Inventor George Sherman Clark incorporated the Clark Patent Blasting Tubing Company in West Virginia in 1922. Clark also obtained a patent for a "blasting[-] paper sheet" featuring printed lines for measuring powder. *Author*

A filled tamping bag is called a dummy, which is a different take on the term, since we normally think of it regarding a replica or prop stick of dynamite.

Dummies were also made in the field with blasting paper and dirt. A dowel (called a mandrel) of the correct length was affixed perpendicularly to a board, and the blasting paper was wound around the dowel a couple of times. The top end of the tube thus formed was folded and the tube was

filled. The open end was then folded shut. Shovel and pick handles were often employed as mandrels.

DuPont sold cork-like tamping plugs made of a wooden, cone-shaped driver surrounded by a shell made of asbestos. The plug was fit into the hole and then driven in, expanding the asbestos and forming a tight seal with the borehole wall without interfering with fuse or wire. DuPont sold its plugs under the name Quick-Seal. Lengths of nearby tree branches of the correct diameter were also used. The Vesey Safety Plug consisted mostly of a dowel and a wedge. The Heitzman Safety Blasting Plug Company of Pennsylvania sold a rubber plug with a wooden driver that

expanded the rubber to fit the borehole. In the 1980s, Austin Powder Company sold concrete plugs for large-diameter holes. A 1997 patent application for a tamping plug noted,

Traditionally, blasters have used the indigenous material to stem the borehole. For example, in rock quarries or other accessible locations, crushed rock is the preferred stemming material. Coal mines use drill cuttings as stemming since crushed rock is not available. However, in wet conditions, drill cuttings are unsatisfactory because they offer little confinement of the blast, blowing out of the borehole in a stream of mud. Mechanical stemming plugs also are used to confine the blast in the borehole. Tamping of boreholes to contain an explosive charge was taught 100 years ago in the "Rudimentary Treatise on the Blasting and Quarrying of Stone" by Maj.-Gen. Sir John Burgoyne, London, 1849. Gen. Burgoyne taught the use of iron tamping plugs in the shape of a barrel, a cone, and a cone with wedges. Generally, then as now, the stemming plug is placed above the charge to enhance and hold the stemming material such as crushed rock. When used correctly, stemming plugs can reduce flyrock, improve fragmentation, reduce airblast, expand borehole patterns and increase crushing. Such improvements allow the use of less explosive powder per ton of finished product.[26]

Tamping bags and plugs are still used today, but the plugs are usually made of plastic of one form or another, and the bags are made of polyethylene more often than paper (and the explosive involved is usually no longer dynamite). Other variations on the

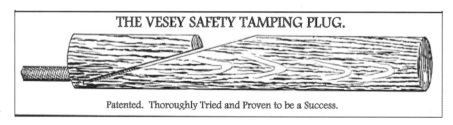

THE VESEY SAFETY TAMPING PLUG.

Patented. Thoroughly Tried and Proven to be a Success.

Richard S. Vesey of Idaho Springs, Colorado, patented this invention in 1898. "The Vesey Safety Tamping Plug," advertisement, *Mining Reporter*, November 1, 1900, 1.

tamping-bag theme include dummies made of wood or foam rubber.

The convenience of tamping bags was offset by the additional cost and the time it took to fill the bags. Larger companies had the luxury of ordering prefilled bags or shells. In smaller operations it was common for the rookies in the mining camp, or even children, to have the honor of filling bags. Preformed, tearable tubes made of blasting paper were also available. Clark's Blasting Tubing featured a patented, feed-from-the-center design that kept the roll intact during use. Atlas Powder Company sold a continuous tube of blasting paper in tube form called Sureshot Blasting Shells. For use for stemming, tubing saved time because longer bags could be fashioned. Thick scrap paper was the cheapest way to "roll your own" dummies. Through the 1890s the term "blasting paper" also referred to gunpowder-coated paper that was rolled into very small-diameter cartridges of less than ½" and detonated via fuse or blasting cap. According to DuPont,

> Tamping bags are filled in the following manner: The mouth of the bag is opened up by pressing the sides, the end is then placed to the mouth and blown into. It readily takes a cylindrical shape. By then placing a hand at the top of the bag the adobe we use for tamping is put in with the other hand, it only taking a few seconds to fill the bag. The adobe is packed into the bag by bumping the end against some object. After the bag is filled to within one inch of the top, the loose ends are folded over and the filled bag laid to one side. The men generally fill from ten to thirty bags at a time, this taking a very short space of time.[27]

Another stemming tool is the blasting mat, which often consisted of a large woven rug made of heavy sisal or steel wire. According to DuPont, its mats were made to order depending on the needs of the customer. The company's steel mats were produced only for large operations, where the machinery was available to move them. Makeshift mats of interlocking tree branches were also used. Sometimes, heavy objects such as railroad ties were placed over the borehole and then covered with the mat.

Commercial blasting mats can be reused until they are no longer viable. Many modern blasting mats are made of old tires. Currently, one can choose among more than a dozen US makers of blasting mats.

Some localities require that shot holes be covered to reduce the potential of damage to nearby structures and roadways from hazardous flying debris. For example, Massachusetts requires the use of a blasting mat for any blast within 100 feet of a building. Federal guidelines recommend the use of blasting mats when necessary.

Some mines used thousands of tamping bags a day. Large, well-financed operations constructed purpose-built structures for filling tamping bags in quantity. The buildings were sometimes called "sand houses."

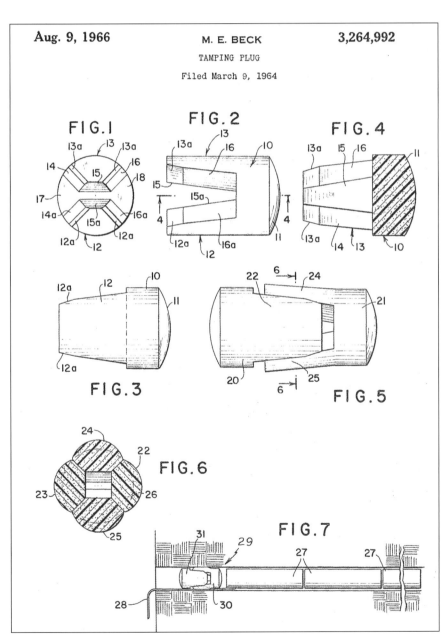

Tamping plugs predate tamping bags and were originally used to close holes filled with stemming from the site. M. E. Beck, Tamping plug, US Patent 3,264,992, filed March 9, 1964, and issued August 9, 1966.

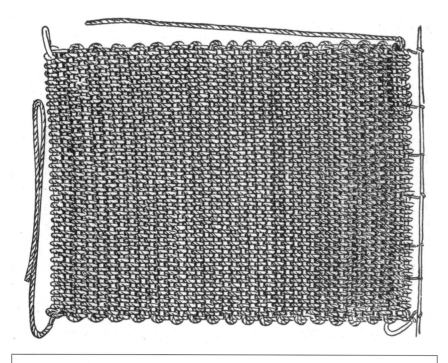

In loading short holes with a tamping stick, many powdermen make a mark or notch on the stick, to indicate the distance from the collar of the hole to the top of the pocket or chamber. By this means they can tell when the pocket is loaded and when the charge begins to rise in the borehole. It is also a good idea to make a mark, or place a rubber band around the tamping stick, opposite the collar or mouth of the hole when the stick rests on the top of the charge when finished. The distance between this mark and the end of the tamping stick will give the amount of tamping in the hole so that in the event of a misfire, it will be possible to remove nearly all of the tamping without danger of digging into the explosives. The tamping stick, of whatever material, should not be a close fit in the borehole, as in loading holes full of water, where the tamping stick fits the borehole snugly, misfires have occurred, in tamping the hole, from the piston effect forcing the water into the electric blasting cap.

In loading deep well-drill holes, where it is not practicable to use a pole, a tamping block of hardwood, attached to a rope, is used. This block should not be weighted unless the holes have water in them. When weighted, a hole about 1 ½ in. diam. is bored in the center of the block. A quantity of molten lead or babbitt metal is then poured in and the opening plugged with a wooden plug in such a way that no metal is exposed. These plugs are usually about an inch smaller than the diameter of the drill bit and from three to four feet long. The rope, for convenience and safety, is passed over an open block or pulley, supported by a tripod seven or eight feet high over the center of the borehole.

Blasting mats should be elevated "several feet above the blast, so when the blast is fired the flying rock is stopped by the underside of the mat," per DuPont. Arthur La Motte, *Blasters' Handbook* (Wilmington, DE: E. I. DuPont de Nemours, 1925), 40.

Some larger operations used tamping-bag-filling machines, which varied in complexity. One simple setup devised by DuPont consisted of a tray of sand with copper tubes attached to the bottom. Empty bags were placed over the tubes, and the sand ran into the bags by gravity. Other contraptions were quite complex and had pneumatic and vibrating elements to ensure a tightly filled bag. In the 1930s the Tamping Bag Company of Illinois sold a machine called the Dummy Maker, which required a crew of two but could fill five hundred bags an hour.

Tamping bags were also used to make cartridges of black powder on-site, although any stiff scrap paper was preferable due to cost.

Despite the wide use of stemming materials in conjunction with gunpowder and blasting powder, some miners believed that the confines of the hole itself were enough to contain and focus the blast of more-powerful dynamite. Experience and formal testing proved that blast efficiency was greatly increased when holes were tightly filled with compacted stemming material.

The primer cartridge was never slit or directly compacted, owing to the danger of damaging the fuse, wires, or blasting cap. Other cartridges were usually slit. Aetna at one point advised completely removing the wrappers

Marking a tamping stick made it easier to retrieve a misfire. Given the propensity of miners to favor convenience over precision, putting this much thought into loading a hole was likely not done in the majority of circumstances. Adapted from "Blasting Hints," *Western Engineering*, October 1, 1915, 170.

Fig. 98.—A wooden tamping stick. No metal parts are permissible.

Even nonsparking copper tamping rods were documented to cause accidents by crushing and detonating blasting caps. Arthur La Motte, *Blasters' Handbook* (Wilmington, DE: E. I. DuPont de Nemours, 1925), 81.

Despite their rough appearances, rigs such as this were designed for the controlled, precise application of tamping pressure, rather than violent ramming action. Arthur La Motte, *Blasters' Handbook* (Wilmington, DE: E. I. DuPont de Nemours, 1925), 76.

"for the perfect charge"[28] and even recommended that the cartridge be "cut into small pieces and rammed into the hole."[29] While it sounds dangerous, cutting and shaping cartridges was common practice. Sticks of gelatin dynamite were even rolled into ropes that were wrapped around tree trunks, beams, and boilers.

Many small tamping poles were improvised, although commercial tools were also available, especially for deeper boreholes. An old shovel or broom handle was ideal, as were straight tree trunks of the appropriate diameter (in 1874, a Canadian inventor patented a shovel with a tamping stick as the handle). In 1949 a 10-foot-long, 1¼" diameter wooden tamping pole cost $1.00.

Tamping poles that were made to be joined together for use in deep holes had connectors made of nonsparking aluminum or brass. For very deep holes, a length of flexible rubber or plastic tubing with an attached wooden tip was fed from a coil into the hole. Sometimes only a weighted block was lowered by rope into deep holes. Such blocks could cause a premature explosion by igniting a fuse if sand particles stuck to a tamping block. For this reason, a piece of burlap was used to set the block on so that dirt would not adhere to it. Although it was obviously dangerous, some miners attached wooden tamping blocks to steel drill bits.

Blasting

Exploding a charge of dynamite is accomplished by six basic methods. A fuse can convey a flame to detonate a blasting cap. An electrical impulse can be sent to explode a blasting cap via a wire. Since the 1970s, a cap can be fired by sending a tiny traveling explosion through a plastic tube. And in the last two decades, wireless systems have evolved.

Another way involves detonating cord. While detonating cord has been produced by other manufacturers, the brand that became synonymous with the product is Primacord. Detonating cord explodes along its entire length when initiated and will in turn detonate any cap-sensitive explosive compound with which it is in contact.

Finally there is sympathetic propagation, which uses the concussion from one charge to detonate another.

If not for the prior invention of safety fuse, originally used to set off blasting powder, the dynamite industry might never have blossomed. *Author*

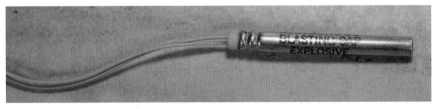

Electric blasting predates fuse caps but did not come into wide use for exploding dynamite until the 1910s. *Author*

Signal tubing is made of 3 mm diameter plastic tubing with a 1 mm diameter hollow core. This core contains a dusting of a high explosive called HMX that almost instantaneously detonates along the entire length. *Author*

Unprimed cartridges within the same hole are always detonated by shock, and entire nonprimed columns of charges can be exploded this way to instantly dig ditches in moist soil.

(In 1974, Hercules Powder Company introduced a system called Hercudet that used two hollow tubes filled with an explosive gas mixture connected to the cap. When fired, the gas exploded along the lengths of the tubes like shock tubing. The system was a bit cumbersome and was around for only twenty years.)

Safety fuse consists of a core of black powder wrapped in cloth threads and paper tapes. The fuse is varnished and waxed for resistance to weather and damage from loading and tamping. Commercial safety fuse burns at a regular rate, either ninety or 120 seconds per yard.

Cold fuse is stiff and must be warmed prior to uncoiling. Moisture resistance is a primary issue, and fuse manufacturers offered a range of products designed for "wet work."

Before a fuse was lit, it was recommended that the tip be slit to expose more of the inner powder to the flame. The Prince cap crimper and a few other designs had a built-in fuse splitter consisting of a hole in the tool with a razor blade inside. The fuse was inserted into the hole, and the blade neatly split the fuse. Many other cap crimpers had built-in fuse cutters, but these tended to dull quickly.

For multiple shots, different lengths of fuse were attached to a trunk line or lit individually. This latter task was sometimes accomplished with special igniters that fit over the end of the fuse that could be pulled to light the fuse.

Quarrycord, Minecord, Ignitacord, Thermalite, and Spittercord were essentially long, thin fuses that were attached to the ends of safety fuses for sequential firing. Short lengths of safety fuse can themselves be used as multiple fuse lighters, especially if all fuse tips are split. Other means of igniting multiple fuses included

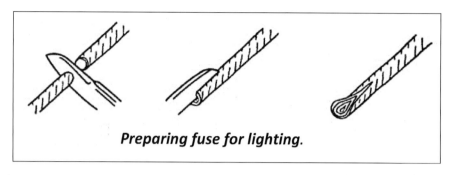

Preparing fuse for lighting.

Above: A knife for trimming and splitting fuse must be razor sharp. Fuse was cut atop a wood box or block. US Bureau of Mines, *Miner's Circular 58* (Washington, DC: USGPO, 1947), 21.

Right: In the 1910s, some experts recommended inserting a small wedge of Giant Powder dynamite into the split end of a fuse to ensure ignition. The dynamite did not detonate but burned "with a bright fierce flame," according to mining engineer John Cosgrove. John Cosgrove, *Rock Excavation and Blasting* (Pittsburgh, PA: National Fireproofing, 1913), 85.

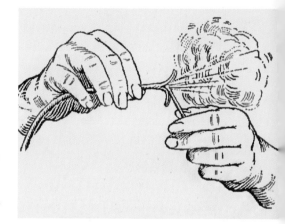

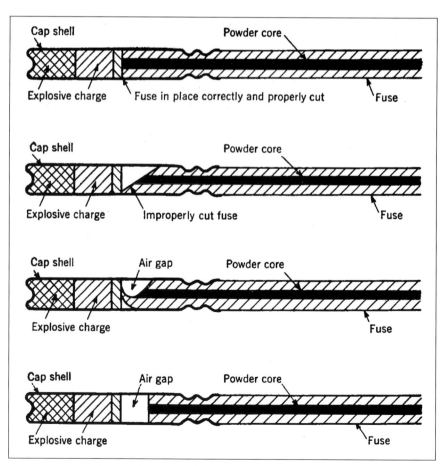

Here are some of the problems associated with incorrectly cut fuse. A ragged end cut with a dull knife might not properly abut the base charge inside a blasting cap. US Bureau of Mines, *Miner's Circular 58* (Washington, DC: USGPO, 1947), 23.

"hot-wire" lighters resembling sparklers or rolls of solder.

Before the invention of safety fuse by William Bickford in 1831, crude methods were used to ignite charges of gunpowder and then blasting powder. Hollow straws, reeds, and even goose quills were filled with powder. Another popular fuse was a long strip of touch paper, which originally was plant-fiber paper or rag paper permeated with potassium nitrate (paper was not made primarily of wood until the 1880s). Astonishingly, touch paper was used in some northeastern US coal mines into the 1920s. Squibs, patented for mining in the 1870s, were used to ignite black powder charges through the 1950s but could not be used to detonate dynamite.

In 1954 the Homestake Mine in Lead, South Dakota, reported using 195 miles of safety fuse over the course of the prior year; 210,142 blasting caps and 29,605 fuse lighters were used to detonate 1,386,815 pounds of dynamite. The company estimated that if the cartridges were laid end to end, the chain of dynamite would measure 525 miles.

Bunching fuses for simultaneous lighting was common practice and safe if done correctly. The US Bureau of Mines advised bunching no more than fifteen lengths of fuse in one bundle. It was essential that all fuses were precisely trimmed to achieve both proper firing order and simultaneous ignition of the fuse tips. For lengths of forty-second-per-foot fuse, 2" increments were typical intervals. This allowed for delays of about three seconds—a far cry from the millisecond delays required for advanced quarry blasting. Still, for a basic wedge cut, blasting with cap and fuse allowed for the inner charges to remove the center of the cut before the outer charges finished the job. The goal of timed blasting is to blast a successive series of faces, each of which is created by the prior blast. This meant that each charge had less material to move.

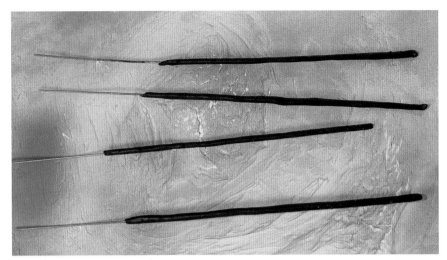

These devices were called hot-wire fuse lighters to distinguish them from pull-wire fuse lighters. The latter were friction-activated sleeves that fit over the ends of fuses. *Author*

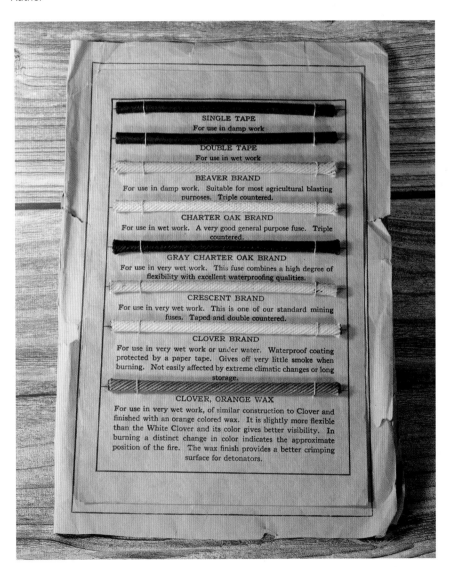

In 1898, Ensign-Bickford offered twenty-seven different brands of fuse, including White Double Tape, Double Wove, Colliery, Treble Wove, Thread Sump, Tape Sump, Double Tape Sump, Double Gutta Percha, Metallic, and Gutta Percha Countered Metallic. All had professed advantages. This sample card is from the 1940s. *Author*

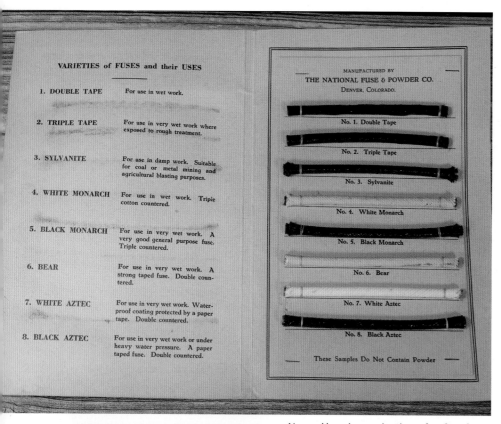

VARIETIES of FUSES and their USES

1. DOUBLE TAPE	For use in wet work.
2. TRIPLE TAPE	For use in very wet work where exposed to rough treatment.
3. SYLVANITE	For use in damp work. Suitable for coal or metal mining and agricultural blasting purposes.
4. WHITE MONARCH	For use in wet work. Triple cotton countered.
5. BLACK MONARCH	For use in very wet work. A very good general purpose fuse. Triple countered.
6. BEAR	For use in very wet work. A strong taped fuse. Double countered.
7. WHITE AZTEC	For use in very wet work. Waterproof coating protected by a paper tape. Double countered.
8. BLACK AZTEC	For use in very wet work or under heavy water pressure. A paper taped fuse. Double countered.

MANUFACTURED BY
THE NATIONAL FUSE & POWDER CO.
DENVER, COLORADO.

No. 1. Double Tape

No. 2. Triple Tape

No. 3. Sylvanite

No. 4. White Monarch

No. 5. Black Monarch

No. 6. Bear

No. 7. White Aztec

No. 8. Black Aztec

These Samples Do Not Contain Powder

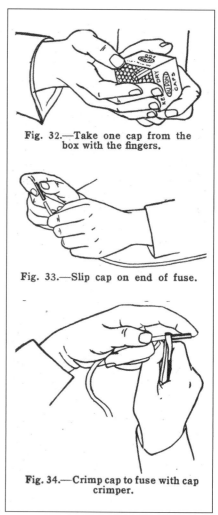

Fig. 32.—Take one cap from the box with the fingers.

Fig. 33.—Slip cap on end of fuse.

Fig. 34.—Crimp cap to fuse with cap crimper.

Above: Here is a selection of safety fuse sold by the National Fuse and Powder Company of Denver, Colorado. The card dates from the 1940s. *Author*

Left: While wire for electric blasting was originally sold by the pound, fuse came by the foot. Cases of fuse usually contained from 1,000' to 6,000' of fuse in rolls containing 100' each. In 1921, a 6,000' case cost about $20.00. Ensign-Bickford also sold fuse in 3,000' lengths wound on metal reels. *Author*

DuPont advised inserting fuse into cap "gently."[30] The Bureau of Mines added, "Dirt or foreign material may lodge in the open end of a cap; this makes it extremely dangerous when inserting the safety fuse, especially if the blasting cap is rotated on the fuse."[31] Arthur La Motte, *Blasters' Handbook* (Wilmington, DE: E. I. DuPont de Nemours, 1925), 71.

A clever method of simultaneous fuse lighting employed by one mine consisted of a wooden tray with holes drilled in the sides to accept fuse. The ends of the fuses were centered in the tray and ignited all at once.

The Ensign-Bickford Company introduced Cordeau in 1913. This was the first commercially successful detonating cord and consisted of a long lead tube filled with TNT. It was sold as Cordeau-Bickford in the United States. Cordeau resembled very thick soldering wire. Ensign-Bickford replaced Cordeau in 1937 with Primacord (Primacord-Bickford in the United States), which was a detonating cord that looked and handled like safety fuse. Primacord is pentaerythritol-tetranitrate (PETN) wrapped in cloth and plastics. It cannot be initiated by lighting the end. Instead, a blasting cap is hooked to the end of the cord with a special connector. The blasting cap is fired with fuse, wire, or signal tubing. Detonating cord can be tied together in a series of trunk and down lines and threaded through cast explosives.

Blasting-cap consumption has steadily decreased with the use of newer explosives and the increased use of detonating cords such as Primacord. One blasting cap initiated by shock tubing can initiate several strands of detonating cord, which can in turn detonate hundreds of explosive charges.

Detonating cord was "a new technique in blasting,"[32] as Ensign-Bickford put it, "acting as the detonating agent for each load and also serving to connect all loads."[33] In other words, all cartridges of a column in a blasthole

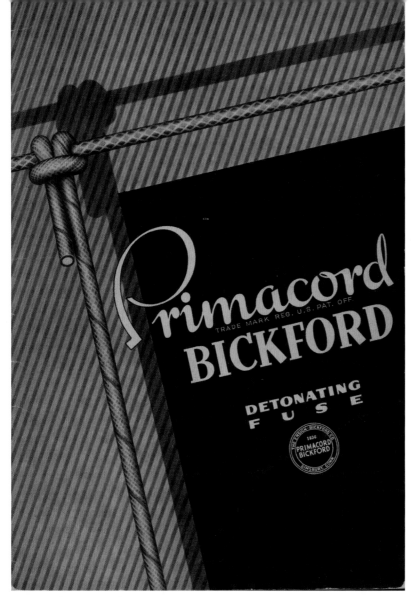

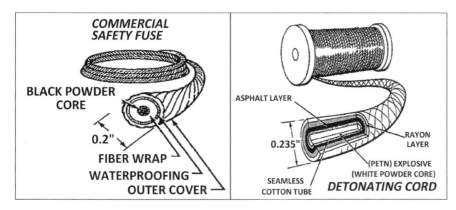

explode as if individually primed. A single blasting cap can initiate the detonation of many very long sections of detonating cord. Lighting detonating cord does not explode it. Detonating cord was used primarily for initiating large-diameter dynamite cartridges. To prime a small cartridge of dynamite with detonating cord, a hole is punched into the cartridge, and the tip of the cord is inserted into the hole. The cord is then taped to the outside of the stick. The cord can also be laced through the sides of the cartridge. With large-diameter cartridges, detonating cord can be inserted and then looped around the bale of the cartridge. Detonating cord was originally sold in spools of 500' and 1,000'. Ensign-Bickford offered at least sixteen different varieties of Primacord categorized by strength, water resistance, and speed of detonation. These included Plain Primacord, Reinforced Primacord, Lo-Temp Primacord, and PETN Plastic Primacord. Diameters ranged from 0.165" to 0.323".

Detacord was a short-lived, small-diameter detonating cord introduced by DuPont in 1957. Other brands of

Large producers such as DuPont, Atlas, and Hercules had their own blasting-cap factories. Smaller companies such as General Explosives Company resold caps that were made by companies such as Metallic Cap Manufacturing Company in tins marked with their own brand. *Author*

detonating cord exist, but none have equaled Primacord in popularity. Modern detonating cord has either four or seven internal strands of explosives.

A blasting cap consists of a small explosive charge enclosed in an aluminum, steel, or copper (actually, a more workable alloy of copper and zinc called gilding metal) shell. Initially the internal charge consisted of mercury fulminate, but many other explosives have been used. The length of blasting caps has also varied, from ½" for the earliest "Number 1" fuse caps to well over 6" for some electric and shock-tube delay detonators. As dynamites became less sensitive for safety reasons, stronger and stronger caps were needed to achieve reliable results. Commercial nonelectric-blasting-cap strengths culminated in #8.

Steel-shelled blasting caps were sometimes used in coal mines because

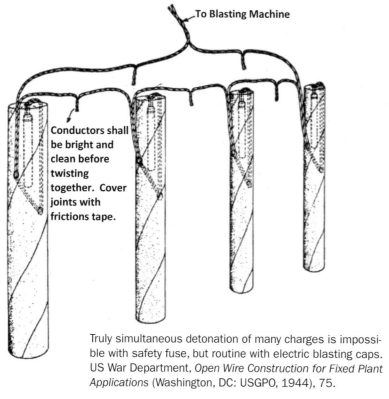

Truly simultaneous detonation of many charges is impossible with safety fuse, but routine with electric blasting caps. US War Department, *Open Wire Construction for Fixed Plant Applications* (Washington, DC: USGPO, 1944), 75.

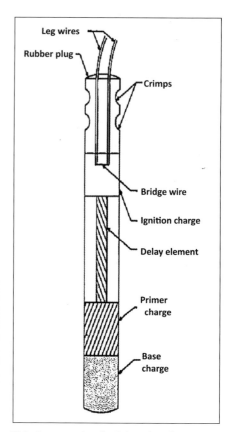

This is a very basic delay electric blasting cap. Modern delay caps have tiny, wafer-like delay elements. US Department of Labor, *Blasting Requirements—Surface Coal* (Washington, DC: USGPO, 1994), 23.

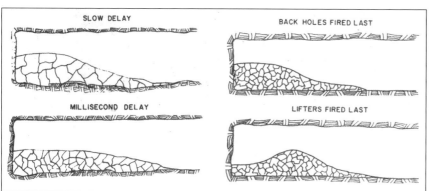

Differing configurations of delay blasting caps produce dramatically different results. US Department of the Interior, *Explosives and Blasting Procedures Manual* (Washington, DC: USGPO, 1983), 69.

the fragments could be removed with magnets during size sorting and screening. Otherwise, copper-shelled caps were used.

Crude timing of blasts was accomplished by trimming fuses to different lengths, but it took the advent of electric caps and delay electric caps to modernize quarry blasting. While early electric caps had a single explosive charge, modern caps have wafers of various compounds and incorporate precise delay elements that enable the coordinated detonation of thousands of charges in a single, highly orchestrated sequence. The leg wires of electric blasting caps are attached to one another if long enough; otherwise, connecting wire is used.

The primary advantage of sequential blasting is that explosions can be engineered so that material is removed or weakened by one blast an instant before the next blast takes over. Generally, slower-delay blasts produce coarser fragments and leave a concentrated muck pile. Millisecond delays make for finer material distributed over a wider area.

The first US patent for a firing device employing an electric current to ignite blasting powder was obtained by Moses Shaw in 1830, after successfully demonstrating the hand-cranked, friction-based machine in 1829 (patent numbers were not issued until 1836). Ben Franklin and others had ignited blasting powder by an electric *spark* as early as the 1750s. Franklin's design of cartridged blasting powder even had an electric wire built in. Of course, a mere spark is useless as a reliable means of exploding a cartridge of dynamite. Oddly, electric blasting was first achieved in 1832, three decades before Nobel devised the nonelectric detonator. (Blasting machines are sometimes erroneously called detonators. Early on, plunger machines were also referred to as blasting batteries, contrasted with the term as used to refer to the storage batteries used for single-shot blasting.)

Electric firing remained primitive until the first commercially successful, plunger-operated blasting machine was patented by prolific inventor and businessman Henry Julius Smith in 1886.

In the interim, a variety of contraptions, many sporting cranks, appeared. These employed static electricity, friction, batteries, and magnetism. None were reliable enough to endure.

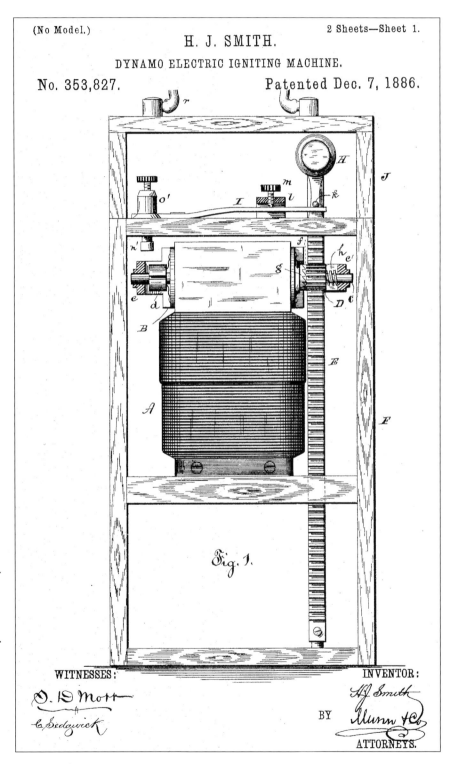

Henry Julius Smith obtained his first patent at age twenty-two. Smith founded several businesses, including the Smith Electric Fuse Works, launched in Pompton Lakes, New Jersey, in 1886. DuPont acquired the business in 1908. H. J. Smith, Dynamo electric igniting machine, US Patent 353,827, issued December 7, 1866.

Many models of the plunger-activated blasting machine exist. Some earlier versions were made so that the plunger was pulled up. All employed an electricity-generating gizmo called a dynamo. Moving the plunger spins a magnet, which imparts a current to a coil. If a complete circuit is run from the ends of the coil, this electricity is discharged through the wire. The number of blasting caps that could be detonated on one circuit depended on the strength of the blasting machine and, for the basic plunger type, ranged from one to 150.

All plunger-activated machines relied on a rapid and complete descension of the rack bar that energized the magneto. Because of the buildup of current, the plunger is more difficult to depress as the rack bar approaches its lowest position. Thus, momentum must be maintained from the start. DuPont stated, "Do not be afraid of pushing the rack bar down too hard. The machine is built to stand it, and this is the only way to use it successfully."[34]

The lightest models of plunger-type blasting machines weighed around 15 pounds. The heaviest, highest-capacity machines weighed as much as 70 pounds.

Above: Blasting equipment such as blasting machines, galvanometers, and rheostats were sold by most dynamite producers. The equipment was usually purchased from outside sources and branded with an appropriate company nameplate. *Author*

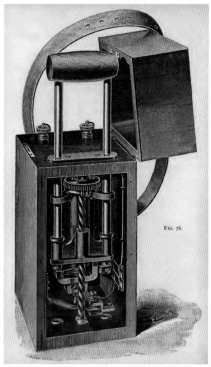

New designs of the plunger-activated blasting machine were patented into the 1940s. This 1890s version has an attached, hinged cover. Zacharias Daw, *The Blasting of Rock in Mines, Quarries, Tunnels, Etc.* (New York: Spon & Chamberlin, 1909), 166.

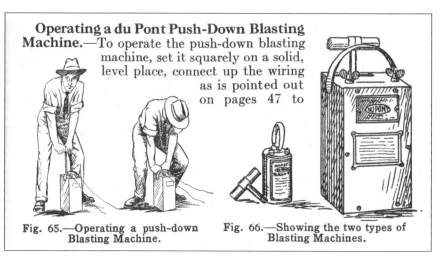

Operating a du Pont Push-Down Blasting Machine.—To operate the push-down blasting machine, set it squarely on a solid, level place, connect up the wiring as is pointed out on pages 47 to

Fig. 65.—Operating a push-down Blasting Machine.

Fig. 66.—Showing the two types of Blasting Machines.

The plunger on a blasting machine must be pushed quickly and forcefully down. The advice frequently given to operators was to "try to knock the bottom out of the machine." Current is not discharged if the plunger is not pushed to its full limit. Arthur La Motte, *Blasters' Handbook* (Wilmington, DE: E. I. DuPont de Nemours, 1925), 71.

According to the 1958 *Blasters' Handbook*, successful electric blasting depends on "(1) selection and laying out of the blasting circuit, (2) connection of the wires and protection of the joints, (3) testing of the circuit, (4) utilization of the electrical energy available, and (5) safeguarding of the blasting circuit from extraneous energy."[35]

While the complexities of electric circuitry are beyond the scope of this book, protection of the joints means waterproofing caps and splices with a sealing compound or taping them for stability. As far as extraneous energy, it is barely possible to accidentally fire circuits with radio waves under the exact right conditions. Lightning is another story, and electric-blasting operations are never conducted in the vicinity of thunderstorms. Electricity from a lightning strike can carry a considerable distance through wet ground or through metal, such as railroad tracks and underground pipes. Buildups of static electricity were also known to cause premature initiation of electric blasting circuits.

One of the smaller blasting machines was the Eveready single-shot blasting unit. The device was modeled on an existing product: the battery-operated flashlight. At the time, Eveready was a brand held by the National Carbon Company. The device measured 8½" × 1½"—just larger than a standard stick of dynamite. Leading wires were attached to the firing pin, and the pin was inserted into the battery housing to fire. The ring was for hanging on a wall or belt. To gain approval from the US Bureau of Mines, this and other blasting machines were tested for reliability, arcing, and other issues relating to air quality and safety in coal mines. The Eveready unit had a safety feature consisting of a spring that prevented the firing pin from engaging unless deliberately pressed home.

Probably the most compact blasting machine was the Miner's Individual Shot-Firing Battery, sold by Mine Safety Appliance Company. The device,

Left and above: The model 2696 Eveready single-shot blasting unit was introduced in 1926, cost $3.00, and was powered by three number 950 Eveready unit cells, now known as D-size batteries (assigning letters to batteries did not start until 1928). *Author*

Below: Electric miner's cap lamps debuted in 1907, and a miner's cap-lamp battery adapted to blasting, called the Edison Shot-Firing Unit, soon followed. Other commercial single-shot-firing attachments for cap-lamp batteries appeared just a few years later. *The Mining Catalog, Metal and Quarry Edition* (Pittsburgh, PA: Keystone Consolidated Publishing, 1921), 461.

Edison Shot-Firing Battery, Showing how special cover for shot-firing is attached to regular Edison Mine Lamp.

The Edison Electric Shot-Firing Battery

Showing cover for M-8 Mine Lamp Battery arranged for shot firing. Shot-firing cable is attached to a key furnished with cover. Key is then inserted in receptacle on cover, and detonation is accomplished by pressing key into contact. Cover is furnished with or without fittings for lamp attachment.

It is simple, light-weight, strong and durable. One charge of the battery is capable of firing a great many shots.

Several variations of the Edison Shot-Firing Unit were produced. With this model, leading wires are inserted into holes in the terminals, and then both buttons are firmly depressed to fire the unit. *Author*

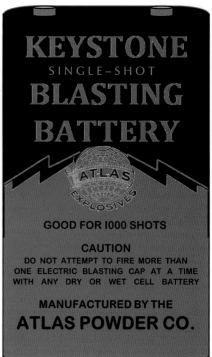

The Keystone Blasting Battery sold by Atlas Powder Company was not approved for blasting in coal mines but would have been ideal for farmers and loggers. *Author*

This battery was one of just five of its kind to earn permissible status from the Bureau of Mines from 1924 to 1938. Only fifteen shot-firing devices or batteries of all types were approved during that same period. *Author*

powered by a 3.6-volt flashlight battery, was slightly bigger than a deck of playing cards.

Portability and simplicity made cap-lamp-battery blasting units immensely popular with coal miners. Leading wires were connected directly to blasting-cap leg wires for one-shot blasting.

Many other single-shot-firing units were powered by batteries. Some stand-alone batteries were marketed as shot-firing batteries. These were ordinary, 3-to-9-volt, two-pole batteries. Such batteries could be used for a long list of other uses. The term "blasting battery" was generally used to refer to blasting machines. "Shot-firing battery" could either mean a blasting device or a stand-alone, conventional electric battery. The Bureau of Mines referred to stand-alone batteries as package-type, dry-cell batteries. Such batteries were not permissible in coal mines unless they had recessed terminals, enclosed terminals, or "plunger-type" terminals. The latter were posts into which wires were inserted. The posts were then depressed simultaneously to fire a shot. A Rayovac brand 3-volt battery labeled as a shot-firing battery sold for a dollar in 1953.

Only single-shot blasting units passed permissibility tests until 1938, when the first ten-shot machine was approved. Whether permissible or not, single- and ten-shot machines were

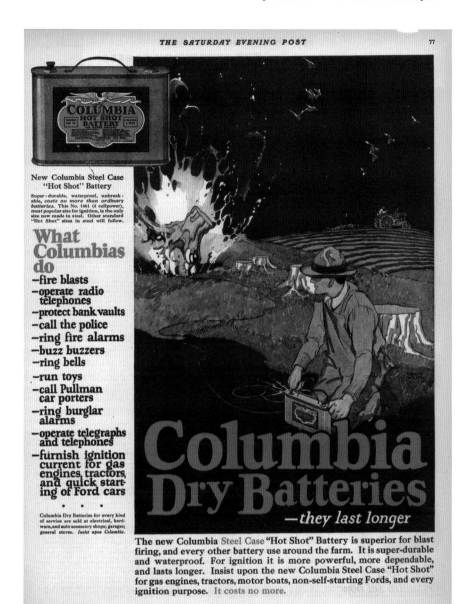

unsuitable for quarry work, where thirty to fifty shots were fired per blast at a minimum.

When available, power lines are also tapped into; large mines and projects have full electrical systems that are linked to the local grid. Portable electric generators normally allocated for lighting and for powering welding machines are used as well, with a shunt installed between the generator and the blasting circuit.

Farmers were known to rig up homemade blasting machines by hooking a series of small, inexpensive batteries together. In 1925 the US Department of the Interior estimated the cost of such a system at less than $2.00, versus $15.00 for a small, push-down blasting machine. The problem was such a device had "live" terminals, while a factory-made blasting machine had terminals that were nonconductive until the plunger was activated or the handle was turned. While more-industrious farmers might engineer a shunt or switch, most simply touched the leading wires to the last pair of battery terminals in the series to fire a shot.

Regardless of power source, blasting circuits were tested with a galvanometer, which detects undesirable resistance on a wire. The Bureau of Mines recommended that the leg

Above: That conventional battery manufacturers would market their product in this way reflects the commonality of dynamite on the farm in the early twentieth century. Nonetheless, calling the battery the "Hot Shot" avoided limiting the battery's perceived uses to blasting. National Carbon, advertisement, *Saturday Evening Post*, December 9, 1922, 4.

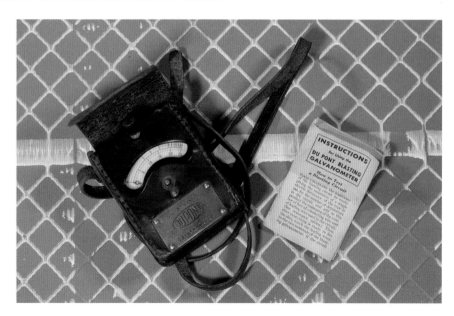

Right: Galvanometers were and are used much more for testing other electric devices than for blasting machines. An ordinary galvanometer is identical to this blasting galvanometer, except for the company nameplate. *Author*

wires of each electric blasting cap be tested as well, but this practice was seldom followed. The fitness of a blasting machine was checked with a rheostat, which measured how many caps could be fired given the length and type of wire, and the complexity of the circuit. Dynamite producers sold company-branded galvanometers and rheostats to ensure high quality and provide a single source of explosives, tools, and supplies for all of one's blasting needs.

In 1915, DuPont introduced its five-cap-capacity Pocket Blaster. Hercules followed suit with its four-cap Midget Blaster in 1916. Giant Powder's version of this small machine was ironically named the Giant Blaster. This style of device uses a scaled-down magneto actuated by the swift twist

of a handle or key. Some models had the handle slot on the bottom or the side of the machine instead of the top.

Slightly larger versions with a capacity of ten caps appeared in 1920. This style of device is so reliable and durable that the identical design is manufactured to this day and was adopted as standard US military equipment starting in World War II. At 4 to 5 pounds in weight, the machines were ideal for farmers, loggers, and subterranean miners.

According to DuPont, when operating a twist-type blasting machine, "the

quicker the twist, the more current is developed."[36] The capacity of the devices, also known as key-twist blasting machines, tops out at twenty shots.

Because twist-handle blasting units have always been a mainstay of military blasting, US Army field manuals provide ample instructions concerning their use. "Exercising the blasting machine"[37] was recommended and consisted of operating the handle several times before hooking up the circuit. The purpose was not to build up current, which was impossible, but simply to test the functionality of the

This machine is a DuPont-labeled, ten-cap model from the 1940s made by the Davis Instrument Manufacturing Company (few explosives producers actually made their own blasting accessories). It cost $30.00 new; a brand-new ten-cap machine, identical in design, currently goes for about $1,000.00. *Author*

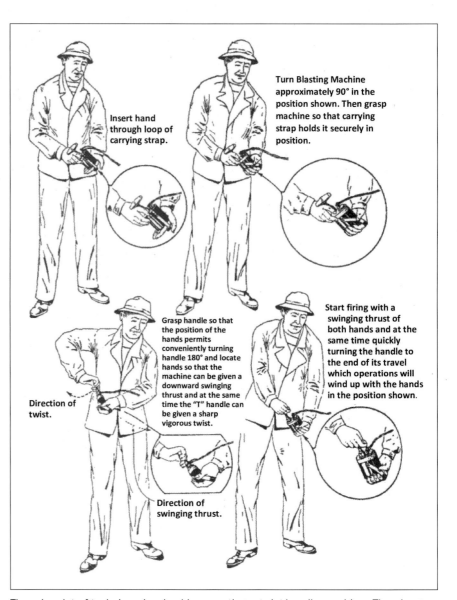

There is a lot of technique involved in operating a twist-handle machine. The plunger-type blasting machine is obviously easier to use, but the twist type is much more portable and simpler in design. US War Department, *Open Wire Construction for Fixed Plant Applications* (Washington, DC: USGPO, 1944), 76.

This DuPont CD-32 blasting machine weighs 22 pounds and measures 9" × 9" × 10½". In 1988, Research Energy Corporation of Ohio sold a one-thousand-cap-capacity machine that weighed 20 pounds and measured 6" × 6" × 10". The company also offered an eighty-cap version measuring 5" × 5" × 2" that weighed 2 pounds. *Author*

machine. In addition, "new operators"[38] were advised to practice with blasting circuits that were hooked up to blasting caps with no other explosives involved. Manuals also recommended that the leg wires of each blasting cap be checked with a galvanometer.

Another type of battery-powered device called the condenser (or capacitor) discharge blasting machine appeared in the 1940s. In 1952, DuPont introduced its CD-12 and CD-24 models, followed by the CD-32 and the CD-48 the following year. Batteries temporarily charge a capacitor, and the stored electricity from the capacitor is used to fire the blasting caps. The CD-32 could fire up to 480 caps, while the CD-48 could fire up to 1,220 caps. Atlas introduced its Shotmaster CD device in 1954, but its condenser was charged with a crank instead of batteries. With most other models, a button was depressed to charge the capacitors by running current from internal batteries. The second button discharged electricity to fire the circuit.

Despite the new technology, Atlas and DuPont still sold wooden-box, plunger-activated blasting machines into the 1960s.

The wire used to make the circuits is classified as leading wire and connecting wire. Leading wire runs from the blasting machine to the blasting-cap circuit array. The wire connecting the

Leading wire was originally provided in coils containing from 200' to 500' of wire, sold by the pound, while connecting wire came in 1-pound spools. Now both types of wire are sold by the foot, in spools of up to 1,000' in length. *Author*

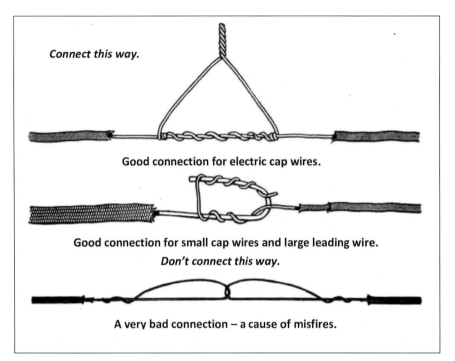

Making secure connections that can withstand the placement of charges is an art in itself, with many recommended procedures and many more inferior practices. Institute of Makers of Explosives, *The Use of Explosives for Agriculture* (New York: IME, 1917), 161.

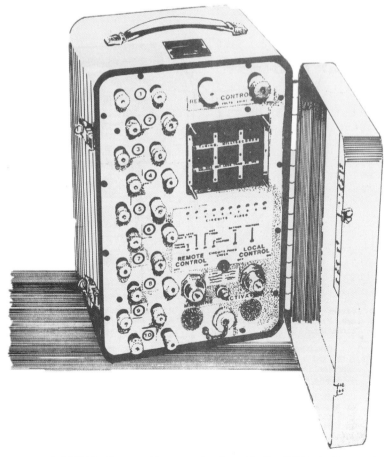

The sequential CD blasting machine was developed in the 1970s by Research Energy Corporation of Ohio. US Department of the Interior, *Explosives and Blasting Procedures Manual* (Washington, DC: USGPO, 1983), 169.

blasting-cap leg wires together is aptly called connecting wire. Leading wire is 12-to-18-gauge, single-strand (not twisted multiple strands), insulated copper wire, either in two separate wires or one easy-to-separate "duplex" cord. When a circuit is being set up, the ends of the two strands of leading wire are twisted together to create a temporary short circuit. This prevents the wires from accidentally carrying current to the blasting caps prematurely.

Connecting wire is 20- or 22-gauge insulated aluminum, iron, or copper wire. Aluminum wire is lighter and cheaper, but has only 60 percent of the conductivity of copper wire. Iron is the strongest and cheapest, but with only about one-fifth the conductivity of copper.

Leading wire can be used multiple times, while connecting wire is consumed in the blast. Specially insulated wires (and blasting caps) are available for submarine work. When necessary, splices are wrapped with waterproof tape.

Early sequential blasting machines were limited to ten delay periods. Modern sequential blasting machines are programmable to produce delays of one to a thousand milliseconds in one-millisecond increments. They can be used either with instantaneous or delay caps to achieve reliable initiation in extremely complex blasting sequences.

Blasting systems now incorporate computers to control the firing order. The designs integrate electronically coordinated, multiple capacitors for extremely precise timing. These systems are currently quite specialized from manufacturer to manufacturer but will no doubt become more standardized, as all innovations do. The systems offered include proprietary wires or shock tube and sophisticated programmable initiating devices. Wireless systems, currently a novelty, will no doubt become ubiquitous.

The shock-tube system has a variety of special connectors, some of which

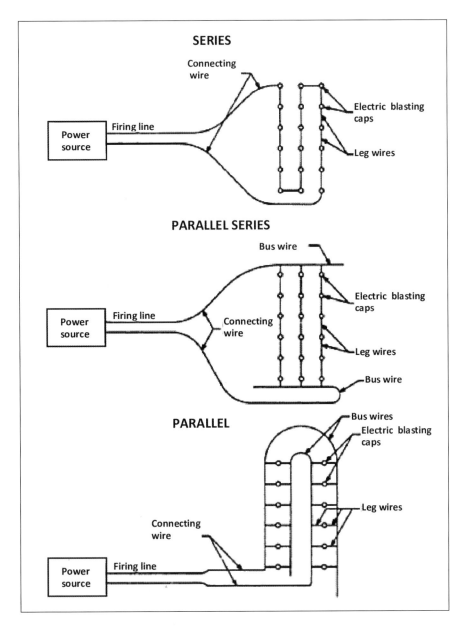

SERIES

Connecting wire

Power source — Firing line — Electric blasting caps — Leg wires

PARALLEL SERIES

Bus wire

Power source — Firing line — Connecting wire — Electric blasting caps — Leg wires — Bus wire

PARALLEL

Bus wires — Electric blasting caps — Leg wires — Connecting wire — Power source — Firing line

These basic circuit schematics can vary practically endlessly to fit the area to be blasted, and could include thousands of precisely coordinated detonations. Adapted from US Department of the Interior, *Explosives and Blasting Procedures Manual* (Washington, DC: USGPO, 1983), 24.

contain a blasting cap. Bunch connectors provide a way to initiate a number of separate shock-tube "trunk" lines via a blasting cap attached to the primary incoming tube. Other connectors attach shock tube to detonating cord, and some connectors are designed to work as delay elements. In sort of a throwback to the era of single-shot firing devices, ordinary shock tube can be fired by handheld percussion-cap strikers that fit over the end of the tube and launch a tiny traveling explosion.

Digging Deeper

Basic Tools

Cap crimpers retailed for around fifty cents in the 1910s and seventy-five cents in the 1920s. Nowadays a basic crimper will set you back about $30.00; a dual crimper that puts in two crimps at once, $109.00; a "deluxe" dual crimper, $149.00; and a superdeluxe, adjustable-depth model that applies a "fourfold crimp," $700.00.[39] Applying two crimps to a cap provides more moisture resistance for the detonating charge inside. And while US Army manuals mandate double crimping, other government publications strongly advised against it, warning of the danger of accidentally crimping into the internal charge. The modern double crimpers are definitely safer than applying two separate crimps.

While the vast majority of blasting is currently accomplished electrically, or with signal tubing, there are still instances where the use of cap and fuse is preferable. Because of the minimum required safety distances for electric and signal-tube blasting,

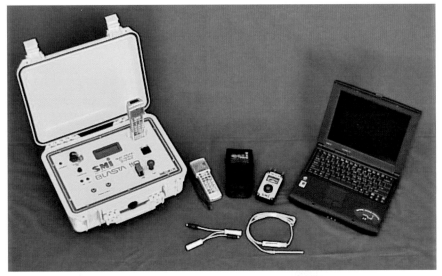

Modern computerized systems use software to determine hole size and spacing, and the amount of explosives per hole. Complex delay sequences are programmed and initiated in the most efficient possible way. US Department of the Interior, *Office of Surface Mining Reclamation and Enforcement, Blasting Module 2* (Washington, DC: USGPO, 2017), 46.

it is sometimes easier to light a fuse (of a safe length, of course) rather than laying out hundreds of feet of wire for a small blast. There are also situations where extraneous electricity is a concern but using shock tube is overkill. Then, ordinary blasting caps are sometimes still employed. Cap and fuse are still used for stumping and boulder blasting in US national parks.

During World War II, civilians gathered scrap metal to be donated and resmelted for war materiel. One such drive, organized by Hercules Powder Company employees in Wilmington, Delaware, gleaned "five large blasting machines, with wooden jackets but valuable metal within . . . single-shot blasting units, shells of various size, midget blasting machines, similar to the larger machines in construction, cap crimpers of steel, used for adjusting blasting caps, galvanometers, stamping machines and rheostats."[40]

In *Explosive Control Tokens*, David Schenkman writes, "Regardless of whether they were purchased or furnished at no charge to the miner, tokens played an important and practical role in the issuance of explosives."[41] However, "Exploder tokens were used by only a very small percentage of coal companies,"[42] and "Paper receipts were sometimes used, as were coupon books with tear-out coupons."[43] (Here Schenkman is using the term "exploder token" as a synonym for "explosive control token" or "explosive checks," as they are also called. More specifically, tokens marked "exploders" were exchanged for blasting caps.) Coincollector.com relates that metal scrip appeared in the early 1900s (Schenkman states that "a good guess would be the 1880s or 1890s."[44]), and that 75 percent of scrip was used in coal mines in Kentucky and the Virginias. Scrip was not limited to coal mines, but the isolation of central Appalachian coal-mining camps in particular meant that merchants outside the mining company could not afford to operate. Obviously, scrip was utterly impractical for use in quarries, where

hundreds if not thousands of rounds were fired a day.

One major downside of scrip was that it allowed miners to become indebted to the company. The cost of room, meals, supplies, and incidental expenses, such as visits to the company doctor, could entirely offset one's paycheck and more. Miners' wages averaged about seventy-five cents to a dollar an hour in the 1930s. Scrip was officially outlawed by Congress in 1967.

Drilling

The *1921 Mining Catalog* gave a chart of statistics on hand drilling. Among the figures: A single jacker could drill a 1⅜" diameter, 2' deep hole in gneiss at the rate of 0.35' an hour, and a ⅞" diameter, 1.5' deep hole in tough sandstone at the rate of 1.19' per hour. A team of double jackers could bore a 1¼" diameter hole through limestone at 1' per hour, and a 1¾" diameter hole through firm red porphyry at 1.48' per hour. Triple jackers could punch a 1¼" diameter hole in mica schist at 1.5' an hour. The table noted that for a 10' deep hole, the starting diameter of the drill steel was 2", and the finishing diameter was 1½". This prevented the latter bits from getting stuck in the hole.

Detachable bits took decades after their introduction to replace bits that were built into the shaft. A study by the Bureau of Mines in 1936 related, "The success or failure of detachable bits at a given property depends not only on the comparative service of the bits but also upon the quality of the fabricated rods or shanks to which the bits are (temporarily) attached,"[45] and "The development of the rods has not kept pace with improvements in the manufacture of drill bits."[46] Problems with shafts and rods "might cause an adverse decision on the adoption of detachable bits even though their drilling performance were [sic] favorable in all other ways."[47] An article in 1941 quoted the manager of a Michigan mine as stating, "In our drilling operation we are now using the comparatively

new 'Jack Bit' which requires the transporting of the bits only to the shops for sharpening. This device saves the transporting of drill steel repeatedly with its attendant hazard of injuries from several handlings of hundreds of tons of that material."[48]

Thawing

One popular way to thaw dynamite was to place it near a running steam engine. The practice was hazardous because the dynamite could overheat and explode. Even more foolhardy was placing frozen dynamite near a running internal-combustion engine. Sparks from engines caused numerous fires and explosions, including one in which "forty-two holes and other injuries were accumulated"[49] by an unfortunate miner. Perhaps the most tragic accident occurred in New York City in 1907, when "several hundred tons of thawing dynamite"[50] exploded, killing fifty.

Dynamite was still being thawed long after it was necessary. In 1957 an article reported on another death caused by thawing dynamite near an engine. In 1961, three men were killed at a quarry thawing house.

There are several reasons why thawing was still practiced, if rarely, decades after it was needed. First, ingrained habits die hard, and if a veteran did things a certain way, it was tough to change. Second, cartridges were no longer marked "Low-freezing." For all the efforts to educate users, not everybody was "in the loop." While we take the dissemination of information for granted today, some areas of the country were all but isolated even into the middle of the twentieth century. Some Appalachian miners in particular clung to antiquated customs, being among the last to use blasting powder, miner's squibs, and other obsolescent technologies.

Priming

Accidents involving serious injury occurred when a cap was crimped "too low," causing the charge of mercury fulminate to be crushed and detonated. However, cap crimpers could be dangerous in other ways, as an article thoughtlessly headlined "Crimped!" related in 1946. According to the story, a pair of cap crimpers "caused injuries" to a man who had fallen on ice. "The crimpers, police said, which were in his pocket, penetrated his back, causing two deep puncture wounds."[51]

The shells of blasting caps are thin enough to be crimped with one's teeth, but the crimp is not as secure as non-biological crimping. The practice is dangerous as well. In 1913 an article warned that "many a miner has blown a hole through his face in biting the cap on the fuse"[52] by chomping the cap "too low." In 1887, accused anarchist bomber Louis Lingg committed suicide in his prison cell by biting a smuggled blasting cap.

Ordinary pliers were not recommended for crimping caps because the powder core of the fuse could be disrupted, resulting in a misfire. Knives were sometimes used, but the crimps produced were flimsy.

A 2020 coin issued by Niue Island features a bust of Alfred Nobel and a bundle of dynamite. Astonishingly, every stick of dynamite in the bundle has its own fuse! We know that's not how dynamite works; one stick in a bundle is primed, and the detonation of that stick sets off all the others. The fuses emanating from the bundle shown on the coin are optimistically tied together in the hope that the short sections of fuse would somehow burn such that all detonated their respective sticks simultaneously, without one exploding a fraction of a second ahead of the others and disrupting all the other carefully arranged fuses.

Loading and Tamping

An 1893 article related,

> For tamping, dry sand was as good as anything, save perhaps pebbles, the worst material of all being wet clay. The tamping material was best placed in a tamping bag of paper

made the diameter of the hole, and about 12" to 30" long, as these bags could be filled with the best tamping material some time before required, instead of having to use otherwise, the nearest stuff that would do. Dry clay, moistened broken brick, and the dust from the drill hole could be used if nothing better was at hand.[53]

In the 1920s it was reported that in a pinch, miners would remove the wrappers from dynamite cartridges and refill the wrappers with sand to use as tamping bags. The unwrapped dynamite was loaded into the hole as per usual.

In 1924, tamping bags cost a quarter cent each. By 1958 the price had doubled. DuPont sold its tamping bags in "bales" of ten bundles of five hundred bags each. The bags were thin, but fairly tough, and weighed 31 pounds per bale of 1¼" × 8" bags. A man could fill seventy-five bags an hour by hand and two hundred bags an hour by machine in 1918.

In 1947 a tamping bag was implicated in a mine explosion that killed 111 men. According to transcripts from a congressional hearing, two tamping bags had been found at the disaster site, one filled with clay and one with coal dust. Some testimony from the hearing:

> "Why would a man put in coal dust?" inquired Senator O'Mahoney.
> "Pure old neglect and carelessness," Scanlan answered. "I found a man using coal dust there sometime back. I reported it and the man was discharged by the company." "If coal dust is used in a tamping bag," Senator Cordon asked, "what is the result?" "It is that much more apt to ignite the dust in the mine," Scanlan told him. "Could that have been the cause of the explosion?" Cordon asked. Scanlan replied, "It could have been."[54]

Following rules for the use of permissible explosives, which prohibit coal dust as stemming, would have prevented this accident.

"Shot tubes" were fashioned out of

mailing tubes and stiff paper and were made to fit snugly over a primed cartridge. The tubes protected the fuse or wires and greatly reduced misfires. Shot tubes were used mainly in the 1920s through the 1940s, and then mostly in regions where borehole walls tended to be rougher than normal. Tamping bags and blasting tubing of the appropriate diameter were also sometimes used to sheath primed cartridges for protection.

In the 1940s, tamping dummies appeared that were made of Tamcot, "a fibrous non-combustible material made up into stemming plugs used in place of ordinary stemming."[55] Another stemming ploy was the Powder Monkey, a machine that blew sand into boreholes.

Tamping sticks were sometimes launched from the borehole if an ill-advised metal pole was used, or if a wooden rod crushed and detonated a blasting cap. Primed cartridges were never to be directly tamped for this reason. On occasion, miners were impaled. In one such incident in 1905, a tamping stick "passed directly through"[56] the head of a Minnesota iron miner, according to a newspaper account. He recovered but lost the sight in his left eye.

Blasting

DuPont estimated that "a fairly industrious blaster should average 150 day's (per original source) employment per year in the North and 250 days per year in the South. In the first instance his earnings at $5.00 per day would amount to $750.00, and if he used an average of fifty pounds of dynamite per day and supplies for same, his profit on the goods would be about $1.00 per day or $150.00 per year. On the same daily basis his net receipts in the South the first year should not be less than $1,600.00."[57] DuPont also was confident that "in the second year[,] receipts should be at least twice those of the first year, owing to greater ease in securing the work, better knowledge of the costs, and the probable employment of assistant blasters."[58] DuPont aided

1	1-1 ½ in. punch bar for making boreholes in earth	$2.40
1	1 ½ in. steel subsoil bar for making holes in hard ground	$2.00
1	sledge or maul	$1.00
1	1-1 ½ in. wood auger for boring holes in stumps	$2.60
2	extensions for punch bar	$1.00
2	extensions for wooden auger	$1.00
6	ft. of 1-inch chain with hook	$1.00
1	wooden tamping bar (home-made)	
1	chambering spoon	.50
200	ft. Duplex leading wire	$2.50
2	lbs. connecting wire	.50
1	No. 2 blasting machine	$15.00
1	home-made magazine, capacity 500 lbs.	$10.00
200	lbs. Red Cross Dynamite (100 lbs. 40%; 100 lbs. 20%)	$28.00
500	DuPont No. 6 caps	$4.50
100	DuPont No. 6 6-ft. electric blasting caps	$3.50
1	DuPont cap crimper	.40
500	ft. of Crescent fuse	$3.00
1	strong box for tools	$15.00
	Total	$93.90

According to DuPont in 1915, to get started as a blaster "you must have" the items listed. Adapted from E. I. DuPont de Nemours, *Agricultural Blasting, A Money-Making Profession* (Wilmington, DE: E. I. DuPont de Nemours, 1915), 2.

these entrepreneurs with such items as free posters, fliers, signage, and business cards to advertise their services. The average yearly per capita income in 1915 was around $700.00 versus about $50,000 today. Small blasting companies still operate, but licensing, permits, and reams of paperwork are now mandated for explosives use of any kind.

Blasting machines can be divided into two categories: generator and discharge. The generator type includes all plunger and twist-handle machines, plus the early versions that used friction and cranks. Generator units are powered by the physical motion necessary to activate the magneto. Discharge units gain their power from batteries, which either temporarily impart their charge to capacitors or directly send a current to the blasting cap. Straddling the line were machines that had to be cranked to impart a charge to a capacitor, which then sent electricity down the leading wires on command via the press of a button or buttons, or the turn of a key.

Plunger-type blasting machines with three connecting posts instead of two were available in the first two decades of the twentieth century. DuPont sold its version only on special order. Although the three-post machines could fire more caps than two-post machines, they were not popular because they were more complicated to use, and they disappeared in the 1920s.

No longer is simply shouting, "Fire in the hole!" sufficient. Now government regulations recommend that a warning system must be implemented, consisting of "hand signals and/or shouts," "using horns or whistles," or "direct radio communication."[59] In addition,

The system must be simple and understood by everyone in the vicinity. If sign text states a signal sequence, then the system must match the system on the sign. Warnings of the impending blast "Blasting" must be provided at least twice: five minutes before detonation, and one minute before detonation. It is recommended that these warning signals be used one at the time of pre-blast inspection, one at the time of blasting machine hookup (at least one minute before detonation), and another ten seconds before detonation. These signals must be different enough to be individually identified by the guards. The system must include an "All Clear" signal, given after the post-blast inspection and distinctly different from the warning signals. The signal system must be a "positive response" system where the guards can effectively communicate to the blaster-in-charge any need to halt the blast prior to instant of detonation, and the blaster-in-charge can effectively acknowledge that communication. The signals must be readily identifiable by the guards, with no risk of confusion about each signal's meaning. For this reason, the use of radio communications is strongly encouraged, while the use of hand signals or voice-only signals are strongly discouraged.[60]

Radio communications cannot be such that there is a danger of initiating the circuits connecting the charges to the blasting machine. If wireless blasting is being conducted, all blasting initiation transmissions must be unique and encrypted.

CHAPTER 3

Dynamite Uses

This chapter delves into both the most-common and most-unusual ways that dynamite was used.

Hard-rock mining was dominated by dynamite. Blasting powder was not nearly strong enough to conquer the toughest rock in an efficient, cost-effective manner. The blasting of softer material, such as coal and clay, was also accomplished with lower-strength dynamites, and quarrying consumed high explosives in prodigious amounts.

Huge quantities of dynamite were also expended in construction projects, land clearing for highways, and blasting for tunnels, waterways, and dams.

It is not common knowledge that dynamite was an ordinary tool on the early-twentieth-century farm. There it was used to remove boulders and stumps that impeded productivity.

Dynamite saw many other applications. Most of us know it is used for the implosion of tall buildings. Some have also heard of dynamite being used to blow out oil fires. But there is much less awareness of the use of dynamite for underwater blasting and seismic prospecting, and its extensive service in rerouting water. And very few have heard of dynamite's most-unusual uses, which include scrapping old machinery,

fighting large metropolitan fires, and even digging graves!

Ore Mining

According to the 1949 edition of the *Blasters' Handbook*, "Blasting is primarily used to break the ore loose from the main body in a form that can be economically handled."[1]

The ancients used fire and cold water to fracture large stones. Gunpowder, a mixture of charcoal, sulfur, and potassium nitrate, was used for mining beginning in the early seventeenth century. Sodium-nitrate blasting powder replaced gunpowder for mining in the 1850s. Gunpowder and blasting powder are low explosives that explode by burning or deflagration. The slower blasts produced were ideal for coal mining, since the relatively soft coal is broken and heaved forward in pieces of the desired size. Blasting powder was also suited for removing packed-dirt overburden and for mining softer materials such as clay. The shattering power of dynamite is so superior to blasting powder that dynamite quickly replaced powder for mining the hard rock in which metallic ores are ensconced.

There are two basic types of ore mining: open work and belowground or closed work. With open work, the ore body is attacked from above with vertical boreholes. Once a pit is established, the faces and slopes are

Blasting gelatin was the most powerful dynamite and was used to blast the hardest materials, such as granite overburden. Because of its noxious fumes, blasting gelatin was not used for underground mining. Blasting gelatin was also used for working oil wells and was sometimes labeled "oil well explosive." *Author*

The myriad uses for dynamite meant opportunities for marketing focused on specific customers such as miners, farmers, and engineers. In addition, general advertisements emphasized the value of explosives by illustrating some of those varied uses. Aetna Powder, advertisement, *National Geographic*, July 1919, 9.

worked. An open cut can also be made on a mountain or hillside. Sometimes, enormous amounts of material, both valuable and not, are removed in a process called strip mining.

Belowground mining is a lot more complicated. The first consideration is how to tunnel through hard rock to get to the valuable ore. Tunnels are blasted to ascertain the extent of an ore deposit, then the deposit is worked, either by more drilling and blasting or with tools ranging from picks and shovels to heavy machinery. Blasting underground is also much more difficult, owing to issues involving moving explosives underground, the potential for cave-ins, poisonous or explosive fumes (or both), and the many mundane hazards associated with large numbers of men in confined spaces.

Ore mining required a broader range of dynamites than other applications. Straight, gelatin, and extra formulations were used to address situations that varied depending on the character of the material being blasted and the phase of mining being conducted.

For tunneling through hard overburden, high-power dynamites of at least 50 percent strength were the rule. DuPont recommended its high-strength gelatin formulations. Gelatin dynamites were particularly recommended for their ease in loading upward-slanting holes. Gelatin mixtures are much more plastic than granular mixtures and more readily stay in place when tamped. DuPont also advised using perforated or preslit cartridges for all ore mining for the same reason: they are easier to tamp firmly in place.

For open mining, DuPont recommended high-strength straight dynamites for hard rock. At 100 percent strength, blasting gelatin was reserved for the hardest overburden. Ammonium nitrate extra dynamites were preferred for blasts where large and sometimes mammoth amounts of explosives were packed into tunnels and chambers

Common metals and their primary ores are (*left to right*) pentlandite, nickel; acanthite in galena, silver; galena, lead; sphalerite, zinc; cassiterite, tin; chalcopyrite, copper; bauxite, aluminum; and hematite, iron. *Author*

hollowed out for the purpose. These methods, called coyote tunneling and springing, were used for ore mining, quarrying, and large construction projects. The techniques are described later in conjunction with quarrying.

Once the operation moved into the production phase, slower dynamites with less fracturing power were used, especially with deposits with lots of seams. As with coal mining, completely pulverizing material is undesirable. DuPont recommended its Red Cross ammonium nitrate mixtures in 20 to 60 percent strengths for most rock. Formulations like these, with their relatively low nitroglycerin content, provided a less violent lifting action to move ore without reducing it to dust.

Underground mining was once more common than surface mining; now the reverse is true. One reason is that most of the known high-grade, near-surface ore deposits have already been mined.

THE USE AND EFFECT OF EXPLOSIVES IN ORE MINING

Ore mining probably offers the most interesting studies in the use of explosives than any other kind of work. In some mines it is desirable not to break the ore too fine on account of the loss in valuable material. On the other hand in most iron and copper mines, the only desire is to get the ore out as cheaply as possible.

In the Cripple Creek district the values are in the soft tellurides of gold and some of the soft metallic sulphides that contain gold in paying quantities. These occur in very narrow veins and stringers which make it necessary to mine large quantities of waste rock. By breaking the material large, most of the waste rock can be picked out by hand, leaving a small amount of rich fines to be treated at minimum cost. Therefore in the Cripple Creek District the comparatively slow acting "Repauno Gelatin" and "Monobel No. 1" are used.

In the Lake Superior copper industry, as it is impossible to shatter the native copper most of the mines use the stronger grades of "Giant Extra" and "Red Cross Extra Dynamites." The nitroglycerine dynamites were discarded because the fumes combined with the heat of the lower levels and had a bad effect on the men. Some of the mines of that district have wide veins varying from 30 to 40 ft., where the copper occurs in irregular shoots. In these mines the vein material is broken down in large pieces by drilling flat holes, having medium burdens, in the stopes and blasting them with "Giant Extra" or "Red Cross 30% Extra Dynamite" which have a slow-heaving action.

The ore in the fines is sent to the mill while the waste remains in the stope for filling. Here also the value of a slow acting explosive is very apparent.

The mining of sulphide ores that require concentration is another class of work where slow explosives having a strong-heaving action are necessary. This applies in particular to sulfide copper ores and arsenical sulphides containing gold or other metals in paying quantities. With the sulphide lead ores this point is not of very great importance because fine lead ores are sintered in roasting. But with the sulphide copper ores the greatest care should be taken not to have the material in any finer condition than is possible for treatment.

Where the ore or rock to be blasted occurs in alternate layers of hard and soft material, a quick-acting explosive must be used. If dynamite having a slow heaving action is used under these circumstances, its strength will be largely wasted by blowing out through the soft streaks.

According to DuPont, "Blasting in ore mines is influenced by a wider variety of conditions arising from the type and distribution of the ore, the nature of the rock in which the ore is embedded, the stage of the mining operation, and the method of mining."[2] Adapted from *Mining and Scientific Press*, February 8, 1913, 266.

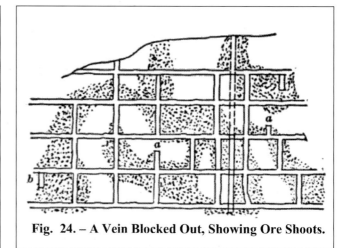

Fig. 24. – A Vein Blocked Out, Showing Ore Shoots.

Underground ore mining often depends on tunnels that must be blasted without first knowing which will cut through valuable material. M. C. Ihlseng and Eugene Wilson, *Manual of Mining* (New York: John Wiley & Sons, 1911), 70.

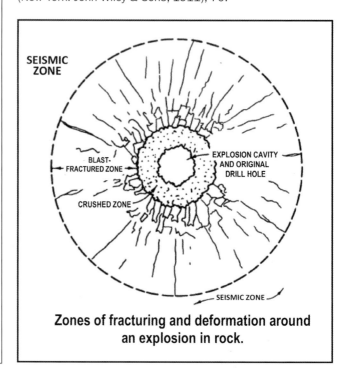

SEISMIC ZONE

BLAST-FRACTURED ZONE

EXPLOSION CAVITY AND ORIGINAL DRILL HOLE

CRUSHED ZONE

SEISMIC ZONE

Zones of fracturing and deformation around an explosion in rock.

The placement of boreholes takes advantages of the fracture zones produced by adjacent boreholes. US Army Corps of Engineers, *Systematic Drilling and Blasting for Surface Excavation* (Washington, DC: USGPO, March 1972), 5-2.

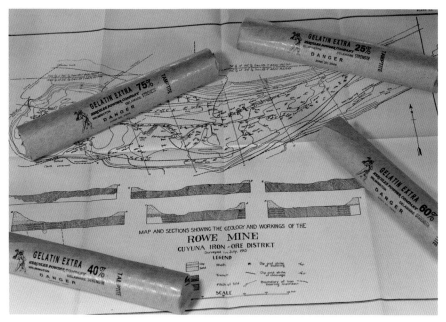

Gelatin dynamites were a mainstay of ore mining. A study by the Bureau of Mines in 1936 revealed that, at that time, well over half of the open-cut metal mines in the United States used nothing but gelatin formulas of one kind or another. *Author*

flooring or steel plates are sometimes put in place for easier scooping. Tunnels must not interfere with one another in terms of stability, and flat floors are needed for laying tracks for ore carts.

Ore miners quickly adopted delay blasting caps when they became available around 1900. An improvement known as millisecond-delay blasting caps debuted in the 1940s. Delays meant that a single round of shots could first detach and then push material. Accomplishing this two-step process with cap and fuse was much more difficult and was restricted as to the number of shots fired per blast. Firing in precise sequence meant that each successive charge had less work to do because the charge next to it had already weakened the rock on that side.

Combination loads were sometimes used for removing overburden in open-cut operations such as iron mines. In 1936, one such mine reported loading its holes with 90 percent strength gelatin dynamite at the bottom of holes. These charges were topped with 80 percent strength gelatin dynamite, atop

Another reason is that technology has allowed for the profitable recovery of valuable minerals from very low-grade ore. Deep, large-diameter boreholes mean that huge quantities of explosives can be detonated in a short amount of time. But as low-grade, near-surface deposits are depleted, even deeper underground mining comes back into play for some minerals.

The 1958 edition of the *Blasters' Handbook* observed that "ore veins vary in thickness from a fraction of an inch in the case of precious metal deposits, to massive veins several hundred feet thick. The ore may be pure and surrounded by country rock, or disseminated throughout the rock. . . . Ores and rocks vary greatly in hardness, toughness and structure, and the pitch of the veins may be anything from horizontal to vertical."[3]

The arrangement of boreholes is key to successful extraction of valuable material. Hole patterns are tailored to shatter and push material forward in successive waves. Explosions are engineered such that ore lands in the most convenient place available for collection and hauling away. Material

is blasted so it tumbles down an incline or falls into a chute. Extraction is done in steps or "stopes" to take advantage of gravity while avoiding steep slopes. Underground, temporary timber

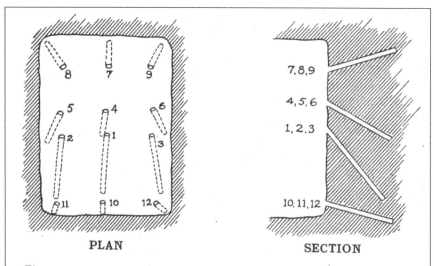

PLAN   SECTION

Fig. 162.—These sketches show the type of round known as the hammer cut. Holes 1, 2 and 3 are fired in the order numbered and blow out the bottom, thus undercutting the breast. Holes 4, 5 and 6 are fired next, and then holes 7, 8 and 9 and then 10, 11 and 12 which break grade and turn the muck back from the face. In small drifts, or in soft, easy-breaking rock, holes 4, 7 and 10 may often be dispensed with.

Drill hole patterns were standardized and given names such as "wedge cut," "hammer cut," and "pyramid cut." The spacing, depth, diameter, and angle of holes can all be adjusted. Arthur La Motte, *Blasters' Handbook* (Wilmington, DE: E. I. DuPont de Nemours, 1925), 130.

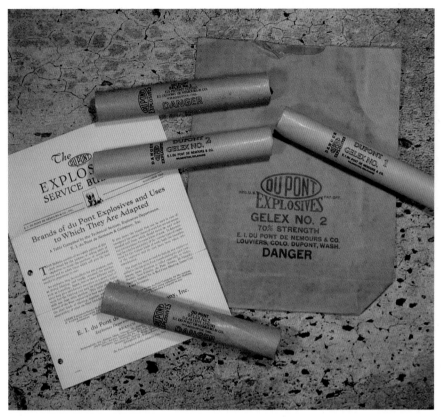

Left: DuPont Gelex was sold in five strengths: No. 1, 60 percent; No. 2, 45 percent; No. 3, 40 percent; No. 4, 35 percent; and No. 5, 30 percent. Gelex was produced from 1929 to 1977, when DuPont quit making dynamite. Later cartridges were marked with their percentage strengths, rather than numbered brands. *Author*

which was loaded 60 percent strength gelatin dynamite, and topped off with 40 percent strength gelatin dynamite. The general theory is that the upper, less powerful charges have less rock to lift.

The impact of dynamite on the productivity of ore mining cannot be overstated. US silver production quadrupled from 1860 to 1880. An article in 1935 noted that dynamite "was the greatest aid to man ever developed in overcoming nature,"[4] because without the enormous amounts of copper and iron that high explosives enabled access to, much of large-scale industry would be impossible to sustain. Another writer noted, "Not a barrel of cement was produced in America until dynamite made big-scale rock blasting practicable. . . . Nickel was a precious metal before dynamite; only the rich could afford silver knives, forks, and spoons. Seemingly impenetrable rock imprisoned the bulk of our common minerals."[5] A 1925 piece in *Industrial Engineering and Chemistry* declared that "future historians, when writing the history of our civilization, will note the sudden rise in industrialization with the application of high explosives during the last half of the nineteenth century."[6] According to the Institute of Makers of Explosives, after the introduction of dynamite, "the extracted tonnage of iron, coal and copper increased a hundred fold."[7]

Farming

In 1902, the US Department of Agriculture opined that while dynamite was "serviceable" in the removal of stumps

EXPERIMENTS WITH DYNAMITE IN FARM OPERATIONS.

A week or two ago a large number of gentlemen interested in the application of dynamite to agricultural purposes met on the farm of Mr. R. Bell, Broxburn, and witnessed some interesting experiments illustrative of the use of this explosive in clearing the ground for farming operations. In many instances, farmers who would otherwise have their land opened up with the steam plow have been unable to employ this implement on account of large boulders lying under the surface, which would effectually prevent its use; and was in order to ascertain to what extent dynamite would be available for removing such obstructions that the following experiments were made. The first subject operated upon was a large boulder of from 3 to 4 tons weight. A hole 1 ¼ inch diameter and 14 inches deep had been previously made, and into this tow or three cartridges of dynamite were dropped, and pressed down with a wooden rammer. Into the uppermost cartridge (or primer) a detonator cap, with fuse attached, was inserted, and the mouth of the hole filled up with clay or sand. As the result of the explosion, the boulder was completely shattered, and the greater part of it left neatly for carting away. In order to shew the immense explosive force of dynamite, a small quantity was then placed on the upper surface of several smaller boulders, which were also shattered, no bore-holes having been required. The company afterward witnessed the effect of a cartridge exploded beneath a tree stump. For uprooting and clearing away of these obstacles dynamite has been found particularly efficacious, and even living trees of large size have been bodily removed from the soil by this means. After witnessing other experiments of a like nature, including a blast in the solid rock, the company partook of refreshments provided by Mr. Bell, Greendykes, who had also prepared the grounds for the day's experiments. Before they parted, one or two cartridges were set fire to with a match and burned in the open air, to illustrate the freedom from danger of explosion in the use of this material. It is said that arrangements are being made by several leading farmers in the district for clearing their lands by the aid of dynamite, with the view of introducing the system of plowing by steam.

Above: Dynamite was doubtless used on the farm soon after its invention. By 1876, as this article attests, the tool was gaining wider acceptance for use in "clearing the ground for farming applications." Adapted from "Experiments with Dynamite in Farm Operations," *Country Gentleman's Magazine*, January–December 1876, 371.

on the farm, it was at that time "too expensive for general use"[8] except for extremely large stumps. But by 1950 the Department of Agriculture estimated that an area of land twice the size of Texas had been improved directly as a result of the use of dynamite. This was in addition to the clearing for the development of housing areas and the expansion of cities. Farmers consumed about 5 percent of the dynamite produced in the early twentieth century, with the balance split between coal and ore mining and road, railway, dam, and other large construction projects.

The removal of tree stumps, called "stumping," was a part of farm life, and dynamite was also used by loggers and highway engineers to blast away woody obstacles. Producers offered numerous stumping dynamites, so stumping gets its own section. Ditching and boulder blasting were also endeavors undertaken on and off the farm, but are covered here.

Stumps and boulders prevented the use of wide plows and required that crops be organized around the obstructions. Cleared land was always valued higher than land with many obstructions to planting. Every obstacle that was removed represented both more space for crops to grow and increased efficiency in tilling, sowing, and harvesting. Large rocks buried just below the surface were especially noisome

because they could dent a plow blade, year after year.

A book called *Farm Knowledge* noted in 1919 that "while dynamite is a high explosive, it can, with reasonable care, be used on the farm with no more danger than attends a great variety of farm operations. It is sent from the manufacturer to the buyer by freight, thus receiving rougher treatment in transit than is ordinarily given it during use."[9] Atlas Powder Company reassured farmers that its farm dynamites were "safe in the hands of ordinary farm help, as long as our plain and brief instructions are followed."[10]

All major explosives producers and some minor ones marketed heavily to farmers. After the divestiture of DuPont in 1912, Atlas and Giant Powder Company, Consolidated, marketed their own farm dynamites. Advertisements for dynamite appeared in newspapers, agricultural trade journals, hardware catalogs, and even magazines such as the *Saturday Evening Post*, *Literary Digest*, *Harper's Monthly*, *Sunset*,

Scientific American, and *Popular Mechanics*. Booklets (see p. 100) were another way to reach potential users. DuPont, Hercules, Atlas, and some smaller firms such as Jefferson Powder Company and Illinois Powder Manufacturing Company published material promoting dynamite on the farm. *DuPont Magazine* and Hercules Powder Company's journal, the *Explosives Engineer*, were also filled with articles directed at farmers. Explosives producers and distributors gave demonstrations, often trading the service of removing a stump or boulder or two in exchange for the space and time on the farm to do it. Although farmers were a relatively small market overall, the profit margin was much greater because farmers typically purchased their dynamite at full retail prices. Farmers purchased their dynamite from smaller wholesalers and distributors who did not qualify for the substantial discounts afforded large consumers such as mines and construction projects.

Du Pont Explosives

ATLAS POWDER   HERCULES POWDER
GIANT POWDER   RED CROSS DYNAMITE

For STUMPS and BOULDERS

Write for FARMERS' CATALOGUE

E.I. DU PONT COMPANY

OFFICES:

New York	Philadelphia	Chicago
Chattanooga	Pittsburg	Kansas City
Scranton, Pa.	Cincinnati	Duluth
Hazleton, Pa.	Terre Haute	Marquette, Mich.
Wilmington, Del.	St. Louis	Houghton, Mich.
Huntington, W. Va.	Joplin, Mo.	Birmingham, Ala.

At the time this ad ran in a farming journal in 1906, DuPont sold Atlas, Hercules, and Giant brands. E. I. DuPont de Nemours, advertisement adapted from *Farming*, August 1906, 32-b.

Most dynamites labeled as farm dynamites were 20 percent strength ammonium nitrate formulas. Because such mixtures are lower in density, cartridge count was high. Farmers, who were mostly blasting stumps and soil, appreciated the economy of bulkier, lower-strength dynamites. *Author*

Relative to miners, farmers were occasional users of dynamite. Much information was passed on from miner to miner or imparted by mining companies. With farmers, the main source of instruction in the use of explosives came from articles in farm journals and newspapers, and from books like these. *Author*

As noted in the introduction, in the 1910s there were 500,000 daily users of explosives. By 1952, with farmers decreasingly using dynamite, only 20,000 "regular"[11] consumers remained, almost all of them miners and engineers, according to DuPont. Dynamite is occasionally still used to dig ditches and remove obstacles on farms, but now the work is done by licensed, professional blasters.

Subsoiling, also known as vertical farming, meant using explosions to loosen hard-packed soil for easier irrigation and tilling. Subsoiling was described by DuPont as follows: "Cut a cake of butter with a sharp knife; the cut surface is left hard and smooth. Just so in subsoiling with a plow. It rains, the water soaks up through the topsoil and then follows the course of the plow, soaking no further. But in subsoiling with dynamite exactly the opposite occurs. The ground is 'heaved,' shaken, and broken many feet deep, and is left so open and porous that all the rainfall is absorbed and retained."[12]

Nutrients in the soil were also freed up via subsoiling. According to DuPont, "Chemical analysis of soils down to a depth of twenty feet show[s] that on the average acre there are tons of plant foods which become available only when the roots can penetrate them, or when ascending moisture brings them up to the roots that cannot get down."[13]

Subsoiling had to be done in the dry months, because wet soil tended to compact even further from blasting, landing in clumps that hardened. Dry clay was pulverized by an explosion. As DuPont put it, "Tight subsoil blasted when *dry* showed increase in crops varying from 25 percent to 300 percent; the average increase was better than 50 percent."[14] Subsoiling in autumn also had the advantage of preparing the cropland to absorb winter moisture.

Holes of a minimum depth of 4 feet were used for subsoiling, with all dynamite producers advising the deeper

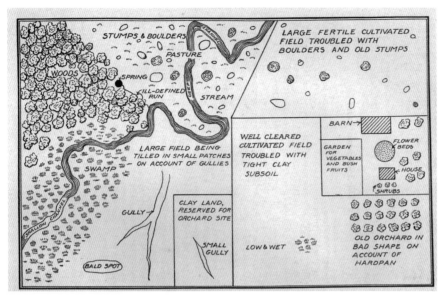

This illustration from DuPont's *Farmers' Handbook* was captioned "The undersized farm, showing how the productive area is reduced by defects that blasting will remedy." E. I. DuPont de Nemours, *DuPont Farmers' Handbook* (Wilmington, DE: E. I. DuPont de Nemours, 1915), 4.

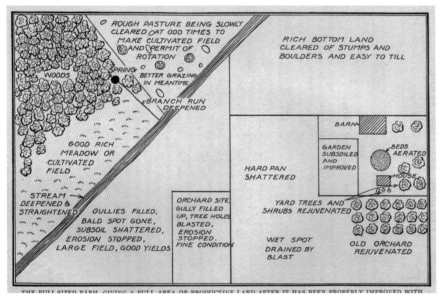

Now we see "the full-sized farm, giving a full area of productive land after it has been properly improved with DuPont Explosives." E. I. DuPont de Nemours, *DuPont Farmers' Handbook* (Wilmington, DE: E. I. DuPont de Nemours, 1915), 6.

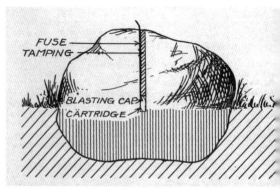

Blockholing is the most efficient way to blast boulders. E. I. DuPont de Nemours, *DuPont Farmers' Handbook* (Wilmington, DE: E. I. DuPont de Nemours, 1915), 153.

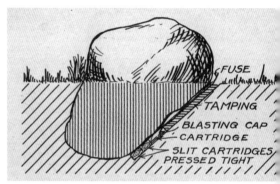

For efficient snake holing, the dynamite should touch the surface of the boulder or, as one engineer put it, there must be "intimate contact."[15] E. I. DuPont de Nemours, *DuPont Farmers' Handbook* (Wilmington, DE: E. I. DuPont de Nemours, 1915), 154.

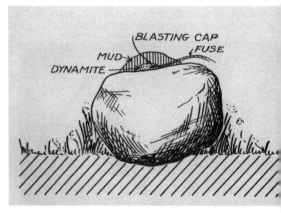

A mudcap shot was also known as an adobe shot, 'dobe shot, poultice shot, bulldozer shot, plaster shot, or blister shot. E. I. DuPont de Nemours, *DuPont Farmers' Handbook* (Wilmington, DE: E. I. DuPont de Nemours, 1915), 155.

the better. Atlas Powder Company recommended drilling down to 20 feet, using a soil auger with multiple extensions.

There are three ways to blow up a pesky boulder. A charge can be snake-holed under the boulder, blockholed in a hole drilled in the boulder, or mud-capped on the surface of the boulder. Stronger dynamite of at least 40 percent strength was recommended for larger boulders. Miners refer to boulder demolition as "secondary blasting" and often rig multiple boulders for simultaneous destruction.

A partially buried rock is the best candidate for a snake-hole shot because the ground provides some measure of confinement. "Snake holing" was also a term used in the context of quarrying, referring to large, sprung holes at the bottom of a vertical rock face. With the right material, blasting the bottom of the formation will bring down the entire side of a rocky hill.

While the most time-consuming of all methods, blockhole shots are the most efficient. It was always vital to properly stem the hole to prevent blowouts. Blockhole shots were also called "pop shots." Only blockhole shots were recommended for underground secondary blasting, particularly in coal mines. A 5' diameter boulder takes twelve 8" × 1¼" cartridges of 50 percent strength dynamite for a mudcap shot, six sticks for a snake-hole shot, and one for a blockhole shot.

Surprisingly, several inches of mud provide enough confinement to shatter a boulder. Tests with completely unconfined cartridges of dynamite produced only "slight cracks"[16] in boulders. Mudcapping was enhanced if cartridges were placed in natural indentations on a rock. Mudcapping a partially buried rock is inefficient because much explosive energy is directed through the boulder and into the ground. DuPont recommended first snake-holing a buried boulder to dislodge it, then mudcapped shots could be used.

According to Atlas Powder Company, "Farming is a business of earth moving. Always there are ditches, grades, holes, stumps, and rocks to dig or move or break. The weak strength of men should not be pitted against this heavy work."[17] Referring to dynamite as "powder," Atlas proclaimed, "Powder is energy with which you can lengthen your arm and strengthen your back a thousand times over. It waits your convenience, eating nothing. It never gets weary. It does its work quickly."[18]

Keystone National Powder Company introduced its farm dynamite around 1912, followed by Hercules and Aetna a few years later. In 1923, Atlas debuted its Non-Freezing Farm Powder with the guarantee that "it can be used at the North Pole without thought of thawing."[19] Atlas Powder introduced its Farmex Boulder brand in 1929,

along with its Farmex Stumping and Farmex Ditching labels. Atlas Farmex Stumping was a low-strength, extra dynamite, while Farmex Ditching was a 50 percent straight dynamite. Giant Powder Company, Consolidated, also sold Farmex dynamites.

Aside from stumping, boulder removal, subsoiling, and ditching, *Better Farming with Atlas Farm Powders* listed these uses in 1923:

breaking up old machinery, concrete wrecking, drainage, gully filling, log splitting, loosening frozen logs and lumber, road straightening and widening, well digging, blasting holes for planting trees, posthole digging, and demolishing old buildings.

Digging ditches by sympathetic propagation was pioneered in California around 1908, per DuPont. The method took advantage of dynamite's

Dynamites used for boulder blasting were 40 to 60 percent in strength and were formulated for a fast, shattering action. *Author*

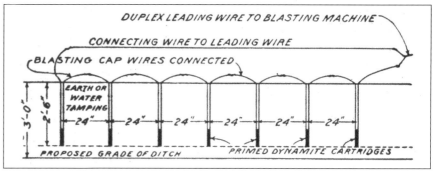

This is a very basic hole configuration for blasting a ditch with electric blasting caps. Two or three parallel rows of charges were used to make wider ditches, and holes were spaced closer together in tougher ground. Arthur La Motte, *Blasters' Handbook* (Wilmington, DE: E. I. DuPont de Nemours, 1925), 121.

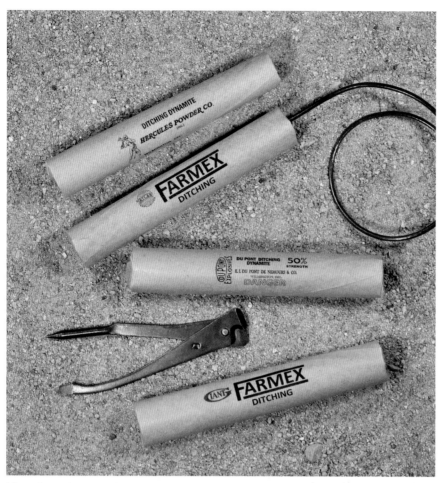

Most dynamites marked as "ditching" were 50 percent straight formulations. Other companies offering ditching dynamites included Austin Powder Company (Red Diamond Ditching), Burton Powder Company (50 Percent Ditching), and Illinois Powder Company (Gold Medal Ditching). *Author*

Dynamites intended solely for farming included Hercules Hercotol and Dupont Agritol. DuPont Dumorite was also billed as suitable for mining. *Author*

sensitivity to shock by placing charges in close-together, evenly spaced holes. The detonation of just one of the charges in one column will instantly detonate the other columns in a line of any contour or curve. Dirt is blasted away, leaving a ditch of uniform dimensions.

Dynamite used for this type of ditching was always straight dynamite, which is more sensitive than other formulations, although all will explode with a strong enough shock. Ammonium nitrate dynamites did not make good ditching dynamites because of their insensitivity.

Straight dynamite is occasionally still used for trenching and ditching; no other explosive can reliably detonate another from a distance. The one caveat is that the soil being blasted must be moist to carry the pulse. Otherwise, each hole must be individually primed and electrically fired. Using cap and fuse for nonsympathetic ditch blasting is impractical because the blasts are impossible to synchronize. The holes used for electric blasting could be spaced much farther apart than those used for sympathetic propagation, and were also loaded with more explosives. Most ditch blasting was done electrically.

According to the US Bureau of Mines, 1 pound of 50 percent strength dynamite was required for each cubic yard of soil to be blasted. This was an average figure, and different kinds of soil took more or fewer explosives.

Coal Mining

US production of coal increased dramatically from 1900 to 1920, leading to a surge in accidents. The worst coal mine disaster in US history occurred on December 6, 1907, at the Monongah Mines in West Virginia, when 350 miners were killed. Two weeks later, 239 lives were lost in another explosion at the Darr Mine in Pennsylvania. From 1901 to 1910, 111 major coal mine explosions left 3,300 dead, with blasting accidents blamed for half.

Other accidents were caused by ignition of concentrated volatile gases or dust by mining lamps, cigarettes, candles, sparks, and even static electricity. This poor safety record accelerated research into reducing the hazards of blasting in coal mines. Many incidents involving dynamite were due to the explosive's large flame and heat output during detonation. Large, hot flames had a much-greater likelihood of detonating mine gases or suspended coal dust.

Fumes produced by the detonation of dynamite were not themselves flammable but could be deadly nevertheless. Miners were known to have been asphyxiated when the amount of blast-produced gases replaced breathable oxygen. This was much more likely to occur if miners returned to work too soon after a blast.

The first low-temperature, low-flame dynamite formulation was patented in Germany in 1887 under the name Carbonite, and a similar mixture called Monobel was developed by Alfred Nobel at a factory he had established in England. A variety of similar explosives were formulated in Europe in the 1890s, with the first roster of English "permitted explosives" published in 1899. Permitted explosives had to pass rigorous tests for fume and flame production.

Testing explosives for their safety in coal mines in the United States was authorized by Congress in May 1908. A testing laboratory was established in Pittsburgh, and exactly one year later the first list of permissible dynamites was published.

DuPont acquired both the Carbonite and Monobel brands for US use and began selling them as DuPont products beginning in 1906. Monobel was one of

The inaugural list of permissible explosives consisted of seventeen dynamites. Seven were DuPont products, including five grades of Carbonite. By 1917 the list contained 153 brands of dynamite. The Carbonite label was produced until 1931. *Author*

List of permissible explosives October 1, 1909.

Brand	Manufacturer
Aetna coal powder A	Aetna Powder Co., Chicago, Ill.
Aetna coal powder AA	"
Aetna coal powder B	"
Aetna coal powder C	"
Bituminite No. 1	Jefferson Powder Co., Birmingham, Ala.
Black Diamond No. 3	Illinois Powder Mfg. Co., St Louis, Mo.
Black Diamond No. 4	"
Carbonite No.	E.I. Du Pont de Nemours Powder Co., Wilmington, Del.
Carbonite No.	"
Carbonite No.	"
Carbonite No.	"
Coalite No. 1	Potts Powder Co., New York City
Coalite No. 2-D	"
Coal Special No. 1	Keystone Powder Mfg. Co., Emporium, Pa.
Collier dynamite No. 2	Sinnamahoning Powder Mfg. Co., Emporium, Pa.
Collier dynamite No. 4	"
Collier dynamite No. 5	"
Giant A low-flame dynamite	Giant Powder Co., (Con.) Giant, Ca.
Giant B low-flame dynamite	"
Giant C low-flame dynamite	"
Masurite M. L. F.	Masurite Explosives Co., Sharon, Pa.
Meteor Dynamite	E.I. Du Pont de Nemours Powder Co., Wilmington, Del.
Mine-ite A	Burton Powder Co., Pittsburgh, Pa.
Mine-ite B	"
Monobel	E.I. Du Pont de Nemours Powder Co., Wilmington, Del.
Tunnelite No. 5	G.R. McAbee Powder & Oil Co., Pittsburgh, Pa.
Tunnelite No. 6	"
Tunnelite No. 7	"
Tunnelite No. 8	"

Initial rosters of permissible explosives listed only dynamites with the brand name and manufacturer. Later lists included an explosive's velocity, cartridge count, and average cartridge weight. Adapted from US Department of the Interior, *A Primer on Explosives for Coal Mines* (Washington, DC: USGPO, 1909), 25.

DuPont operated an explosives testing gallery at the Repauno works in New Jersey. E. I. DuPont de Nemours, advertisement, *Colliery Engineer*, April 1914, 9.

DuPont's most popular brands and was produced until the 1960s. Duobel and Gelobel rounded out DuPont's initial line of "safety explosives," soon to be called permissible dynamites.

A good coal-mining dynamite had two main features. First and foremost was the control of hazardous fumes and flames. Second, an explosive was needed that did not shatter the coal to the point of being unsellable. Dynamites were formulated to explode at a lower temperature, decreasing the likelihood of igniting gases or fine coal dust. As for the dust itself, the US Geological Survey issued a two-hundred-page report called *The Explosibility of Coal Dust* in 1911. The incredibly detailed bulletin covered every aspect of the potential dangers of fine coal dust, from "a definition of dust" to "factors that affect explosivity,"[20] which included density, moisture, ash content, structure, and size.

While the slowest-acting dynamites were not as gentle as blasting powder,

their judicious use could achieve similar results with fewer risks.

There are four basic types of coal: anthracite, bituminous, subbituminous, and lignite. Anthracite is the hardest and has the highest carbon content at more than 86 percent. Bituminous coal is softer and sootier and has a carbon content of between 69 and 86 percent. Subbituminous coal has less than 69 percent carbon, and lignite or brown coal is a very soft, low-carbon coal.

Coal forms over millions of years as plant matter breaks down and becomes compacted by the pressure of the rock above. High-carbon anthracite is the most deeply buried and most difficult to get to.

The most widely used coal for energy is bituminous coal, which is sold in many grades and sizes. Lump coal is generally larger, from 3" to 6" in diameter. Smaller sizes include egg, stove, stoker, and even smaller pea, nut, and rice sizes. The finest of all is called slack. The smaller

Bituminous coal is burned to generate electricity and is used for making steel. Subbituminous coal and lignite are also used to produce electricity. Although lower in energy, low-carbon coals are more environmentally friendly. Anthracite coal is used to fuel some stoves and furnaces and to make products such as charcoal briquettes. *Author*

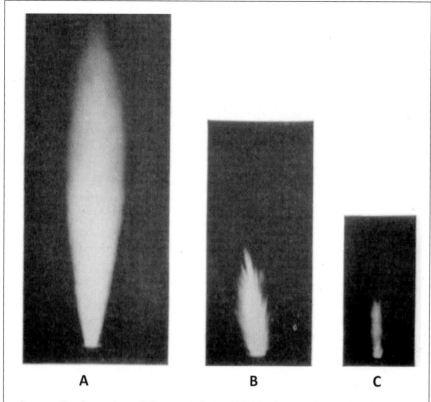

A   B   C

Comparative flame sizes of blown-out shots of (A) blasting powder, (B) 40% dynamite and (C) permissible dynamite.

Reducing the size of the flame during an explosion also included proper tamping with the correct stemming materials. Improperly loaded holes could blow out the top with a large flame, and improper stemming materials such as coal dust could ignite during the blast. US Bureau of Mines, *Explosives Accidents in Bituminous Coal Mines* (Washington, DC: USGPO, 1954), 11.

the furnace, stoker, or other burning apparatus, the smaller the coal needs to be. The ad below shows that coal producers also added other adjectives as a matter of branding.

Coal is separated as to size by running it through a series of screens from coarse to fine. A freight-car load of lump bituminous coal can come in dozens of different variations, depending on the carbon content, upper and lower size limits, whether the coal has been dry- or wet-washed, how it has been dried, and whether it has been double-screened or further sorted.

The largest US coal mines today are surface mines, with the largest producing ten times more coal than the largest underground mine. Explosives are used to loosen the overburden, which is then hauled away. Large machines such as draglines and excavators then harvest the coal.

Monobel, Duobel, and Gelobel were DuPont's most widely sold permissible brands. According to one ad in 1922, the DuPont Company produced the first "permissible" dynamite for use in coal mines. "If you are using this type of explosives, it will pay you to investigate the qualities of Du Pont Duobels, Monobels, and Carbonites. Where Blasting Powder can be used, more Du Pont is being shot today than any other kind. A letter to our nearest

DuPont Monobel and Duobel were the only brands of explosive to be included on both the first and last government-issued printed lists of permissible explosives. *Author*

Pittsburgh Coal Company operated more than sixty mines throughout Pennsylvania. Pittsburgh Coal, advertisement adapted from *Coal Trade Bulletin*, October 1, 1915, 74.

branch office will bring a Service Representative who will be glad to demonstrate any of these explosives right on your own work."[21] Blasting powder was never permissible by the Bureau of Mines, yet it was still used more than permissibles for coal mining up until the 1940s, mainly in mines where gases and dust were not as prevalent. Slow acting and ideal for delicate coal, blasting powder was seldom used for twentieth-century hard-rock mining.

The first list of explosives deemed to be permissible in US coal mines was released as *Explosives Circular No. 1* on May 15, 1909, by the Technology Bureau of the US Geological Survey. The US Bureau of Mines was established in 1910 and assumed the helm, publishing lists of permissibles until 1976. In 1988 the Mine Safety Health Administration issued its final parameters. The last government tests for permissibles were performed in 1999. In 2018 the Institute of Makers of Explosives lamented that "the government has not conducted tests for over

nineteen years, and the equipment is in disrepair and the corporate knowledge needed to conduct the tests is slipping away into retirement."[22]

To evaluate prospective permissible dynamites, the Bureau of Mines launched its Experimental Mine 13 miles south of its main testing laboratory in Pittsburgh. The bureau first leased and then purchased 138 acres of land with existing infrastructure from the Pittsburgh Coal Company. The bureau also focused on other safety aspects of mining, including ventilation, first aid, accident prevention, mine rescue, and "post-disaster survival."

There were no penalties (other than injury and death!) for violation of the bureau's recommendations until 1969, when the Coal Mine Health and Safety Act was passed (other legislation such as the 1952 Federal Coal Mining Safety Act had no provisions for enforcement). In 1973 the Bureau of Mines' regulatory purview was transferred to the Mining Enforcement and Safety Administration, which became the Mine Safety and Health Administration in 1978. The

US Bureau of Mines closed in 1996, and its remaining functions were taken over by the National Institute for Occupational Safety and Health (NIOSH). (The United States Bureau of Mines operated under the US Department of the Interior from 1910 until 1925, when it was placed under the US Department of Commerce. The Bureau of Mines was "retransferred" to the Department of the Interior in 1934.)

In 1913, Atlas Powder Company purchased the US rights to the Coalite brand of permissible explosives from the Potts Powder Company of Pennsylvania. At the time, white wrappers were extremely unusual. After the acquisition of Giant Powder Company by Atlas, Giant-branded Coalite was also produced. The brand label was popular into the 1980s and was even recycled for nonnitroglycerin emulsion explosives (water gels were accepted for use in underground mines in 1970, and emulsions were approved in 1981).

Coalite came in ten grades. According to the Bureau of Mines, Coalite was formulated exactly like DuPont

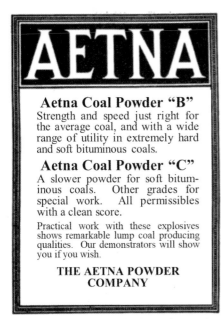

Four grades of Aetna Coal Powder made the first list of permissible explosives in 1909. By 1917, Aetna offered more than thirty different brands and grades of coal-mining dynamite. Aetna Powder, advertisement, *Colliery Engineer*, July 1911, 40.

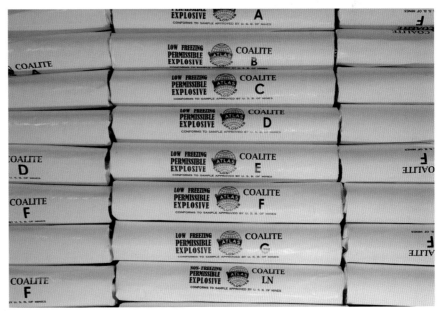

In 1920, Atlas wrote of the various grades of Coalite: "There are grades that are quick, that are slow, that are strong, that are weak, that are dense, and that are bulky."[23] Strong, quick-acting Coalite A was recommend for anthracite, while slower, less powerful Coalite D was for softer, bituminous coal. *Author*

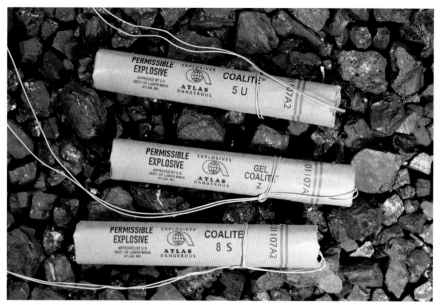

Atlas produced its Coalite permissibles until 1995. The practice of putting either a red or black stripe or stripes on permissible cartridges began in the 1940s but was never a legal requirement. *Author*

Detachable rotary coal bits such as this were introduced in the 1890s. *Author*

Monobel. The same was true of Hercules's original Red H line. After the breakup of DuPont, the three resultant companies continued to share expertise and product knowledge.

The boom in the number of permissible brands from seventeen on the initial list attests to their popularity. The figure reached 160 products in the 1930s and peaked in the 1950s with more than two hundred explosives listed. A gradual decline then ensued, and the last list in 1976, which also included nondynamite explosives, contained only seventy-two products.

Because percussion drills produce more dust, rotary drills were preferred for underground coal mining. Bituminous and subbituminous coal are soft enough that hand augers were used. Pickaxes, and later a variety of saws and other tools, were employed for "undercutting," which involved carving out a channel in the very bottom of a coal face, parallel with the floor of the mine. This weakened the face and dramatically enhanced the efficiency of blasting. Current handheld coal-cutting devices resemble chain saws but are seldom needed, because blasting coal underground is only rarely done today. A process called longwall mining was perfected in the 1970s and uses a huge cutting machine to shear coal from seams and walls. Another newer technology is the continuous mining machine, which is similar to the longwall apparatus and uses rotating drums with protruding picks that claw coal from the deposit.

In the 1890s, other countries tried sheathing cartridges in special "flame-extinguishing envelopes." Cartridges with built-in sheaths made of inert materials were popular in Europe and Canada in the early twentieth century but saw very limited use in US mines. Austin Powder's Rockbuster was a later (1980s) example of a US-produced sheathed permissible. Rockbuster was a bag within a bag, rather than a cartridge, and was used for secondary boulder blasting underground.

Other efforts to reduce the heat of the flame included bottles of water or chemicals placed over the opening of the borehole. Ultimately, reformulating the dynamite mixtures was necessary to produce reliable, low-temperature explosions. Initially this was done by adding sodium chloride or ammonium chloride, which reduced the size and heat of the flame produced by a blast. Later permissibles contained a variety of flame-resistant additives.

It was important that all recommended practices regarding permissible dynamites be carefully followed. The Bureau of Mines advised the following in 1965:

The continued permissibility of specific lots of different brands of explosives as approved by the Bureau of Mines is contingent upon the following: (1) The explosive must conform with the basic specifications, within limits of tolerances prescribed by the Bureau of Mines; the cartridges must be of diameters that have been approved. (2) The explosive must be stored in surface magazines under conditions that help maintain the original product character; it must be used within forty-eight hours after being taken underground. (3) The explosive must remain in its original cartridge wrapper throughout storage and use, without admixture of other

Hercules Red H permissible dynamites first made the permissibles list in 1913 and were produced by Hercules until 1985. The brand was then taken over by IRECO and passed on to Dyno Nobel. The label was produced until 2020 and was one of only a few dynamite brands to ever achieve centenarian status. *Author*

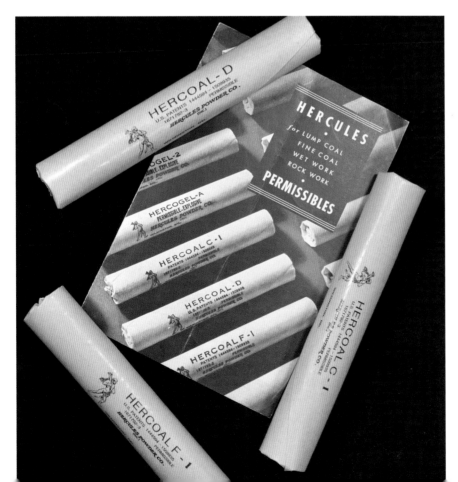

substances. (4) The explosive must be initiated with a copper or copper-based alloy shell, commercial electric detonator (not cap and fuse) of not less than #6 strength. (5) The explosive must be used in conformance with all the applicable provisions of the most recent edition of the Federal Mine Safety Code.[24]

Aluminum-shelled blasting caps were prohibited in coal mines because aluminum burns with a large, hot flame. Explosives on the permissibles list were deemed permissible only if used exactly as prescribed. For example, no frozen dynamite was considered permissible, whether on the list or not.

Permissible explosives achieved their objective: no fatal accident has ever occurred where permissible explosives were used as recommended. Coal-mining deaths in general showed a steady decrease from 1909 onward.

Blasting with cap and fuse was deemed permissible only until the 1930s, when electric firing was required, and then only with single-shot blasting devices. These blasting machines and many other pieces of equipment were also evaluated by the Bureau of Mines and had to have features that prevented arcing, which could ignite coal mine gases and dust. Other requirements included nonleaking batteries where applicable, and that the units generally perform their function adequately, so as not to induce the use of less safe fallback procedures.

In 1923, T. W. Bacchus, whose career in explosives spanned forty-seven years with DuPont and Hercules, reported that thanks to dynamite, US coal production "rose from 33,000,000 tons in 1870 to 650,000,000 tons in 1920."[25]

Sales of permissibles peaked in 1947 and then declined as surface

Hercules offered other permissible formulas, including its Hercoal, introduced in 1928, and Hercogel, which debuted in 1930. Hercogel was the first semigelatin permissible dynamite and was higher in strength for use in tougher deposits. *Author*

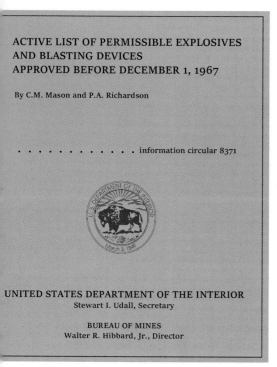

ACTIVE LIST OF PERMISSIBLE EXPLOSIVES
AND BLASTING DEVICES
APPROVED BEFORE DECEMBER 1, 1967

By C.M. Mason and P.A. Richardson

. . . . information circular 8371

UNITED STATES DEPARTMENT OF THE INTERIOR
Stewart I. Udall, Secretary

BUREAU OF MINES
Walter R. Hibbard, Jr., Director

With a few gaps here and there, per-missibles lists were published yearly until 1968. Lists were then published in 1970, 1973, and 1976, with 1976 being the final formal list. US Bureau of Mines, *Active List of Permissible Explosives and Blasting Devices* (Washington, DC: USGPO, 1967), front cover.

To blast a typical stump, a hole was drilled into the roots at a 45° angle with a soil punch or long-shafted hand drill. A primed cartridge was loaded into the hole, which was then compacted with clay or dirt to confine the charge. *Library of Congress Prints and Photographs Online Catalog*, LC USF34-021971-C.

coal operations replaced underground mines. Currently, two-thirds of coal is extracted from surface mines. The rest is usually harvested with modern machinery rather than explosives.

Stumping

Removing a large tree stump in the early twentieth century was a staggering physical undertaking. Without modern grinding machines, an enormous amount of chopping, sawing, and digging were required. Stumps were also burned out; pulled out with horses, mules, and tractors; and, starting in the 1920s, bulldozed.

Stumps come in three kinds. Old, rotten stumps are the easiest to remove because their root systems have relin-quished their grip. Dead stumps that are only a few years old are much

Either cap and fuse or electric blasting was used to remove small stumps requiring only a single hole and charge. Electric blasting was better for large stumps requir-ing multiple holes. *Library of Congress Prints and Photographs Online Catalog*, LC USF34-021963-C.

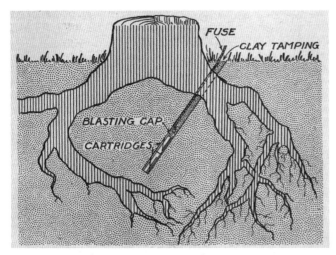

The setup for blasting a small stump. E. I. DuPont de Nemours, *DuPont Farmers' Handbook* (Wilmington, DE: E. I. DuPont de Nemours, 1915), 4.

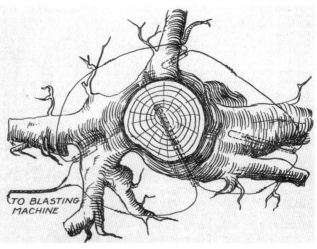

Here, a large stump is rigged with multiple charges. E. I. DuPont de Nemours, *DuPont Farmers' Handbook* (Wilmington, DE: E. I. DuPont de Nemours, 1915), 4.

more difficult, and freshly cut, green stumps are the toughest of all. Trees with large lateral root systems are more difficult to remove than those with single vertical taproots.

According to DuPont, "Several factors very materially influence the blasting of stumps, notable of which are: The character of the root, whether tap or fibrous. The nature of the soil with regard to the resistance it offers the explosive. The moisture content of the soil. The state of preservation of the stump, whether sound or partially decayed."[26] Stumps in moist soil need less dynamite because the extra weight of the soil helps confine the blast to the roots, rather than dissipating through the ground. The amount of moisture in the soil is also an indicator of the extent of lateral root structures. Trees growing in very moist soil tend to put down shallow, wider roots. Small, tap-rooted stumps were taken out with a charge placed in a hole drilled into the tap root. For larger stumps, the charges were placed alongside the tap root to avoid drilling intersecting holes. For huge stumps, many holes were drilled into the root structure.

Dynamite had early competition from another kind of Hercules, a brand of stump puller made by Hercules Manufacturing Company of Centerville, Iowa. Naturally, this company

In 1923, Atlas Powder wrote that "stumps break your tools—an added expense every year. Stumps drive boys to the city. They make you swear. They destroy your pride in your home, and your family's interest in farm life."[27] A good stumping dynamite could solve all of one's problems. *Author*

An advertising poster stamp. The stamps are larger than government-issued ones, and many resemble small posters. Because the stamps get little respect in the philatelic world, they are also called Cinderella stamps. *Author*

Giant Powder Company estimated that one or two sticks of 20 percent strength dynamite would take out a 10" diameter stump. A 40" diameter stump took twenty-two sticks, and a 72" diameter stump took forty sticks. *Author*

and others advertised that mechanical stump pulling was cheaper and easier than dynamite. DuPont admitted that "while the use of explosives for stumping has been found to be a great money-saver in clearing land, there are some classes of stumps that can be more cheaply disposed of by combining the use of explosives with a good steam, horse, or hand puller."[28] Large areas of trees with widespread root systems were blasted with small charges to split the roots and loosen them from the soil. Ideally, the sections of the root structure could then be mechanically pulled out with relative ease.

Stumping dynamites were low in nitroglycerin content and strength, generally 20 percent or less. DuPont's Pacific Stumping Dynamite was 95 percent ammonium nitrate and was rated at 17 percent strength. Gigantic stumps required as many as forty sticks in strategically drilled holes. Less powerful blasting powder or free-running dynamite was sometimes poured into a natural split in the wood that had been widened by a metal wedge.

Another rule of thumb provided by DuPont was to use 1½ pounds of 20 percent strength dynamite for each foot of diameter of stump, up to 4', and 2½ pounds per foot of diameter after that. *Author*

Farmers benefited enormously from the labor that dynamite saved. Expanding one's fields and pastures necessitated reshaping the land in a variety of ways. Flat land was better than sloping land, and obstacles to plows such as boulders and stumps impeded productivity. Farmers who were unwilling to blast stumps themselves could avail themselves of the services of "professional" blasters. Early on, these were usually one- or two-man crews who were just as likely to be farmers themselves, who simply had more experience with explosives and were willing to travel. Dedicated blasters were more prevalent in the 1930s, and by the 1960s, stump blasting was done almost exclusively by licensed contractors or employees of bona fide companies.

Loggers used stumping dynamite to clear land for access roads and reforestation. Construction managers used it to ready a wooded area for development. And highway and railway engineers could not have completed many projects without high explosives to remove stumps. DuPont devoted eight pages to stumping in its 127-page book *Road Construction and Maintenance* in 1915. Stumping strategies specific to road building included avoiding blasting large holes that would require filling, and simply burying stumps in the banks on the sides of roads instead of removing them.

Quarrying

Quarrying is generally understood to be the mining of rocks as construction materials. This is a distinction from surface ore mining, which produces metallic ores and industrial minerals. While a limestone quarry and an open-cut iron mine might look similar, especially from the air, one is a quarry and one is a mine.

Quarrying broadly falls into two main types. The first involves the reduction of large masses of material into economically useful or easily transportable fragments. Aggregate materials such as gravel, limestone, and clay are extracted via wholesale blasting involving enormous numbers of large-diameter drill holes and vast quantities of explosives.

The second kind of quarrying is the production of precisely cut slabs, blocks, and sheets called dimension stone. This is a much more delicate task and involves fewer explosives and fewer, smaller-diameter boreholes placed in tight rows to hew square chunks of rock free from the side of a deposit. Cutting dimension stone is rarely done with nitroglycerin-based explosives because of the risk of fracturing the stone. However, Dyno Nobel's Stonecutter dynamite was specifically formulated for blasting dimension stone.

"Presplitting" is the blasting of very large slabs of stone. Dyno Nobel's Dynosplit brands, which come in 16" and 24" lengths and ⅞" and 1¼" diameters, are intended for delicate presplitting operations. The long, thin cartridges are meant to be connected together with supplied cardboard or plastic couplers to make very long sticks that can be inserted easily into very deep holes. Hercules is said to have pioneered presplitting dynamite with its connectable Hercosplit cartridges. Atlas Powder sold its version as Kleen-Kut.

Quarries are of two general arrangements: pit and hillside. Pit operations are developed on flat terrain and work down either in a circular or stairstep pattern. Hillside cuts are either mined in progressive steps or blasted wholesale to bring down large sections of a vertical formation. The steps are called benches and must accommodate miners and the heavy equipment used to load and haul away blasted material. The heights of the faces in bench or spiral quarries tend to be from 30' to 50' high.

There are many ways to configure a quarry, but the most widely used are

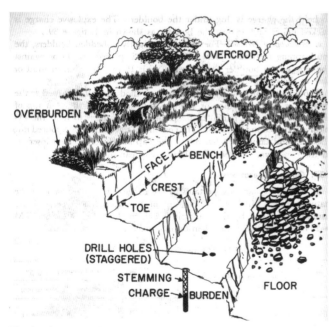

The basic parts of a classic bench-cut quarry. We can add the term "muck pile" near the material piled on the "floor." US Department of the Army, *Explosives and Demolitions* (Washington, DC: USGPO, May 1959), 140.

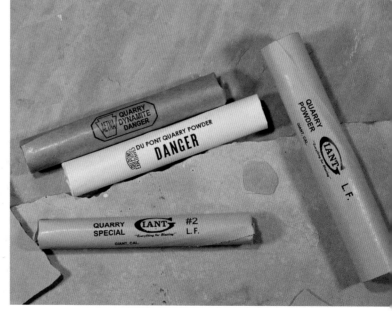

Quarry powders were low-cost, high-ammonium nitrate formulations that were meant to be loaded in quantity for blasting aggregates. *Author*

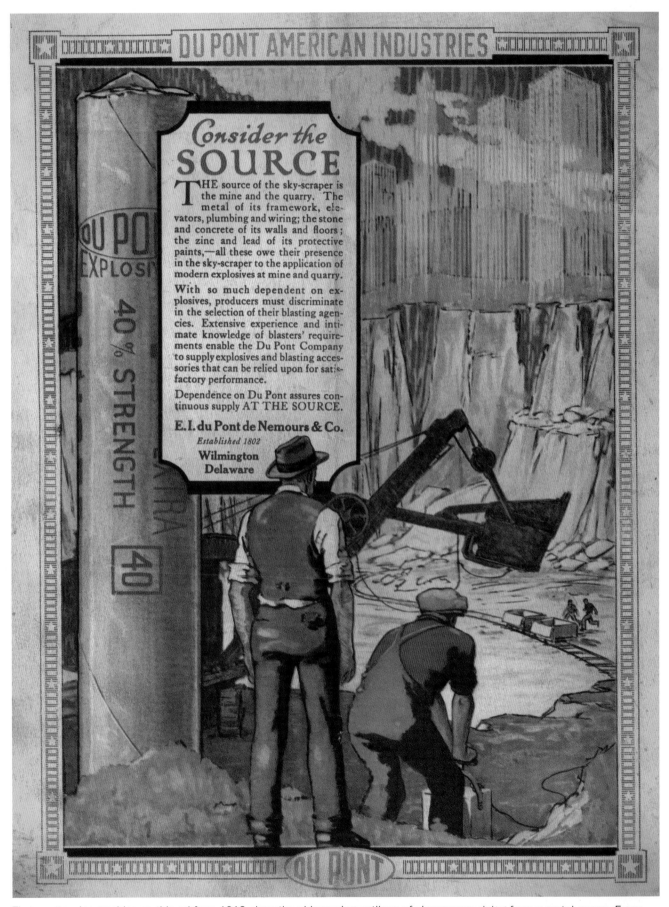

The spectacular graphics on this ad from 1919 show the shimmering outlines of skyscrapers rising from a metal quarry. Every word regarding the need for metals in construction is as true today. E. I. DuPont de Nemours, advertisement, *DuPont Magazine*, July 1919, back cover.

Test No.	Kind of Material	Height of Face	No. of Holes	Average Depth of Holes	Average Distance from Face	Average Distance Apart	Kind of Explosives	Total Lbs. of Explosives	Weight of Material per cu. ft.	Tons Blasted per lb. of Explosives
11	Hard Blue Slate	47'	8	18'	18'	12'	L.F. Extra 40% and Quarry Powder	1,700	165 lbs.	3
12	Limestone	41'	13	45'	16'	16'	L.F. Extra 60%, 50%, 30%	5,900	165 lbs.	2.002
13	Limestone	44'	5	50'	20'	18'	L.F. Extra 50%, 30%	1,250	170 lbs.	6.8
14	Limestone	36'	5	44'	32'	16'	N.G. 60%, 40%, Judson Improved	1,650	165 lbs.	4.8
15	Cement Rock	53'	26	53'	23'	18'	L.F. Extra 50%, 30%	7,950	165 lbs.	5.033
16	Limestone	36'	3	38'	19'	15'	L.F. Extra 40%	1,325	170 lbs.	5.3517
17	Trap Rock	200'	28	136'	43 ½'	20 ½'	L.F. Gel. 60%, 50%, 30%	37,500	180 lbs.	8.14

Table showing the results secured in actual quarrying work. These figures will be found valuable as a guide for estimating purposes.

QUARRYING

This chart shows how much dynamite was used to tackle rock faces of various heights and types. These figures are for single, coordinated blasts and were likely repeated many times for each face. Giant Powder, *Explosives and Blasting Supplies* (San Francisco: Giant Powder, 1924), 48.

the spiral cut and bench cut. The spiral quarry configuration, also called the circular bench, is ideal for large deposits on flat, well-drained land. The pitch, width, and height of each bench vary depending on the material being mined.

Bench-cut quarries are better for attacking a hillside, or for removing material after a pit has been excavated next to a formation. Travel is typically one way, with a switchback at the end for turning around for the journey back.

Beyond cost considerations, the type of dynamite used for quarrying depended on the material being mined. Straight dynamites of 50 percent or more were used for harder materials, while lesser strengths were used for material such as clay. Slower-acting dynamites were used for softer rock so as not to pulverize it. A rule of thumb offered by the Bureau of Mines is that 1 pound of the appropriate dynamite will shatter from 1 to 3 tons of hard rock and from 3 to 5 tons of softer rock.

According to DuPont, factors influencing the selection of a dynamite for quarrying included "the nature of the material, whether it be hard or soft, tough or brittle, limestone or trap; the direction of the strata; the thickness of the ledges; the height of the face or the benches; the method of drilling; the method of loading; the size of the crusher; the purpose for which the stone is used; and whether the work is wet or dry."[29] The "size of the crusher" refers to the in-pit ore crushers employed by quarries to roughly process material after blasting. The larger the crusher, the larger the initial fragments can be, which means that fewer explosives are necessary. Successively smaller crushers are used to further process material.

DuPont recommended its Forcite gelatin and blasting gelatin for shattering very hard rock. Blasting gelatin was known in Britain as gelatine and was rated at 100 percent strength relative to straight dynamite. However, blasting gelatin did not have more than 75 percent nitroglycerin content, the rest of the power being generated by ammonium nitrate and nitrocellulose. For a slower, pushing action, lower-strength straight and gelatin dynamites were recommended. Railroad powder, with strength between blasting powder and dynamite, was employed for the most-delicate work.

The borehole patterns used for quarrying were not nearly as elaborate as those used for underground mining, but quarry blasts had many more holes of much-larger diameter. A simple row was used to blast a small face, and a staggered row of holes were used for bringing down more material at one time. The size, depth, and spacing of holes were calibrated to the material. According to the 1958 *Blasters' Handbook*, "Typical patterns for 5" to 6½" holes in limestone are 20' × 14' for 30' to 50' faces."[30]

Once an area has been blasted, unmanageable boulders are given a "secondary blast." Either a hole is drilled (blockholing), the charge is placed under the boulder (snake holing), or the technique of mudcapping is

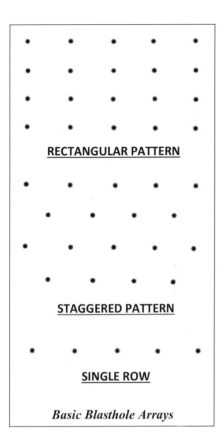

RECTANGULAR PATTERN

STAGGERED PATTERN

SINGLE ROW

Basic Blasthole Arrays

Left: These basic patterns can be expanded to include hundreds of bore-holes. Adapted from US Department of the Army, Corps of Army Engineers, *Systematic Drilling and Blasting for Surface Excavations* (Washington, DC: USGPO, March 1972), 5-1.

employed. The latter method involves simply positioning a charge on the surface of the boulder and gets its name from the practice of using mud or plaster to keep the charge in place. While an essentially unconfined charge wastes a lot of energy, sometimes it is still the most efficacious option. The three methods are illustrated on (see p. 101).

In 1957, Atlas Powder Company offered its quarrying dynamite in high-, medium-, and low-velocity formulations. According to the company, at that time, "some quarry operators continue to spend over a third of their total explosives cost on secondary blasting."[31] By correctly matching the velocity of the dynamite to the type of rock being harvested, secondary blasting could be minimized. Change was in the air, with Atlas noting that "many quarry operators are getting truly efficient, low-cost blasting through the use of one or more of the new Atlas Blasting Agents."[32] At the time, these consisted of various free-running ammonium nitrate formulas.

Before the development of modern pourable emulsion and slurry blasting agents, free-running dynamites were widely used for quarrying, particularly for filling large sprung holes in softer rock.

There are trade-offs among water resistance, strength, density, and cost of explosives. Density is important because the denser the explosive, the more explosive power per foot of borehole. Conversely, less dense

Lower-density extra formulations were used for blasting dry hillsides and very soft rock. The more malleable, gelatin-extra mixtures were better for firm tamping into upward holes. *Author*

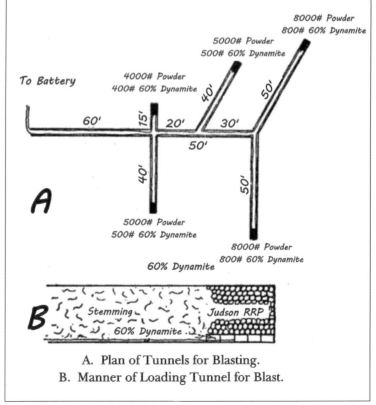

A. Plan of Tunnels for Blasting.
B. Manner of Loading Tunnel for Blast.

Some of the largest networks of tunnels excavated for coyote blasting were more than a mile in total length and were packed with more than 1,000,000 pounds of explosives. Halbert P. Gillette, *Handbook of Rock Excavation Methods and Costs*, US Bureau of Mines (New York: McGraw Hill, 1916), 485.

(bulky) dynamites were needed for engineering enormous, heaving blasts that produce correctly sized material. Using very powerful, fast-acting explosives can reduce salable gravel to useless powder, and profitable lump coal to dust.

Larger boreholes are more efficient because more explosives can be used with less drilling and loading. Where many small-diameter holes might be filled with hundreds of pounds of dynamite, thousands of pounds could be loaded into fewer, wider holes. Another way of maximizing the power of explosives was to pack large amounts of them into a preblasted chamber or tunnel. The two primary techniques to accomplish this are called coyote tunneling (or coyoteing) and springing.

Coyote or gopher tunnels are branching arrangements of shafts excavated into the exposed side of a deposit. The tunnels are filled with explosives and detonated, bringing down the side of the formation. This technique of attacking from the side is repeated for each newly exposed face of material. According to the Bureau of Mines:

"Coyote blasting" is breaking rock with explosives loaded in small adits run into banks or hillsides. The explosive is generally concentrated at desired points in the adit working, in crosscuts, and in cavities or chambers excavated to hold the charges. After the explosive has been placed and the detonators connected underground, the openings are filled with stemming, usually the material removed in running the adits. The charges are then shot with electric detonators, Cordeau-Bickford, or a combination of both. Relatively large charges are used for this form of blasting, and under some conditions rock can apparently be broken more cheaply by this method than by blasting with drill holes.[33]

"Relatively large charges" is something of an understatement. Staggering amounts of explosives, routinely running into the hundreds of thousands of pounds, were loaded into coyote

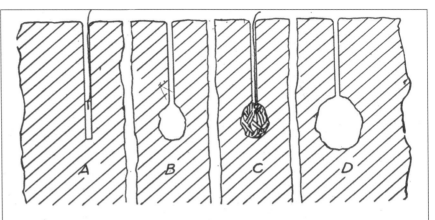

Fig. 85.—Process of springing a bore hole. "A" shows a single cartridge (2 or 3 can be used) in position for the first springing shot in a bore hole of small diameter. "B" shows the result of the first shot. "C" shows the second charge in place. "D" shows the result of the second shot. Additional shots are made until the chamber is enlarged to the desired size.

According to DuPont, "It is seldom necessary to spring more than five times."[34] Even so, hundreds of pounds of dynamite could be loaded into a large, sprung hole. Arthur La Motte, *Blasters' Handbook* (Wilmington, DE: E. I. DuPont de Nemours, 1925), 71.

tunnels. The largest were even laid with temporary track to make transportation easier. Ironically, when free-flowing dynamites were used in coyote tunnels, they were left in the bag. Coyote tunnels were sometimes augmented with charges placed in drill holes to better move material down and out from the formation.

Another way of setting up large blasts is called chambering or springing. While the blasts were nowhere near as large as those produced with coyote tunnels, springing was practiced more frequently because it was easier and more versatile. The method entails the successive blasting and scooping out of the bottoms of vertical boreholes. Small charges of ever-increasing size are exploded at the bottom of a deep borehole again and again until a sizable cavity is carved from the bottom. The enlarged void is finally filled tightly with explosives for the actual blast. According to DuPont, it was advisable to wait one hour after the first springing blast, two hours after the second, and so on. The heat of the hole and smoldering pieces of unfired explosives caused many tragic accidents, and DuPont observed that "it requires considerable judgement on

Hercules used the term "bullet-nosed" to describe its tapered-end, large-diameter cartridges. Atlas Powder Company called its tapered-end cartridges "domed." Atlas also sold large cartridges with fluted ends, which had one end twisted and closed with a wire tie. *Author*

the part of the blaster to spring holes successfully."[35] Not only was safety a consideration, but underloading or overloading could either do nothing or collapse the chamber.

Coyote tunneling and chambering fell out of practice as large-diameter boreholes became easier to drill.

Large-diameter cartridges came into wide use in the 1920s as the technology to drill large-diameter boreholes became more economical. An ideal mix consisted of 6" holes and cartridges, but even-larger cartridges (up to 12" in diameter) were used until the 1970s, when larger-diameter dynamites were relegated to use as primers for slurries, ANFO, and emulsions.

Some large dynamite cartridges came with bales or handles and tapered ends for easier loading. DuPont and other companies supplied loading tongs for large cartridges. The tongs were attached to a rope to slowly lower dynamite into a hole. *Author*

Gelatin-extra dynamites were the work-horse explosive for many construction projects, owing to their water resistance, plasticity, and economy, but blasting for dams and tunnels in hard rock required stronger dynamites. *Author*

While large-diameter holes are generally more cost effective, smaller holes and charges are still used for some quarrying applications. These include situations when finer fragmentation is desired, and for "presplitting" hard dimension stone. Many closely spaced, small-diameter holes produce finer pieces of softer materials that are easier to haul away. With hard rock, a controlled cutting action occurs. The goal is to maximize primary breakage of correctly sized material and minimize secondary blasting, thus reducing the overall quantity of explosives employed.

Construction Engineering

The use of dynamite in construction falls into four categories: land clearing/leveling, ditching/drainage, demolition, and tunneling.

Like farmers, construction engineers must clear land by removing boulders and stumps, and the same techniques are used. Huge trees were also felled and blown into manageable pieces with dynamite. As for leveling a plot of land, the use of explosives was adapted to the terrain. Swampy land was blasted forward and then replaced with viable fill dirt. Sometimes the fill material was loaded on top of the wet ground, and the muddy water was blasted out from underneath, with the fill settling in place. Nuisance hills, rock banks, and slabs of upended bedrock were removed to make room for foundations, piers, and all manners of infrastructure. The dynamites used in engineering span the gamut from straight formulations for ditching and demolition, to extra dynamites for wholesale blasting, to gelatin dynamites for wet and submarine work.

Rerouting water was accomplished with ditching dynamite when practical, and with techniques similar to aboveground mining in the rockiest terrain. Blasted culverts divert marsh water and channel irrigation water, and blasted reservoirs store water. "Pipelining" is the blasting of trenches for laying water pipe.

Sites for new construction often have to be cleared of old structures. Demolishing obsolete bridges, overpasses, and buildings involves placing many small charges at the base of a structure. Arrays of single sticks are exploded in precisely timed sequences to weaken load-bearing portions of a building, and then the weight of the building does the rest. Controlled implosion is one realm of blasting where dynamite is still the tool of choice. Such demolition accounts for only 1 percent of all building demolition.

Tunneling was accomplished with dynamite, as in mining: a wedge of rock was blasted out of a rock face, then the resultant cavity was repeatedly blasted and the debris transported out. Many tunnels of the New York City subway system were originally blasted as open cuts that were then capped off to support the streets above, in a process called cut and cover. Tens of millions of pounds of mostly small charges of dynamite were used, often with blasting occurring right next to the foundations of skyscrapers.

Digging tunnels with blasting powder was a slow and arduous task, particularly in hard rock. Dynamite revolutionized tunnel excavation in the late 1800s, in the same way that giant boring machines have transformed the modern process of tunneling.

Dynamite was used extensively in public-works projects such as dams, irrigation networks, highway construction, and drainage canals. To make the Colorado River flow around the Hoover Dam construction site, four 4,000' long, 56' diameter tunnels were blasted through the Black Canyon from 1931 to 1932. The ducts were used until 1935, when the last one was plugged with concrete. According to the US Department of the Interior, 3,561,000 pounds of dynamite were used to blast the four tunnels.

Steam power and electricity receive much of the credit for early industrial progress. But without high explosives, the metal for the engines, wire for electrical transmission, and material for railway construction would not have been available in the quantities necessary to meet demand. An article in the *American Magazine* in 1924 related the following:

Our road builders use enormous quantities of dynamite, the average being 1,000 pounds to every mile of concrete highway. Farmers clear their land of stumps and boulders with it and drain marshy fields. In the Florida Everglades 3,000,000 acres of swamp are being drained in a single project. The Government estimates that of 113,537,000 acres of land in the United States at present too wet for agriculture, 91,543,000 acres will be of value after draining. Virtually all railroad grading and tunneling, subway work in our cities, under-

According to DuPont, Red Cross Extra dynamite was ideal for blasting dry, soft soil. Such nongelatin, ammonium nitrate formulas were ideal for blasting away dirt hillsides where stronger dynamites were too much and water resistance was not an issue. *Author*

Data regarding drill holes used in New York rapid-transit tunnel (subway).

BENCH HOLES.

Order of firing.	Number and kind of hole.a	Depth.	Size of charge.	Explosive used.
		Feet.	*Pounds.*	
A........	7 grading......................	3–5	50	"40 per cent" dynamite.
B........	5 bench......................	9½	45	Do.

HEADING HOLES.

B........	6 trimming......................	3–9	42	"40 per cent" dynamite.
C........	8 center cut......................	9	56	"60 per cent" dynamite.
D........	8 side......................	8	48	"40 per cent" dynamite.
E........	8 dry......................	8	36	Do.

a All holes tapered from 3 to 2¼ inches in diameter.

Blasting for the first New York City subway line began in 1900, and the initial system opened in October 1904. Expansion has taken place ever since, but tunnel-boring machines are now used for underground work. US Department of the Interior, *Primer on Explosives for Metal Mines and Quarries* (Washington, DC: USGPO), 78.

and lagged and the top is blasted down on these timbers by means of light explosives charges. The muck from these shots can be drawn out by gravity into cars below, either from chutes or through holes made by moving some of the lagging. Where this method can be carried out, it will effect a very great saving in the loading costs.

Sequence of Shots.—The sequence of blasting holes in a heading is governed largely by the method of loading in use. In any case, however, it is generally best to arrange the shots so that a cut is made clear across the heading as soon as possible. This will make the blasting of the remaining holes used to square up the face easier and cleaner and will help to avoid the wasteful practice of shooting against an arch. As regards the sequence of the remaining holes, when the muck is loaded by hand, the best practice, as a rule, is to load the lifters heavy and shoot them last. By doing this, the grade of the bottom is maintained and the muck is thrown back from the face so that a minimum amount of material has to be shoveled back to enable the drillers to set up their machines and get to work.

Figure 194 shows a top heading and bench round. In this delays are used only in the trim holes. If the ground is not too tight, the face can be squared up by blasting sections 2 and 3 with one application of the current, with first delays used in the holes marked D. If the ground is tight, it will be better to blast section 3 by itself, using instantaneous caps in the two middle side or trim holes and first delays in the corner lifters and the remaining three trim holes at the top. In very tight ground, the slant of the cut holes can be changed to bring the bottoms closer together.

Where the full face is blasted and mechanical loaders are used to clean out the face before drilling is started, these loaders will work most effectively if the muck is left in a compact pile against the face. When this method is used, the shots should be arranged so that the bottom holes or lifters are blasted as early as possible in the sequence of shots and the top or back holes are shot last.

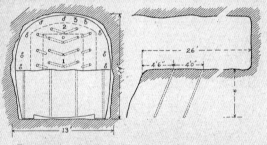

Fig. 194.—A top heading and bench round for tunnel driving.

This will reduce the time the loader wastes in cleaning up the approach to the muck pile and enable it to spend more time on the real digging operation in the pile of broken material.

Figure 195 shows a typical round of holes in a tunnel where a full face is carried in hard tight rock and mechanical loaders are used to clean up the muck before starting the drilling of the next round. The sequence of blasting operations is shown by the numbered blocks. In this, instantaneous and first delay electric blasting caps are used. The holes having the delays are marked D. It will be noted that the blasting of this round is designed to leave the muck piled up against the face to allow the most efficient operation of the loader.

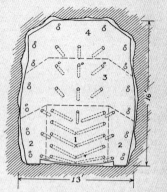

Fig. 195.—A typical round of holes in a tunnel where a full face is carried in hard, tight rock, and mechanical loaders are used to clean up the muck before starting drilling of the next round.

A heading and bench round which has been especially developed to take the place of the full-face round in an 8 by 8 foot or 9 by 9 foot tunnel shown in Figure 192, when the simultaneous blasting of the cut holes throws the broken rock so far down the drift as to slow down the loading and make this operation too costly, is illustrated in Figure 196.

The bench, which is shot with one or more lines of lifter holes, is kept one cut behind the face. By firing the three middle lifter holes with instantaneous electric blasting caps at the same time that the cut holes are fired, the rock blasted upward by the lifter shots meets that thrown out by the cut holes and prevents most of it from being hurled a long distance down the drift. This reduces the cleaning up along the drift to a minimum, for the

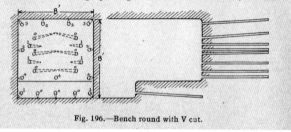

Fig. 196.—Bench round with V cut.

Borehole patterns for tunnel blasting were complex and almost always included angled holes, which directed material toward the center of the blast. The technique helped prevent fracturing too far into the rock on the sides of the tunnel. Arthur La Motte, *Blasters' Handbook* (Wilmington, DE: E. I. DuPont de Nemours, 1930, 169, 170.

river tubes, and harbor dredging is carried on with dynamite. In the South, abandoned rice plantations are being reclaimed for other crops through dynamite ditching. Almost all of our naval stores—turpentine, resin, pitch, tar, and various pine oils—are obtained by dynamiting stumps in the pine swamps and then distilling the fragments. One American company I know of is employing dynamite to fell big trees in Dutch Guiana; the tropical wood is imported and used in veneering furniture. During the past century, through the use of dynamite and its fellow explosives,

over a quarter of a million miles of railroads and more than 2,000,000 miles of highways have been built in this country, more than 400,000,000 acres of land have been cleared and improved, and in one single year over 600,000,000 tons of coal have been mined. The Panama Canal has been excavated; the Moffat Tunnel is being cut through the Great Divide; the famous Roosevelt Dam is a reality; the new Erie Canal, and the Los Angeles and New York Aqueducts built; and in New York alone millions of tons of stone have been removed for the building of tubes and subways. These

achievements in engineering would have been literally impossible without dynamite.[37]

Van Gelder and Schlatter noted that the sprawling Transcontinental Railroad relied on explosives to "blast down mountains and fill up valleys, so that today we can ride in comfort over and under, through and round, this great country and transport merchandise at a fraction of the average cost before the days of modern explosives."[38] The same can be said of the Interstate Highway System. In the 1950s, about

Left: An article in 1911 observed that "stone for bridges and retaining walls is quarried with dynamite, and carloads of dynamite are used by manufacturers of cement, which is also used to a great extent in bridges and retaining walls." E. I. DuPont de Nemours, advertisement, *Good Roads*, February 3, 1917, 16.

one-fifth of the dynamite sold in the United States was used for construction applications.

Seismic Prospecting

Seismic prospecting was first conducted with gunpowder in 1850 by Irish engineer Robert Mallet, who also had invented the seismograph in 1846 and had in fact introduced the term "seismology." The process involves detonating an explosion on or near the surface of the earth and recording and measuring the reflection of the shock waves. Different strata of different densities echo back at different speeds, and a subterranean portrait is developed. The technique reveals where water, bedrock, and oil and gas deposits are located.

Characteristics of a good seismic dynamite are high velocity of explosion, high density, and water resistance. Straight gelatin formulations are ideal. DuPont recommended its 60 percent strength Hi-Velocity Gelatin, as well as several prospecting-specific branded products, both dynamite and not. One DuPont brand called Seismogel began life as a dynamite but then was retrademarked as a water-gel explosive in 1981. Seismogel also went through numerous

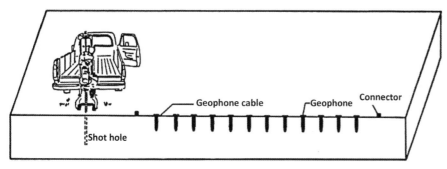

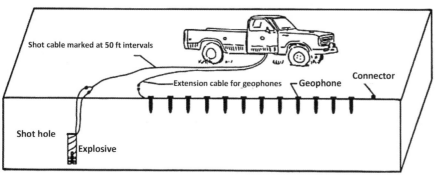

Left: This basic setup for seismic prospecting shows the positioning of multiple geophones to collect data. The truck would be moved and the process repeated many times to complete the survey. US Department of the Interior, *Technology of Water-Resources Investigations of the US Geological Survey* (Washington, DC: USGPO, 1988), 64.

Dynamite used to generate seismic waves is usually loaded into boreholes from 10' to 50' deep. Where this is impossible due to terrain, dynamite is exploded aboveground. A large-diameter cartridge is mounted on a pole with a wooden block atop it to direct the explosion downward. *Author*

packaging changes, going from a rigid cardboard tube to a rigid plastic one in 1966. Resistance to water pressure is crucial, and hard-plastic cartridges were used for deep underwater work.

Another important factor in obtaining uniform readings is using an uninterrupted column of charges. For this reason, tube couplers are employed to prevent gaps between cartridges. Hercules branded its connectable cartridges as Spiralok, while Atlas called its version Twistite. Dyno Nobel currently offers its Vibrogel brand of high-velocity dynamite either in stiff paper or plastic shells. Vibrogel was originally a Hercules property and was acquired by IRECO in 1985, when the company purchased Hercules. Dyno Nobel acquired IRECO in 1984 and operated it as a subsidiary until 1994.

Dynamite is no longer used for submarine seismic prospecting in US waters, owing to environmental concerns. On dry land, explosives are being replaced by seismic air guns and vibroseis vibration systems. The latter consist of large, truck-mounted metal plates that are slammed on the ground to send pulses deep into the earth.

Submarine Blasting

Submarine blasting is needed for the construction of channels, for widening and deepening existing channels, and for removing submerged hazards to shipping. Hull-threatening features such as underwater rock ledges and columns are removed with strategically placed charges. Shipwrecks that pose risks are dealt with in similar fashion. Explosives are also used for excavation, for building underwater foundations and piers, and for digging underwater trenches for laying pipe and cable.

Submarine blasting was accomplished with blasting powder as early as the 1840s and was initially limited to depths of around 20 feet. The fuse of the day would not ignite charges at lower depths. Early electric blasting caps and dynamite were good for another 10 feet. By 1905, special submarine blasting caps were developed, doubling the practical depth for placing charges. The technology has advanced ever since, and modern submarine blasting caps are initiated with signal tubing or purpose-engineered firing systems.

The three main difficulties of submarine blasting are placing the charge, overcoming the effects of moisture, and overcoming the effects of water pressure. Strong currents, boat traffic, and the safety of nearby structures are other considerations.

Bundles, bags, and even entire cases of dynamite were primed, submerged, and detonated (full primed cases were also used for coyote blasting). These instructions from DuPont outline the procedure for priming a case of dynamite for underwater blasting:

In making the charges, the case is opened, five or six cartridges of the 75 percent Gelatin are removed[,] and the same number of 60 percent straight dynamite cartridges substituted. One of the 60 percent straight cartridges should contain the water-proof electric blasting cap well taped and sealed in to prevent the entrance of water, the wires being carried through a small notch cut in the case. The case cover is then nailed on. The primed case is then lowered by means of strong wire or rope, put in place, and a buoy attached to the top end of the rope. The electric blasting[-]cap wires should be attached to the buoy, but all strain and pulling of buoy should be carried by the rope or heavy wire, as the small electric wires are too delicate to stand any strain. Where the tide or undercurrent is strong, No. 14 leading wires are spliced to very short electric blasting[-]cap wires and carried from the charge at the bottom to the buoys. This makes a strong connection.[39]

Drilling boreholes deep underwater poses special challenges, necessitating a stable, floating platform and special drills. For that reason, especially early on, much submarine blasting was done by simply placing charges underneath rock ledges or tying explosives to outcroppings.

Because of their high water resistance, gelatin dynamites revolutionized underwater blasting, enabling more-complex firing patterns and exponentially larger charges. The first gelatin dynamite in America was produced by the American Forcite Company of New Jersey in 1884. The brand was acquired by DuPont under the auspices of the Eastern Dynamite Company and was then produced by DuPont into the late 1920s. Canadian Explosives Limited, which had heavy DuPont ties, also sold gelatin dynamite under the Forcite label.

Ordinary gelatin dynamites in extra-thick shells were recommended by all producers for deep underwater work and had to be special ordered.

Straight dynamites were still used, but mainly for detonation by propagation in closely spaced holes, and for use as primers for gelatin explosives. DuPont billed its 60 percent strength straight dynamite in extrathick, heavily waxed shells as "Submarine Dynamite" and related that its 60 percent strength Hi-Velocity Gelatin was originally developed for submarine work.

Because water pressure reduces the efficiency of explosives, more and stronger charges are used, with boreholes spaced closer together.

Demolition

Demolition requires more preparatory work than any other type of blasting. "Blasting curtains" are placed on nearby buildings to prevent damage, and "blasting blankets" are hung over horizontal boreholes both to help confine the power of the charges and contain flying debris. The US Department of Labor advised that

> several days are required to prepare for dynamiting. The building is first completely stripped to the base shell of concrete, frame, or brick exterior walls, and all salvageable equipment is removed. Holes are drilled at the base of the building's supporting columns and filled with timed dynamite charges which are detonated electrically. Radio silence is coordinated with professional and amateur radio stations in the area, as there have been cases where wave frequencies have set off dynamite. After the charges are detonated, the

---

Submarine Blasting – Deepening harbors and channels

For extended work of this kind, a scow or boat upon which the drilling equipment is mounted is anchored and steadied by means of spuds resting upon the bottom. Each boat carries from one to four large piston drills, or well drills operated by steam or compressed air, mounted in standards somewhat like pile drivers along the side of the boat. The standards can be moved sideways by means of tracks and the drills raised and lowered in the standards.

The bore holes are usually from 2 ½ to 3 ½ inches in diameter, or when well drills are used 5 to 6 inches and where sand, gravel or other material lies above the rock a weighted pipe of sufficient diameter to accommodate the drill bit with a cone-shaped header above the water surface is used. This serves as a guide in starting the hole and sinks down to the rock surface, thus keeping loose debris out of the bore hole.

The work is usually carried forward in the same manner as a single bench in quarry work, the holes being spaced apart and back a distance equal to the depth of hole, except that holes are seldom spaced farther than six feet apart, no matter what the depth. The pressure and resistance of the water preclude, as the rock is not broken into small enough pieces for the dredges to handle easily. Holes are drilled as a rule from five to eight feet below grade to insure thorough breaking to full grade depth. As soon as the hole is drilled to the proper depth it is thoroughly cleaned out by means of a jet of water, and immediately loaded.

The loading in smaller diameter holes is done by means of a loading tube, which is a tube of brass, having a longitudinal slot on one side for the electric blasting cap wires to slide in. Of a proper diameter to slide easily to the bottom of the bore hole, this tube is usually from two to six feet long and is screwed to a smaller piece of pipe of sufficient length to extend above the water surface. The dynamite cartridges, of sufficient size to slide easily into the brass tube, are pushed up into the tube from the bottom in the same manner as in loading a bore hole, the electric blasting cap wires being carried out through the slit. The entire charge is loaded into the tube and held there by wedging the bottom cartridge with a wooden wedge. The tube is then lowered into the bottom of the hole. A long wooden rod or tamping stick is inserted through the pipe and the dynamite charge is held down while the tube is withdrawn. Where the hole is to be fired immediately, a spring clip on the top cartridge is often used to hold the dynamite in the hole, and where the hole is to be left for some time before firing, the clip is placed on the wooden rod.

Underwater blasting can be so difficult that it is sometimes better to drain an area of water first. According to DuPont, "This is quite expensive and only feasible for comparatively small areas, but is practical for foundation work of all kinds underwater."[40] Arthur La Motte, *Blasters' Handbook* (Wilmington, DE: E. I. DuPont de Nemours, 1925), 177.

High-velocity dynamite was used for underwater blasting to overcome the effects of water pressure. Gelatin formulations are virtually waterproof but should still be fired within a few hours after placement. *Author*

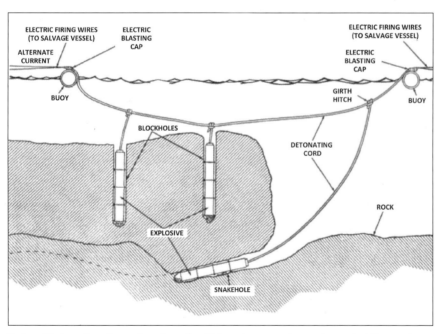

Detonating cord works well underwater. This setup uses buoys to control the position of the detonating cord and leading wires. US Department of the Navy, *Uses of Explosives in Underwater Salvage* (Washington, DC: USGPO, 1956), 45.

are more closely spaced with more explosives per charge. On average, about 300 pounds of dynamite will take down a ten-story edifice. The load-bearing columns are drilled and charges placed and timed such that the innermost columns fail first and the building implodes. The taller the structure, the more floors that need to be weakened. The bottom two stories are sufficient for a ten-story building.

Because tall buildings are often in populated places, another layer of preparation is needed. Permits must be acquired, the proper authorities notified, and safety precautions taken beyond those for normal blasting. Dust mitigation, vibration damage to nearby structures, traffic control, gas and power lines, a plan for toxic material debris to be hauled away, and a safe place for gawkers are other considerations. Local media are often enlisted to spread the word and to cover what is always a spectacular event.

Only 1 percent of buildings are demolished with explosives, which are reserved for very tall buildings, buildings with no room for heavy machinery to attack them, and buildings that must be removed on a shorter schedule than mechanical demolition would allow.

The enormous fires after the 1906 San Francisco earthquake caused far more devastation than the quake itself. Dynamite was used to create breaks between the raging fire and unburned zones of the city. Buildings in the path of the fire were razed, in hopes that the fire would not pass beyond. Dynamite was used in a number of other large metropolitan conflagrations, most notably in the devastating Atlanta, Georgia, inferno of May 17, 1917. DuPont even published a booklet in 1918 titled *Fighting Fires with Dynamite*, advising large municipalities on how to prepare. In 1925 the Institute of Makers of Explosives reported that "it is becoming increasingly frequent for fire departments and property owners to resort to dynamite in fighting large fires."[42] The IME also noted that

floors drop, causing the outer walls to fall inward toward the center of the building. Dynamite is also used to raze structures other than buildings, such as chimneys, towers, bridges, etc. Dynamite charges are strategically placed at the base of the structure, and with proper timing in detonating the charges, the structure can be made to fall in any direction desired.[41]

Dynamite is still used to raze concrete buildings; other explosives and means are needed for buildings with major steel structural elements. Reinforced concrete walls take more holes that

CONCRETE WALL

LEAD WIRE TO BLASTING MACHINE

Left: When large charges are needed, the holes are drilled high enough so that the ground is not cratered by the explosion. Arthur La Motte, *Blasters' Handbook* (Wilmington, DE: E. I. DuPont de Nemours, 1922), 96.

dynamite was used to blast trenches as firebreaks for wildfires.

The successful use of dynamite for firefighting in a few instances was far offset by the dangers of using explosives in such high-pressure situations. One account of the San Francisco fire reported that "many men"[43] were lost during the blasting operations that were frantically conducted amid the roaring flames. Dynamite exploded prematurely due to proximity to intense heat. Collateral damage to infrastructure such as gas lines and electrical systems posed

Left: For taking down concrete structures, 40 to 60 percent strength gelatin and gelatin-extra dynamites were preferred for their high velocity. While water resistance is usually not an issue for demolition, the plasticity of gelatin mixtures makes them easy to tamp to fill boreholes tightly. *Author*

Below: This article details the use of dynamite in the effort to control the catastrophic fires that resulted from the San Francisco earthquake of April 18, 1906. Dynamite was more extensively and successfully used to demolish damaged buildings after the disaster. Adapted from "Fighting Fires with Dynamite," *Fire and Water Engineering*, October 24, 1917, 322.

FIGHTING FIRES WITH DYNAMITE

Some Valuable Suggestions as to Methods of Procedure in the Staying of Large Conflagrations, by the

It is generally known that San Francisco was visited by a serious earthquake on April 18, 1906, at 5:15 a.m. The damage done to the city, while considerable, did not in dollars and cents amount to within 1 per cent of the damage that was afterwards done by the conflagration. Included in the damage done by the earthquake, and the most important of all (so far as after events were concerned) was the breaking of the large Spring Valley waterworks main, which

allow no one to take the balance, since he fully realized that the amount of powder which each wagon could carry, approximately 50 boxes to a wagon, would not last long. With the full authority granted to him by the authorities, and with six men who were skilled in the handling of explosives, he proceeded to a point in front of the fire by a circuitous route around the fire's flank, and set to work to blow down buildings in the path of the flame, with the hope of

in charge remaining nearby in order to count the shots.

8th. The officer in charge will then view the result of the shot and judge the amount necessary for the balance of the square of similar buildings and the same routine to be carried out in future shots.

In the San Francisco conflagration the above organization was effective, and during the 72 hours spent in actual labor not a misshot occurred nor was any one injured. Five buildings were usually

further hazards. The IME's assessment of the ascendancy of explosives as a firefighting tool turned out to be premature. Like the dynamite projectiles of the 1880s, the interest in the use of explosives to fight fires in cities died out in just a few years. Municipal fire chiefs were particularly reluctant to train ordinary firefighters in the use of such a dangerous and specialized tool.

Other Uses for Dynamite

Dynamite was and is used in many ways beyond the most common.

Thick steel plates were blasted into smaller pieces by placing charges along their riveted or welded joints. A large steel rod or cable can be cut with charges on either side. Dynamite can also be used to weld two pieces of metal together. A plate of steel is covered with an inch of free-running dynamite. Another plate of steel is put on top, and an electric blasting cap is inserted into the powdered dynamite. The heat of the blast instantly and permanently fuses the plates together.

A boiler can be cut in half by unwrapping gelatin dynamite and making a rope around the boiler. Although the whole procedure sounds alarming, cutting, shaping, and compacting dynamite cartridges was an integral part of this kind of blasting. The Giant Powder Company advised, "For breaking a boiler the explosive should be removed from the cartridge and pressed in a train in the line of the proposed cut. The train of explosive should not be flattened or thinned out, but should be at least ⅞" thick. Well-tempered stiff clay or mud should be used to cover the charge."[45] Other heavy machinery was also scrapped in this way. In addition, boilers can be cleaned of the fused deposits of ash that accumulate over time. Holes are drilled in the material, and small charges knock it away without damaging the boiler. Removing boiler deposits with dynamite is one of the few uses for dynamite that is currently becoming more frequent, because other methods are so time consuming and labor intensive.

Furnaces for making steel were tapped or drained by blasting a hole with a small charge of dynamite. This was not a new hole in the furnace, but rather in a hard clay plug, closing an existing hole that was used over and over. Tapping allowed the fresh molten steel to flow into railroad ladle cars. Inserting a cartridge of dynamite into a furnace of boiling steel is dangerous. One expert advised loading a copper pipe with a primed cartridge inside for temporary insulation. Regardless, the charge must be fired very soon after loading.

It was recommended that material stuck in a railroad car be blasted out with a partial stick of dynamite. According to DuPont, "Coal, ore, and similar material occasionally freezes so hard in railroad cars that it is next to impossible to dislodge it. Charges consisting of a quarter or a half of a cartridge of Red Cross 40 percent Extra Dynamite, if located with good judgement, will not injure the car in any way and will provide great assistance in breaking up the material so that it can be easily handled. The charges should be pressed into crevices and fired with blasting cap and fuse."[46]

Large chunks of ice, up to and including icebergs, were blasted with dynamite to reduce hazards to ships and infrastructure such as bridges, docks, and piers. The charges were placed underneath or on top of the mass or were lowered through holes drilled in the ice. Whole cases were primed and placed or submerged, with DuPont noting that "where the ice gorge runs up to twenty feet and thirty feet in depth, it may require 1,000 or 1,500 pounds of dynamite fired under it to obtain any results."[47]

Attacking an ice flow from the top was accomplished by mudcapping

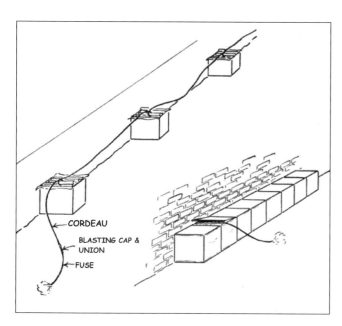

CORDEAU
BLASTING CAP & UNION
FUSE

Fig. 112.—An open vessel is filled with water and a charge of dynamite, properly primed, lowered to a point near the thickest or strongest metal. This charge must not be actually in contact with the metal. For large vessels, two or more charges are used and fired electrically. The dynamite causes the water to transmit a powerful blow to the sides.

Above: DuPont warned that when blowing up an old boiler, "pieces can be expected to travel several hundred feet with the chance of severe injury to persons or damage to property." Encasing the boiler in blasting mats was recommended. Arthur La Motte, *Blasters' Handbook* (Wilmington, DE: E. I. DuPont de Nemours, 1922), 92.

Left: Since the blasting of structures to create firebreaks must obviously be done in a hurry, the Institute of Makers of Explosives suggested lining up full cases of dynamite next to a building. H. E. Davis, "Dynamite in Fire Fighting," *Quarterly of the National Fire Protection Association*, July 1925, 29.

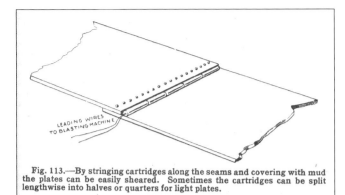

Fig. 113.—By stringing cartridges along the seams and covering with mud the plates can be easily sheared. Sometimes the cartridges can be split lengthwise into halves or quarters for light plates.

DuPont recommended its 80 percent strength Hi-Velocity gelatin dynamite or 100 percent strength blasting gelatin for shearing steel plates. A ¼" thick plate took 1.5 pounds of dynamite per foot along the line of the intended cut. Arthur La Motte, *Blasters' Handbook* (Wilmington, DE: E. I. DuPont de Nemours, 1922), 93.

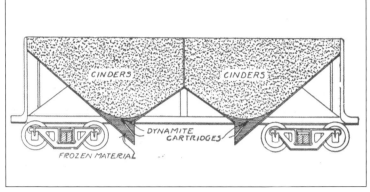

This diagram shows how to release frozen material so that it could pour out of chutes built into a railcar. Arthur La Motte, *Blasters' Handbook* (Wilmington, DE: E. I. DuPont de Nemours, 1922), 95.

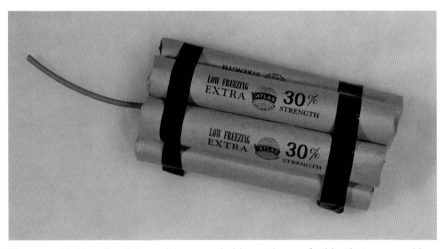

The proverbial bundle of dynamite was probably used more for blasting snow and ice than for any other application. Getting to and from where the blast was to occur often had to be done quickly, with no time to drill holes. *Author*

closely spaced charges. Ideally, 40 percent strength gelatin dynamite was used for blasting ice. However, according to DuPont, "As this work must, as a rule, be done on short notice, little time is afforded to obtain the correct explosive, so the recommendation is made for any low-freezing dynamite."[48] Ice blasting is still a routine part of life in some regions, but TNT and emulsion explosives are usually used instead of dynamite.

In 1914, Atlas Powder Company suggested, "Tie together three or four cartridges of explosives. Prime one of them with a blasting cap and fuse. Tie the bundle to a block of wood and

This image was originally captioned "A bundle of cartridges primed with electric blasting cap floating under an ice floe. When fired, this breaks up the ice and allows it to float away." Arthur La Motte, *Blasters' Handbook* (Wilmington, DE: E. I. DuPont de Nemours, 1922), 98.

This illustration was originally captioned "Showing bundle of cartridges tied together with fuse and cap, used to mudcap floating pieces of ice." Here, "mudcap" is being loosely employed to refer to an unconfined charge. Arthur La Motte, *Blasters' Handbook* (Wilmington, DE: E. I. DuPont de Nemours, 1922), 97.

Right: This study delved into questions such as charge placement, different kinds of explosives, and different snow conditions. Alfred Fuchs, *Effects of Explosives on Snow* (Washington, DC: USGPO, July 1957), front cover.

immediately after igniting the fuse, throw it upon the cake of ice, as near as possible to center. The operation may require another and, perhaps, a larger charge. A nearby bridge may supply a position from which to drop the charge upon the floating ice cake."[49]

Logjams are another waterborne nuisance solvable with high explosives. Giant Powder Company recommended "a fairly heavy charge of explosives, in the form of cartridges tied together,"[50] of 40 percent strength gelatin dynamite placed at a "central point where one or more of the logs have been wedged."[51] According to DuPont,

Log rafts or jams, on careful examination, are usually found to be tied together or held by a log or several logs that act as a key or pivot. It is against this point that attention should be directed. The use of heavy charges of dynamite is usually necessary. The loading should be done as quickly as possible, as it is usually dangerous to remain long on the jam. The dynamite can be loaded into a bag or box, primed and placed in the water as near the key logs as possible. Firing should be by means of electric blasting caps, as there is danger of the loader having difficulty in reaching a point of safety when using blasting caps and fuse.[52]

Explosives are still used to break large logjams. Because of environmental concerns, cranes and heavy machinery, combined with handwork

Sometimes, wooden towers are quickly erected at safe distances on either side of the fire. A cable strung between the towers is used to lower the charge into position for snuffing out the blaze. "Extinguishing an Oil Well Fire," *Popular Mechanics*, January 1925, 68.

Special Report 23

JULY, 1957

Effects of Explosives On Snow

by Albert Fuchs

U. S. ARMY SNOW ICE AND PERMAFROST
RESEARCH ESTABLISHMENT
Corps of Engineers
Wimette, Illinois

EXTINGUISHING AN OIL WELL FIRE
Men in Asbestos Suits Lower Chair Wrapped in Same Material on which Dynamite is Strapped into Flames and Electrically Detonated Blast Puts Out Fire

Asbestos and dynamite both figured in the extinguishment of a California gas well fire recently. Natural Gas Well Martin no. 1., of the Royal Dutch Shell Company, Signal Hill, near Long Beach, Cal., caught fire, it was thought, through a spark of static electricity, caused by friction of gas on the well casing. The well has an estimated daily flow of 100,000,000 cubic feet and when the gas ignited the flames mounted to a huge column many hundreds of feet high.

It was seen that heroic measures would have to be taken to extinguish the blaze before the terrific heat communicated the fire to the surrounding wells. Ordinary methods would have had no effect. Dynamite, it was concluded, would be the most effective means to extinguish the fire, but the problem arose as to how the explosive was to be placed near enough to be of any service, without it being affected by the terrific heat.

with chain saws, are employed whenever possible.

Water itself is directed by explosives in numerous ways. Rivers and streams are rerouted by blasting new channels. Swamps are drained and lakes, ponds, and reservoirs are excavated. Wells are sprung to increase flow, using the same springing techniques employed for mining. Huge quantities of explosives are used to optimally shape sites for building large dams. Small dams can be entirely made by blasting fill material to block water flow.

Finally, water in the form of snow is blasted to produce small, controlled avalanches that reduce the chances for large, destructive slides. The trickiest part is "placing" the charges. Dynamite sticks, primed with fuse lit, used to be hurled from a ledge, ski lift, or helicopter. Nowadays, charges are more likely to consist of a TNT cannister launched with a special air cannon and detonated via remote control. On especially slippery slopes, a charge may be belayed on a rope anchored on a higher point on the mountain.

This drawing facetiously illustrates the dangers of not coordinating dynamite drops from ski lifts. Accompanying advice included "All chairs should be vacant except for the blasting team"[53] and "Charges should be tossed downslope and away from chairs that are moving uphill."[54] US Department of the Interior, *Avalanche Handbook* (Washington, DC: USGPO, July 1976), 128.

Dangerous snow and ice overhangs called cornices are also blasted.

Using liquid nitroglycerin to blow out an oil well fire was first successfully accomplished in 1913. Liquid nitroglycerin was last used in the 1940s, but 100 percent strength blasting gelatin is still used. In 1922, *Fire and Water Engineer* related,

Exploding dynamite in the midst of raging fires to extinguish them sounds like an absurdity, but that is what is sometimes done in the oil fields when wells catch fire. Not long ago, a gasser in Colorado which had blazed steadily for twenty days, in spite of efforts to quell it with a battery of thirteen steam boilers and with pumps throwing water, mud and chemicals, was quickly extinguished with a charge of dynamite that literally blew out the flame and stopped the flow of gas. The action, in the case of an oil well, is much the same as when dynamite is fired in the quarry. Just as the explosion shatters the rock, when the charge is set off in the flaming torch of a gas well, it breaks the blaze into fragments and at the same time, a tremendous volume of rapidly expanding gas is liberated in the midst of the flames with velocity sufficient to snuff out the fire.[55]

Using dynamite to extinguish oil well fires was first done in the late 1910s, but the most famous practitioner was Texan Red Adair, who operated from 1959 to 1992 (a movie about Adair's exploits, titled *Hellfighters*, was released in 1968). Dynamite was placed in an insulated drum and lowered as near to the flame as possible from a horizontal crane. The charge was then pulled directly over the flame and detonated electrically. The force of the blast removes the fuel for the fire and temporarily interrupts the flow of gas, allowing just enough time to quickly cap the well.

Another use for dynamite regarding oil wells was for shooting, which in this context meant removing deposits from a well that impeded oil flow. Initially, a heavily weighted liquid nitroglycerin "torpedo" was lowered to the bottom of a well and exploded, but high-strength dynamite was preferred from the 1890s to the 1960s.

Atlas Powder Company sold three different grades of special dynamite for oil wells. Oil Well Explosive was a 100 percent strength gelatin dynamite. Oil Well Explosive #2 was an 80 percent strength gelatin dynamite and was available both in cartridges and in 25- and 50-pound blocks. Finally, Oil Well Explosive #3 was a less dense, 80 percent strength formulation "for use where conditions are not too severe."[56] Apparently, when blasting oil wells, severity is in the eye of the beholder.

In a book produced for the Atlas Powder Company's twenty-five-year anniversary in 1937, a very unusual use for dynamite is illustrated. The first photograph is captioned "A huge whale ashore. Explosives needed for emergency burial."[57] The second and third photos show the blast (thankfully from a distance) and resultant crater. "High tide finished the job and restored the level beach."[58]

A more famous whale detonation occurred in Oregon in 1970. A beached whale that had perished was discovered near the small coastal resort town of Florence. The state highway department used dynamite to try to blast the leviathan cetacean to tiny bits for the birds to eat. Instead, large chunks of blubber rained down in a radius of a quarter mile, with one giant slab destroying a car. The ex-whale became a local celebrity, and it is an easy matter to find footage of the blast. In 2020, the residents of Florence established Exploding Whale Memorial Park to celebrate the event's fiftieth anniversary.

Dynamite has been used for so-called blast fishing, where an explosion is used to kill fish, which are then scooped up from the surface. Foolish on its face, the practice is illegal in the US and most other countries.

In 1965 the Wisconsin Conservation Department published a study titled *Pothole Blasting for Wildlife*. According to the authors, "Wildlife use of many blasted potholes has been good. A group of forty-four potholes blasted in a dry portion of Horicon marsh showed excellent use by mallards and blue-winged teal in the spring of 1964."[59] The blasted holes, which fill with ground or rainwater, were also associated with "considerable deer activity."[60] The holes were the size of small ponds and provided breeding

When Atlas oil well explosives were first introduced, they were sold by the quart, even though they were packed in cartridges. A weight of 2.85 pounds of oil well explosive was equivalent to 1 quart of liquid nitroglycerin. This made it convenient for well blasters who were experienced in the use of liquid nitroglycerin. *Author*

United States Department of Agriculture

Creating Snags With Explosives
Evelyn L. Bull, Arthur D. Partridge, and Wayne G. Williams

The tops of ponderosa pine (Pinus ponderosa) trees were blown off with dynamite to create nest sites for cavity nesting wildlife. The procedure included: drilling a hole through the trunk, inserting the dynamite, and setting the charge with Primacord and fuse. Trees were simultaneously inoculated with a decay organism. The average cost was $30.00 per tree.

An increasing awareness of the importance of snags (dead trees) to wildlife, particularly cavity nesters (Jackman 1974, Beebe 1974, Balda 1975, Conner et al. 1979, Mannan et al. 1980), has resulted in an interest in how to create snags. Trees can be killed by silvicide, girdling, and burning; however, we wanted a quick and safe method that would provide a facsimile of the dead trees woodpeckers use for nesting. In conifer forests woodpeckers favored snags with broken tops and decayed wood (McClelland and Frissel 1975, Bull and Meslow, 1977, Scott, 1978, Raphael, 1980).

It's not enough to blow up stumps. The tops of trees need exploding now and then, for quite an unexpected reason. Certain birds need cavities or "snags" in dead trees to nest, and dynamite does the trick. Evelyn L. Bull, Arthur D. Partridge, and Wayne G. Williams, *Creating Snags with Explosives* (Washington, DC: USGPO, 1981), 1.

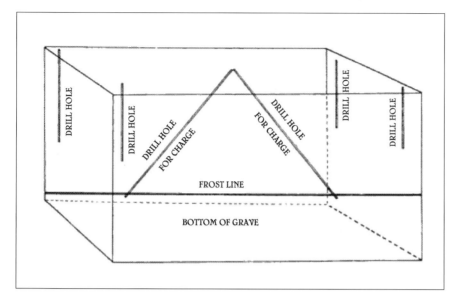

When necessary, frozen ground in cemeteries was excavated with dynamite. Obstacles such as buried boulders that would not succumb to other methods were blasted to kingdom come. "Methods of Digging Graves with Explosives," *Park and Cemetery and Landscape Gardening*, March 1920, 11.

territory for waterfowl and drinking water for mammals.

Finally, digging graves with dynamite was once "common practice,"[61] according to *Park and Cemetery and Landscape Gardening*. In 1920 the journal wrote that "advocates of blasting claim that explosives can be used in cemeteries without danger to damage to neighboring monuments."[62] Another journal, *Modern Cemetery*, noted that dynamite was also used in cemeteries to plant trees, improve drainage, and remove old stumps. After the influenza outbreak of 1918–19, *Park and Cemetery and Landscape Gardening* quoted a cemetery superintendent as saying, "An abnormally large funeral list and a shortage of labor so crippled us that it was impossible to dig the required number of graves per day"[63] without the use of dynamite.

In 2004, Sequoyah County, Oklahoma, officials announced that dynamite would no longer be used to dig graves, citing liability considerations. According to one county commissioner, "A prime concern is the safety of our people. When you go out to shoot a grave, you don't know how that dynamite is going to go off."[64]

Digging Deeper

Ore Mining

As a result of the decreasing amount of accessible, high-grade copper ore, copper mining shifted from underground to open-pit mining between 1960 and 1980. Now huge quantities of very low-grade ore are blasted and processed in a quarry-like operation. A proposed underground mine in Arizona called the Resolution Copper Mine would have shafts descending more than 7,000 feet. The deeply buried copper deposits near the small town of Superior in that state are considered the fourth richest in the world ("rich" by copper ore standards still means that 98 percent of the ore hauled up is waste!). Some ores such as molybdenite often occur in such thin veins that the ore is sometimes still worked entirely by hand in the production phase, using nothing more than picks and sledgehammers.

In 1922, explosives expert T. W. Bacchus related that US production of copper rose from 7,200 tons in 1869 to 539,760 tons in 1920. As for iron, less than 3,000,000 tons were mined in 1860, while nearly 68,000,000 tons were produced in 1920.

Farming

Evidently marketing to farmers worked. DuPont reported that it sold 500,000 pounds of dynamite to farmers in 1908, 750,000 pounds in 1909, 1,500,000 pounds in 1910, and 3,000,000 pounds in 1911.

While ditching and trenching are essentially synonyms, no producer ever made "trenching dynamite." If you want to be technical, "trench" is used more often to refer to a temporary cutout for the laying of pipe, pouring of concrete, etc. "Ditch" usually indicates a more permanent drainage canal.

An article in 1915 stated, "A single row of (24" deep) holes can usually be depended upon to excavate a ditch from seven to nine feet wide and about thirty to forty inches deep. Where larger ditches are required, the holes can be made deeper and loaded heavier, or two or more lines of holes, spaced from 3' to 4' apart, can be used."[65] This assumes a 50 percent strength straight dynamite and moist soil. Because of the limitations of sympathetic propagation, most ditching and trenching were done with electric-blasting-cap circuits rigged to each individual hole. Toward that end, in the 1980s Atlas offered Power Ditch dynamite, which was deliberately formulated to be insensitive to hole-to-hole sympathetic propagation. Hole-to-hole propagation is undesirable when it would disrupt precise delay sequences.

Another on-farm application is the controlled demolition of silos. While using explosives to raze a tall silo is the safest method, sometimes things do not go as planned. In 2018 a blasted silo in Denmark fell the wrong way and crushed a nearby library and music school. Fortunately no one was injured.

Snakes brumate—the reptilian equivalent of hibernate—in tightly packed, cross-species-filled cavities, often under large rocks that retain warmth. This is doubtless the origin of the term "snake holing." In 1912, miners in Harrisburg, Pennsylvania, helped prove this theory. They were blasting a boulder when their snake-hole shot "blew out a solid ball of copperheads, rattlers, [and] garter, water, black, and house snakes as large as a half-bushel measure."[66]

Coal Mining

In 1987, Atlas reported that at that time, the coal industry used 60 percent of all explosives used for mining, and that 60 percent of coal was surface mined.

Atlas also described the two basic underground coal-mining methods. The first, "shooting off the solid," entailed the blasting of holes in a "V" or fan-shaped pattern at the front of the cut. More effective was "shooting an undercut," which involved using a giant sawing machine to create a 6' to 12' deep slot at the bottom of the coal face. Undercutting was accomplished largely by hand until the 1910s, when reliable mechanical coal saws debuted. Still, one source estimated that the job was done by pick in about 20 percent of mines through the 1920s.

The secondary blasting of boulders sometimes has to be performed deep underground in coal mines. Mudcapping a shot was never allowed due to the extremely large flame of an essentially unconfined blast. Multiple secondary blasting is also prohibited in coal mines; each boulder must be drilled, loaded, and blasted individually.

The US Bureau of Mines reported that in 1930, of the 60,000 accidents that occurred in bituminous coal mines, only 475 were due to explosives. Of those, most were related to blasting powder, and not one occurred when permissible dynamites were used correctly.

Stumping

Aetna Powder Company wasted no time getting started marketing to farmers, advertising its first stump-blasting demonstrations in Indiana in October 1881, just a few months after having begun production.

A 1913 ad by Jefferson Powder Company summed up the stumping process: "An auger hole into the stump, a charge of Jefferson Dynamite, a spark, a flash, and it's over. Two days work in just a few minutes, and a cost of a few cents a stump."[67] Of course it wasn't always that easy. Jefferson Powder also employed itinerant salesmen to demonstrate a motorized stumping drill. Often a combination of explosives and machinery worked best.

The US Department of Agriculture provided this formula in 1919: "Roughly, the number of pounds of dynamite required to shoot a stump clear of the ground is the same as the square of the number of feet in the diameter of the stump at the cutoff. For example, a 2' stump will require four pounds, and one 6' in diameter will require thirty-six pounds."[68] However, the article also noted that "all factors, such as kind and soundness of the stumps, and kind and condition of the soil, influence the amount of explosive required for a stump of given size."[69] By the 1970s, stumping dynamite had been largely replaced by bulldozers.

Aetna Powder Company advised using stump fragments as fuel "for your kindling, for kettlewood, for your heating furnace and your fireplace,"[70] noting that the wood "is already partially chopped."[71] Certain stumps were actually much more valuable. In 1922 a Hercules Powder Company representative stated that "quite a lot of explosives are now being used in salvaging pitch pine stumps from cutover pine lands. These pine stumps are very rich in rosin turpentine and pine tar and other products. They are blasted from the ground with powder or dynamite as the case may be, shipped to the reduction plant, ground up, and the naval stores extracted."[72] Hercules operated numerous reduction facilities for this purpose.

Quarrying

By the 1980s the majority of all coal and all metallic and nonmetallic ores were obtained via open-pit surface mining. While collapsing tunnels are obviously not a danger, the stability of slopes and benches is. For this reason, blasting operations pay careful attention to "overbreak," which is fragmentation or cracking of rock beyond the intended area. Holes that are overloaded with too much or too strong an explosive, wrongly positioned holes, and mistimed blasting sequences could weaken surrounding rock structure, making hauling or further blasting operations treacherous or impossible.

DuPont noted a transition from horizontal to vertical boreholes by the 1970s, writing,

> For many years horizontal drilling and blasting were considered to be the most economical and efficient method of fracturing and displacing the overburden. The loading of horizontal holes is usually more difficult and time consuming than with vertical holes. For many years, the holes were loaded with cartridged explosives by loading poles. This required many cycles of shoving the cartridges to the back of the hole, normally 100 pounds or less per each cycle. This was not only slow, but frequently resulted in separation of the charges.[73]

A column of anything other than straight dynamite could not have large gaps between cartridges and still guarantee in-column sympathetic propagation. Even so, some miners used wood spacers to lessen the power of a blast in soft rock. Another technique for accomplishing the same goal involved leaving some boreholes completely empty. The proper use of a lower-strength dynamite obviated the need for such tactics.

Construction Engineering

Atlas summed up a main use of dynamite in construction thus: "When rock becomes an obstacle in the excavation phases of construction projects, controlled blasting techniques are used to fragment and displace the rock."[74] The primary considerations are similar to those involved with demolition: proximity to other structures, acceptable noise and vibration limits, and public safety concerns. In highly congested areas, smaller charges are used to more gradually complete a given excavation project, while in more-remote areas, large-diameter cartridges and modern explosives accomplish the job in a fraction of the time.

Trenching for utility projects was another important construction application for dynamite. Wider trenches were fired with elaborate delay circuits, enabling alternating blasts of center and outer holes. Such strategies resulted in "less flyrock, reduced overbreak, better fragmentation and bottom breakage, and reduced vibration and air blast levels,"[75] per Atlas.

Submarine Blasting

The smallest form of submarine blasting was in a small-diameter hole that immediately filled with groundwater upon withdrawal of the drill bit. Such holes were extremely common, and all the products designed specifically for "wet work" were engineered to make blasting possible in such conditions. For that reason, "sinkability" was another factor that went into the choice of explosives. Some brands of gelatin dynamite were advertised as being formulated to be extra dense, making them more readily sink to the bottom of a water-filled hole instead of floating. Regardless, gelatin dynamites could be submerged indefinitely, owing to their rubbery constituency.

Holes dug in outright mud collapse upon the withdrawal of the punch or bit and so are reinforced with expendable cardboard tubes into which explosives are loaded.

Aside from rigid shells with special waterproofing sealants, submarine dynamites obtain resistance to water pressure via the addition of incompressible fillers to the explosive mixture.

Seismic Prospecting

Oil and gas reserves are the primary focus of seismic prospecting. As DuPont put it, "Seismic prospecting is based on the segregation of the earth by geological processes into layers of varying density and elastic behavior. The purpose of seismic prospecting is to detect irregularities in the layering, such as faults and folds, which may serve as reservoirs for the accumulation of oil and gas."[76]

For marine prospecting, explosives were fired about 30 feet below the surface of the water, according to DuPont. In 1949, California became the first state to start placing restrictions on underwater blasting for prospecting. Currently, less than half of seismic prospecting (also called "reflective seismology") on dry land is accomplished with explosives. Vibration plates, air guns, and computerized sonic devices are instead used to create a pulse that can be bounced off underground strata.

There are actually two basic types of surveying: reflection and refraction. Reflection surveying involves placing seismographs near the source of the pulse, while refraction uses sensors placed over a wider area. Both have advantages and disadvantages.

Among Atlas's offering of seismic explosives in the 1980s were Petrogel and Seis-Prime. Typical of products of the era, both were packaged in easy-to-couple plastic cartridges. While most seismic dynamite was packed close in the hole, Seis-Prime required special spacers between charges for best results.

The underwater seismic sound device that was introduced in 1968 by Esso Production Research Corporation had many advantages over dynamite. The machine, which exploded a mixture of oxygen and propane inside a sonically engineered drum, cost one cent per pulse produced, as opposed

to $15.00 a shot for dynamite. This type of device, called a gas exploder, is also less dangerous to marine life, is safer to use, and is more accurate at greater depths. Other sonic devices such as water and air cannons are also safer than explosives but have recently come under fire for their long-distance impact on sound-dependent creatures such as whales and dolphins.

The members of a crew of seismic prospectors are called "doodlebuggers" because of the scribble-like lines their instruments inscribe on paper to record the depths and densities of various underground structures. According to DuPont, a typical land crew of doodlebuggers consisted of a manager, an observer, a computer analyzer, a shooter, several surveyors (who man the "geophones" that pick up the sound waves), and drillers. The recording equipment is usually housed in a truck or van.

Demolition

The largest and most successful US practitioner of demolition with explosives is Controlled Demolition Inc. of Maryland. The business was launched in a small way in Baltimore in 1947 by Jack Loizeaux, who first blasted stumps and then graduated to chimneys, bridges, silos, and buildings. The family-owned firm still delivers on its promise to fell tall buildings within their own footprint and accurately claims that the company has performed more large building demolitions than all other competitors combined. (A building can also be caused to fall in any direction by setting charges in structural columns near one or two outside walls.) One of Controlled Demolition's biggest projects was taking down the roof of the Seattle Kingdome in 2000. In total, 4,450 pounds of dynamite was loaded in 5,905 "carefully sited holes"[77] and initiated with more than 21 miles of detonating cord. The explosions took 16.8 seconds, and the overall project cost $9,000,000. Doing the math shows that the charges were very small: about 1.3 pounds average per hole.

The sixteen-story Sears Warehouse building in Philadelphia was taken down with 13,000 pounds of dynamite by another company in 1994 without breaking a single bottle at the Canada Dry bottling plant across the street.

Other Uses for Dynamite

One does not normally think of dynamite as a tool for controlling the weather, but in a 2002 company-produced history of DuPont, the author relates that "during times of drought, some Texans exploded dynamite suspended from balloons in attempts to blast rain out of the clouds."[78] And an article in 2007 outlined a proposal to fill a retired submarine with dynamite and detonate it by remote control "directly below"[79] the eye of a hurricane. Supposedly, by "disturbing the physics"[80] of the storm, it would dissipate. The tactic has so far never been tried.

Real dynamite was routinely used to create special effects for movies.

One five-minute re-creation of the D-day landings on Omaha Beach in the 1964 film *The Americanization of Emily* consumed 5,000 pounds of explosives, including 200 pounds of dynamite exploded underwater. The creature in John Carpenter's 1982 *The Thing* was dispatched with a real stick of dynamite. The exploding shark in the 1974 movie *Jaws* was also courtesy of a cartridge of dynamite. In addition to dynamite, black powder and Primacord were used extensively until computer-generated images began to replace many real explosions in the 1990s.

On a related note, another kind of inert or dummy dynamite is prop dynamite for movies and television. Collectors pay hundreds and even thousands of dollars for sticks and bundles seen on screen in movies such as *Inglourious Basterds*, *Sahara*, *Last Action Hero*, *Django Unchained*, *The Specialist*, and *The Patriot*.

Dynamite-filled projectiles were invented in the 1880s. The problem was engineering a gun that did not detonate the dynamite inside the metal shell. The pneumatic launchers that were developed saw limited use and were rapidly replaced by more-efficacious explosive weaponry.

Perhaps the most unusual "use" for dynamite ever was outlined in an article in 1899 describing the experience of a New York quarry supervisor. "One day a professor from Steven's Institute of Technology came to the quarry and asked me to perform a certain experiment for him. We printed with dynamite direct from a newspaper onto a block of iron."[81] The professor "carried three round blocks of iron, 6" in diameter and 3" thick. Then he took a newspaper from his pocket and spread one sheet over the face of the block. I placed half a stick of dynamite on top and covered it with a little heap of sand and touched her off."[82] The result was a legible, reversed image of the writing on the newspaper etched into the metal. The experiment was repeated with an oak leaf, and the two embossed pieces of iron were placed on local display.

CHAPTER 4

Dynamite Companies

This chapter offers detailed histories of seventeen of the primary producers of dynamite in the United States. The largest three—Atlas, Hercules, and DuPont—held most of the market from the late 1910s to the 1970s. Many of the others were absorbed by these three behemoths.

As with most products, advertising was considered crucial to increasing sales of dynamite. Many of the promotional pieces shown in this chapter are marvelous combinations of eye-catching graphics and carefully crafted claims. Others are purely utilitarian but reflect the needs of prospective customers. (It is a statistical curiosity that the list of names of major dynamite producers ends with the letter "N." The smaller companies outlined in chapter 6 have a "normal" alphabetic distribution.)

Aetna Powder Company

Aetna Powder Company was incorporated in 1880 by Horatio Pratt, James O. Parker, and Michael Tearney. Pratt was treasurer and Parker secretary of the new firm. Pratt was also secretary of the Eureka Coal Corporation in Chicago and was a partner in the mining firm of Pratt & Parker Company, along with James O. Parker. Both men had numerous mining-related business interests in Illinois and Indiana.

Tearney was a Chicago justice of the peace and former owner of Chicago

Brick Company. He had worked at the Oriental Powder Company in Cincinnati as a youth.

The three men enlisted Addison Orville Fay to be president of Aetna. Fay was already president and owner of Miami Powder Company in Xenia, Ohio, and had co-owned American Powder Mills in Boston, Massachusetts,

with his father, Addison Grant Fay. The elder Fay was killed in an explosion at American Powder Mills in 1873.

Aetna Powder Company built a large plant on 240 acres of remote wilderness in northwestern Indiana between Miller and Tolleston, in Lake County. The plant began production in 1881. Nearby, a classic company

Aetna Powder Company inherited the Bituminite brand from Jefferson Powder Company and the Collier brand from Sinnamahoning Powder Manufacturing Company. In turn, Hercules Powder Company continued to carry the Coal Powder label for a short time after acquiring Aetna. *Author*

This 1895 poster by Edward Penfield depicts a miner shouting, "Fire in the hole!" to warn of the imminent detonation of explosives. Penfield was one of the most critically acclaimed illustrators of the era and is widely considered the Father of the American Poster. *Library of Congress Prints and Photographs Online Catalog*, LC-USZ4-13310.

town emerged, consisting of living quarters and mercantile outlets for the mostly male populace. The town of Aetna was formally incorporated in 1907. The company reported sales of 18,000,000 pounds of dynamite in 1912, per Van Gelder and Schlatter, who also noted that "Aetna Powder was one of the companies organized at the beginning of the eighties when the increased demand for explosives and the upsetting of the Nobel patents gave a new impetus to the dynamite industry."[1] (The "upsetting of the Nobel patents" began with the Vigorite Powder Company successfully fending off a suit by Nobel patent holder Giant Powder Company in 1879.) Aetna opened another dynamite plant in Fayville, Illinois (named for Addison Orville Fay), in 1911.

Aetna Powder Company was the first dynamite maker in the United States to operate its own electric-blasting-cap factory, beginning in 1889. Aetna bought a small plant that had been launched in Newport, Rhode Island, "in the early '80s,"[2] according to Van Gelder and Schlatter.

An 1883 history of Lake County related that:

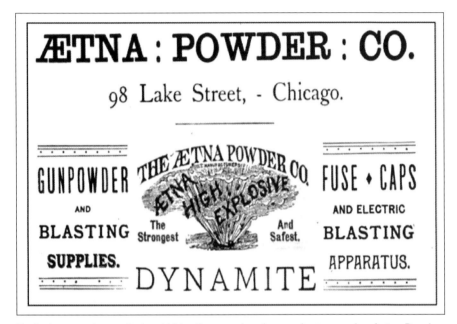

Explosives producers that sold blasting powder also made gunpowder. Aetna Powder, advertisement, *Railroad, Telegraph and Steamship Builders*, 1888–89, 246.

> the works of the Aetna Powder Company are situated one and a quarter mile west of Miller's Station, on the south side of the railroad. The surroundings are attractive, and the company seems to have found a favored spot in this desert region. Although this is called a "powder works," no common powder is made here. It is all "high explosive powder," and nitroglycerine is the active agent in the compound. Here it is manufactured in large quantities and absorbed into substances for shipment and use. It is only fourteen months since the company began here; now they have twenty-six buildings, employ forty-five men, and have a capacity of 60,000 pounds of powder a day. They are at present building another works and twenty workman's cottages. When the new building is completed, the capacity will be 100,000 pounds a day.[4]

The company changed its name to Aetna Explosives Company in 1914, after consolidating with the Miami Powder Company, Sinnemahoning Powder Manufacturing Company, Keystone National Powder Company, Pluto Powder Company, and Jefferson Powder Company. The goal was to combine some of the principal competitors of DuPont for a stronger market presence and to prepare for the output of military explosives for World War I. An article in 1915 reported that "the capacity of the Birmingham plant will be doubled in the near future" and

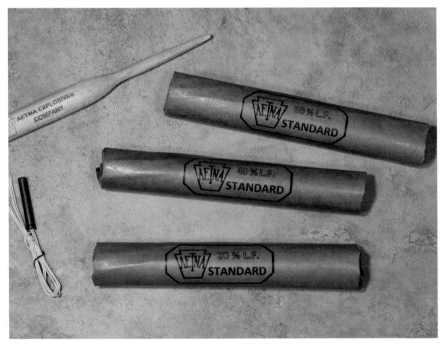

According to Aetna, its Aetna Standard brand was "the best dynamite for blowing stumps, and is very effective for blasting rock that is seamy and rather soft, where a heaving rather than a sharp shattering force is needed."[3] This is another way of saying the explosive was mostly ammonium nitrate. *Author*

"This is the original 'Aetna Dynamite' and, with minor improvements, the same that we have made for over thirty-five years," per Aetna in 1913. The explosive was a straight nitroglycerin formula recommended for work in hard rock and ditch digging by sympathetic propagation. *Author*

According to Van Gelder and Schlatter, "Aetna powder was very carefully put up in neat cartridges, cartons, and boxes[,] which pleased the consumers and for a time set the standard in this respect for other American manufacturers."[6] Aetna Powder, *Handling Explosives* (Chicago: Aetna Powder, 1913), 40.

would be "one of the largest producers of munitions in the United States."[5]

The merger gave the new company dynamite production plants in Birmingham, Alabama; Emporium, Pennsylvania; Sinnemahoning, Pennsylvania; and Ishpeming, Michigan. And like DuPont, Atlas, and Hercules, Aetna now had dozens of sales offices and hundreds of authorized dealers throughout the United States. However, Aetna Explosives Company overspent to meet obligations for munitions during World War I and became insolvent in 1919. Enormous government contracts beginning in 1914 necessitated construction of dedicated plants, which required taking on more debt than the company could bear. Despite its tribulations, Aetna produced a lavish catalog in 1919 and published it in 1920. The company and its facilities were sold to Hercules Powder Company in 1921.

Aetna the place was annexed by the City of Gary in 1928 and is now an eastern suburb. The dynamite sold by King Powder Company was first produced by Aetna Powder Company, then by Illinois Powder Manufacturing Company.

Henry P. Hall was an employee at Aetna in the early 1890s and began experimenting with shell-making machines, building an early model for the company. In 1898, Hall patented his device, which was used by smaller producers into the 1930s.

Larger companies that purchased smaller concerns usually kept viable production facilities in operation, as Aetna did with the Jefferson Powder Company plant in Birmingham, Alabama, and the Keystone National Powder Company plant in Emporium, Pennsylvania. Aetna continued to carry the Pluto Powder Company and Keystone brands after acquiring the two companies in 1914. The Collier label was inherited from Sinnamahoning Powder Manufacturing Company via Keystone, while Bituminite was a Jefferson Powder Company brand. This was common practice with popular brands and was exemplified by early DuPont-branded dynamites also bearing Atlas, Giant, Judson, and Hercules labels.

Apache Powder Company

Apache Powder Company was incorporated in 1920 by Charles E. Mills, Donald E. Fogg, Walter W. Edwards, and Joseph E. Curry. Company president Mills was an Arizona banker and mining engineer who had worked for the Copper Queen Mine near Bisbee,

Arizona. He and Fogg were charged with building a plant southeast of Tucson, and Fogg became its assistant manager. Edwards was a resident of Aetna, Indiana, who had served as the superintendent of Aetna's home plant there. Curry, a mining executive, had worked for the Calumet & Arizona Mining Company and the New Cornelia Copper Company.

The Pluto label was acquired from Pluto Powder Company, and the Keystone label from Keystone National Powder Company. Titan Powder was originally a product of the Dittmar Powder Works (see chapter 6). *Author*

Apache Powder's Amogel was a semigelatin dynamite offered in five strengths. No. 1 was 60 percent strength; No. 2, 50 percent; No. 3, 45 percent; No. 4, 40 percent; and No. 5, 30 percent. *Author*

Construction began in June 1919. The February 8, 1922, *Arizona Daily Star* reported:

The completing touches are being put on the plant of the Apache Powder Company near St. David and it is expected that the company will begin the turning out of products shortly after the first of March. The plant has a capacity of 1,000,000 pounds of powder a month which can be increased within a very short time to 1,500,000 a month should the demand arise. It occupies more than 600 acres and is within about seven miles of Benson and four miles from St. David. The water is furnished to the plant by a 16" artesian well of a capacity of several hundred gallons per minute. The plant will give employment to about 150 skilled workmen when it is run at full capacity. It is in the charge of W. W. Edwards, an experienced powder manufacturer, who is the manager. He will be assisted by D. E. Fogg. Both will be under the general supervision of C. E. Mills, managing director.[8]

All the major shareholders had interests in copper mines. The rationale for the new company was for the state's copper-producing concerns to have control over the production of their explosives, and, most importantly, to reduce the high freight costs associated with shipping from other states. Apache Powder was the only explosives producer in the Southwest.

Apache Powder Company began producing ammonium nitrate fertilizer in the 1940s. The company stopped making dynamite in 1983, changed its

> Benson, Arizona August 24, ,1925
> Acknowledging your Order No. 39109 Shipment
> will be made on or about August 30, ,1925
>
> Thanking you, we are
> Yours very truly,
> APACHE POWDER COMPANY

Postcards like this were a courtesy to let customers know that their order was on the way. *Author*

name to Apache Nitrogen Products in 1990, and thrives to this day.

An article in the *Arizona Daily Star* related that in 1941, Apache Powder Company was "the largest independent producer of nitroglycerin explosives in the United States."[9] "Independent" was intended to refer to companies other than DuPont, Atlas, and Hercules. While the big three producers were not formally affiliated, they dominated the dynamite market for most of the twentieth century.

A 1927 article gave these details about production at Apache: Box components "which have already been stenciled . . . are put through a lock corner machine, and through the joiner where the sides and ends are fitted together and glued. The box then travels to the nailing machine and then on to the sander[,] which smooths up the box."[10] As for shells, "The particular markings of the powder and the date are stamped with a steel die. The roll is then cut into individual lengths for each cartridge, after which the paper is rolled to form the end fold. The cartridge is then ejected onto a belt conveyor. The output is eighty or ninety shells every minute."[11] Here, "stenciled" is being used to refer to printing with metal dies. Only very early cases were actually stenciled.

Apache Powder Company focused mostly on selling to mines in Arizona and New Mexico. The company did very little advertising and published catalogs only occasionally.

Atlas Powder Company

Atlas Powder dynamite was originally made by the Repauno Chemical Company as the company's first product, debuting in 1881. In 1882, Repauno Chemical Company, owned by DuPont, produced ten different lettered grades of Atlas Powder dynamite. The strengths ranged from 15 to 75 percent. "Powder" was used by all explosives companies to refer to dynamite, as well as black blasting

Amogel was introduced in July 1933 but was not formally trademarked until 1974. That same year, the company's distinctive logo was also trademarked. It had seen first use in 1934, according to trademark registration documents. *Author*

The Atlas Powder dynamite offered by Atlas Powder Company was its straight nitroglycerin formula. Atlas Extra was mostly ammonium nitrate, and Atlas Gelatin was a straight gelatin dynamite. *Author*

powder. This usage was encouraged by early dynamite producers, since it capitalized on the established acceptance of blasting powder by miners. Powder companies kept the "Powder" in their names long after high explosives replaced black powder for most uses. Austin Powder Company, DuPont, and Hercules Powder Company also made sporting powders for guns. Plus, early granular dynamite did resemble greasy powder and was often likened to brown sugar in appearance.

Further blurring the distinction, Judson Powder's cartridged, nitroglycerin-coated black powder was

These cartridges would date from the early 1900s. Atlas Powder was sold as a DuPont-branded product from 1906 to 1912. *Author*

When this ad ran in 1914, "100 different kinds" included various lettered grades of brands such as Vigorite and Coalite. Ten years later, the choices had ballooned into the thousands. Atlas Powder, advertisement adapted from *Proceedings of the Coal Mining Institute of America*, 1914–16, 771.

technically classified as dynamite because of the nitroglycerin. And free-running dynamite powder (actually ammonium nitrate sensitized with nitroglycerin) became popular in the 1920s. The linguistic convention continued well into the twentieth century, with explosive-check tokens being valued at one or more "sticks of powder."

Atlas Powder dynamite was produced both by the Repauno plant in New Jersey and DuPont's factory in Kenvil, New Jersey (Repauno Chemical Company was also headquartered

in Delaware). Laflin & Rand Powder Company, which never produced dynamite under its own name, held 25 percent stock in Repauno and sold Atlas Powder through its distributors. Atlas Powder was a product and not a company until 1912.

The court-ordered breakup of DuPont yielded three companies: DuPont, Hercules, and Atlas. Atlas Powder Company was incorporated in Wilmington, Delaware, in 1912 and began production in 1913. William J. Webster, head of DuPont sales in San Francisco, became the new company's president. DuPont salesmen John Findlay Van Lear and Walter A. Layfield were the first vice presidents.

Atlas was assigned four DuPont black-powder plants and four DuPont dynamite plants. The dynamite plants were the Forcite plant near Landing, New Jersey; plants at Senter, Michigan, and Joplin, Missouri; and the Vigorite plant in California. The same year, Atlas acquired the Reynolds Works of the Potts Powder Company near Tamaqua, Pennsylvania. The site of the Joplin plant became known as "Atlas," Missouri. In 1915 Atlas purchased Giant Powder Company, Consolidated, taking over its vaunted California plant and a facility in British Columbia.

Atlas supplied ammonium nitrate for explosives used in World War I and became the world's largest producer of the chemical. G. R. McAbee Powder and Oil Company was acquired in 1922, and its facilities were kept running until 1928. Atlas again supplied explosives for the war effort, beginning in 1939. The company ceased production of black powder in 1947 and got into the ANFO business in the mid-1950s.

Atlas changed its name from Atlas Powder Company to Atlas Chemical Industries in 1961. The company was purchased by Imperial Chemical Industries, Limited, of England in 1971 and became ICI America until 1973. Then ICI America was sold to Tyler Corporation of Dallas, Texas, becoming Atlas Powder Company, a wholly owned subsidiary of Tyler Corporation. Tyler Corporation was a manufacturing and transport business. Atlas was purchased back from Tyler by ICI in 1990 and renamed ICI Explosives, USA, in 1992. In 2001 the company was renamed E-One Holdings.

The Vigorite plant was closed in 1913, but Atlas kept the brand active until 1925. The New Jersey location was shut down in 1933. The Senter, Michigan, plant operated until 1960. The Reynolds plant at Tamaqua, Pennsylvania, was acquired by Copperhead Chemical Company in 1997 and now

manufactures chemicals. The facility near Joplin, Missouri, was acquired by EBV Explosives Environmental Company and still operates, making nonnitroglycerin explosives and disposing of old explosives.

After acquiring Giant Powder, Consolidated, Atlas divided the country into two sales territories. Products sold in western states would have the Giant logo, and those sold in eastern states would have the Atlas logo. Atlas

Printed product brochures showcased an explosive producer's wares and provided a means to espouse the benefits of their brand over the competition. Catalogs of company offerings trace their roots to lists of seeds that were first published in the late 1600s. *Author*

Atlas Gelodyn was introduced in 1936. The semigelatin formula was billed as a bulkier, lower-cost alternative to gelatin dynamite. Gelmax was first sold in 1982. *Author*

A "spinner coin," poster stamp, and employee badge. The reverse of the spinner coin has arrows to determine who paid for a round of drinks. *Author*

The acquisition of Giant Powder, Consolidated, by Atlas led to some cartridges being stamped with both company's names. *Author*

brand Apcodyn was sold as Gicodyn in western states, and Giant Gelatin, sold in eastern states, carried the Atlas logo. Regardless of sales point, cases of Giant products were marked "Giant Powder Co. Consolidated Controlled by Atlas Powder Company." The scheme capitalized on the popularity of the Giant name and its long history.

Atlas began diversifying into other products in 1917, adding "CHEMICALS" to the company logo in 1937. Industrial finishing compounds in the forms of specialty lacquers were the province of the company's Zappon division. Coated fabrics, industrial coatings, and chemicals for purifying and decoloring water were among its product lines.

By the 1910s, many dynamite producers had already instituted their own in-house safety protocol, born of hard experience. Nevertheless, disastrous explosions still occurred, either when rules were ignored or when something truly unforeseeable happened. Many an accident was triggered by something as simple as a container of chemicals being carelessly handled or dropped.

In 1916, the US Department of Labor issued some of the first formal regulations regarding clothes in explosives factories. The rules required that "no pockets be allowed in the clothing worn for such work, except one skeleton pocket in either the coat or trousers."[12] This style of pocket, prominent on the back of the pictured Atlas Powder Company coveralls, is also called a lattice pocket and ensured that no matches or pocketknives could be carried. In addition, no cuffs were allowed on pant legs or shirt sleeves, lest explosive granules or dust collect inside. Also, "Neither iron nor steel buttons, nor other metal attachments, shall be allowed on such clothes, nor shall metal objects, such as knives, keys etc., be allowed to be carried in the pockets."[13] Garments were made of light-colored fabric to show contamination by chemicals or powder and were treated with a flame-retardant chemical. Rubber-soled shoes with no metal parts other than fully enclosed steel toes were mandatory. All these precautions were intended to avoid a spark that could initiate a chain reaction of devastation.

The Labor Department's rules also stated that employees of dynamite factories were not allowed to take uniforms home, and those who worked directly with nitroglycerin and other explosive compounds were required to shower before leaving work.

Booster cartridges are used to detonate insensitive explosives such as ANFO. Detonating cord alone won't detonate most blasting agents, so the cord is strung through purpose-made tunnels in a booster cartridge. Most boosters are detonated by detonating cord or a blasting cap. A few, such as Atlas Hi-Prime, contained small amounts of nitroglycerin for increased sensitivity. Aside from pentolite, other mixtures of RDX and PETN are used.

So a hole might be half filled with an ANFO mixture, onto which is lowered a primed stick of gelatin dynamite or a booster cartridge on detonating cord. The hole is topped off with ANFO.

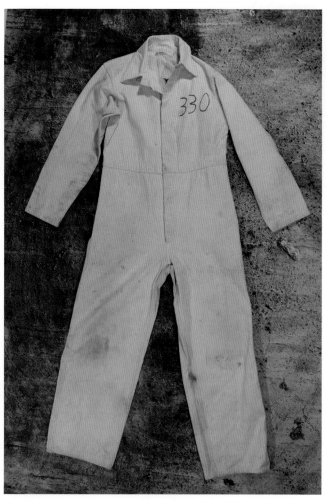

These 1940s-era coveralls were made by the A. Rifkin Company of Wilkes-Barre, Pennsylvania. Founded in 1891 by Russian immigrant Abraham Rifkin, the company initially made work uniforms and sold dry goods. One of Abraham's sons patented the "cover-all" in 1922. *Author*

Coveralls became enormously popular for mechanics and other workers who wore them to protect their civilian clothing. A. Rifkin branched out into bank bags and quit making garments in the 1950s. The company still makes bank bags and related items. *Author*

The blasting cap or detonating cord detonates the dynamite or booster, which detonates the ANFO. Many emulsions and slurries can be detonated only in this manner and are classified as "booster-sensitive." A booster added to a charge of explosives also increases the overall strength of the blast.

Large charges of blasting powder were also initiated with dynamite primers. Atlas also recommended its Power Primer dynamite for use as a main charge for hard-rock mining. The

Left: Atlas Power Primer was a 75 percent strength, high-velocity gelatin dynamite introduced in 1965. "Primer" in this context is an initiating explosive for cap-insensitive blasting agents. *Author*

most extreme example of a primer/ booster is the small fission explosion used to initiate the larger fusion reaction of a hydrogen bomb.

Because of the popularity of mining collectibles, some reproduction and fantasy blasting items have entered the market. Colorful signs, either faithfully reproducing a genuine item or designed as a tribute, are the most-common modern-made collectibles. Very few are intended to deceive. Stencils for dynamite boxes are also available and some are nicely done, acknowledging the fact that they don't produce embossed images.

In the late 1970s, Westclox made a fantasy Atlas Powder Company pocket watch along with other watches commemorating Harley-Davidson motorcycles and John Deere tractors. The

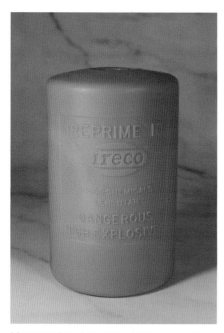

Most modern booster cartridges, such as this IRECO cast booster, are made of an even mixture of TNT and PETN called pentolite. "Cast" refers to the plastic shells in which booster formulations are packed. *Author*

Boosters can also have metal or thick cardboard shells. US Bureau of Alcohol, Tobacco and Firearms, *Modular Explosives Training Program* (Washington, DC: USGPO, 1976), 45.

This pocket watch might be mistaken for something very old if it weren't for one telling detail (*see below*). Regardless, the graphics are attractive. *Author*

According to Atlas, Amodyn offered "a balance of strength and speed which can increase production and save money." *Author*

problem is there was never an "Atlas Powder Co." in Repauno, New Jersey. As we have learned, Atlas Powder Dynamite was made by the Repauno Chemical Company in Repauno, New Jersey. The Atlas Powder Company was in Wilmington, Delaware.

Austin Powder Company

Austin Powder Company was founded in Cleveland, Ohio, in 1833 by brothers Lorenzo, Henry, Daniel, Alvin, and Linus Austin Jr. of Wilmington, Vermont. The original plant was situated on 45 acres in Cuyahoga County, near the Old Forge region, named for the wrought-iron industry that had been established two decades earlier. The location was on the banks of the Cuyahoga River, enabling the installation of waterwheels for power. The site also allowed for transportation by barge, and the first blasting powder (at that time potassium-nitrate gunpowder) manufactured by the company was used in the construction of the Ohio & Erie Canal.

In 1866 the Austin brothers purchased the Cleveland Powder Mills near Brooklyn Station, in Cuyahoga County. In 1870 the original plant was shut down, and in 1907 the Cleveland plant was deemed too close to the city of Cleveland.

Anticipating the encroachment, in 1891 Austin Powder acquired land 25 miles southeast of Cleveland, giving rise

Austin Powder Company's Ex Gel was a line of gelatin-extra dynamite. Austin Powder currently sells two dynamite labels: Apcogel extra gelatin dynamite and 60 Percent Extra Gelatin. Both are manufactured by Dyno Nobel. *Author*

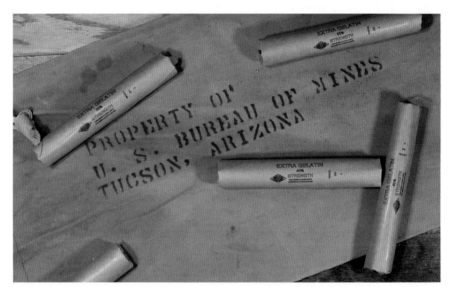

Currently, Austin Powder offers its Extra Gelatin dynamite in only 60 percent strength. This is in sharp contrast to the thousands of different brand/strength/formulation combinations that were sold by the company over the years. *Author*

Hard-hat stickers were and are a popular advertising specialty item. Hard hats debuted in 1919, and self-adhesive stickers became widely available in the late 1930s. Stickers also promote projects, tout safety achievements, and indicate the completion of training. *Author*

to a company town that became known as Glenwillow. The Glenwillow plant began production of black powder in April 1892 and ceased in 1972.

Some of Austin Powder's most popular products were its gunpowder

Austin Powder Company did not make dynamite for the first one hundred years of its existence. Austin Powder, advertisement adapted from *American Coal Miner*, May 15, 1920, 6.

for rifles, which include vaunted brands such as Austin Crackshot, Club Sporting, and Eagle Ducking.

All dynamite sold by Austin Powder Company was made by Illinois Powder Company until September 1931, when Austin Powder opened its first dynamite plant on 920 acres 2 miles northeast of McArthur, Ohio, 50 miles southeast of Columbus. With an initial annual capacity of 18,000,000 pounds, the facility underwent constant improvements and in 1963 was described as one of the "most modern"[15] dynamite plants in the country. The location stopped making dynamite in 1984.

Austin Powder is the oldest continuously operating explosives company in the country, with plants in McArthur, Ohio; Camden, Arkansas; Brownsville, Texas; and Valle Hermosa, Mexico. There are also sixty-five company-owned distribution locations.

Austin Powder became known for its dynamite primers for ammonium nitrate and other cap-insensitive explosives, which it started selling in the

1950s. Atlas, Hercules, Trojan, and DuPont also began producing boosters and primers around the same time. Austin Powder began manufacturing detonating cord in 1950 and electric blasting caps in 1953. When Austin closed the Glenwillow location in 1972, production moved to McArthur. In addition, a new plant was built in Camden, Arkansas, to produce cast boosters. In 1977, Austin acquired the Southwestern Explosives Company. The Texas firm focused on seismic exploration, and Austin expanded into this area.

Austin Powder is currently a subsidiary of Austin Powder Holdings Company of Cleveland, which was formed in 1995. Another subsidiary, Austin International, was launched in 1985. Two additional subsidiaries, Austin Detonator and Austin Star Detonator, are devoted to the manufacture and distribution of electric and nonelectric detonators (here, "nonelectric" means both shock-tubing components and fuse caps).

DuPont

The original DuPont company was launched in 1800 as a real estate development and trading firm under the name DuPont de Nemours et Pere et Fils et Cie (DuPont of Nemours and Father and Sons and Company) by Pierre Samuel DuPont and son Victor Marie. DuPont had added "de Nemours" to his name, à la Leonardo da Vinci, to distinguish himself from another family member living near his hometown of Nemours, France. "Du pont" is French for "by the bridge" and was spelled by the company variously as "du Pont," "Du Pont," "Dupont," and "DuPont." The company currently uses "DuPont" for references and "E. I. du Pont de Nemours and Company" for the formal company name.

The DuPonts' initial business dealings were not fruitful, and the first iteration of the firm failed in 1805. In the meantime, Pierre's other son,

Austin Powder's Red Diamond brand was introduced as a line of permissible explosives in 1917. The Red Diamond label would eventually grow to more than thirty trademarked varieties of dynamites of all types. The MacArthur factory now makes other explosive products but is still known as the Red Diamond plant. *Author*

After acquiring the Atlantic Giant Powder Company and rechristening it the Atlantic Dynamite Company, Dupont sold Judson- and Giant-labeled products. *Author*

ATLANTIC DYNAMITE COMPANY

SUCCESSORS TO

The Atlantic Giant Powder Co. and Judson Powder Co.,
Small & Schrader, Gen. Agts. Offices, 245 Broadway, N.Y.

GIANT POWDER
THE HIGH EXPLOSIVE

SAFE, RELIABLE, POWERFUL, AND EFFECTIVE
ALSO MANUFACTURERS OF

JUDSON POWDER

A SUBSTITUTE FOR BLACK POWDER
Guaranteed 50 per cent stronger than the best Black Powder, and far safer to use. Adapted for all kinds of blasting where Black Powder has heretofore been used. Also dealers in Blasting Caps, Fuse, Electrical Batteries, etc. For sale in all sections of the country by our authorized agents.

Éleuthère Irénée, had established the Eleutherian Powder Mills 3 miles north of Wilmington, Delaware, in 1802. The plant began production in 1804 and was renamed E. I. du Pont de Nemours and Company the following year. The endeavor began as a single gunpowder mill on the bank of the Brandywine River but enjoyed immediate success and expanded nonstop throughout the nineteenth century. Most of the sprawling DuPont family initially resisted the idea of producing dynamite, seeing it as unnecessary competition to blasting powder and dangerous as well.

Éleuthère Irénée's grandson Lammot had begun investigating the potential of dynamite in the 1870s and by 1880 had obtained enough backing to form a new company to manufacture and sell high explosives. The Repauno Chemical Company was founded near Gibbstown, New Jersey, in 1880 and began production of dynamite in 1881. The location was near the Repaupo River, but Lammot DuPont had decided that "Repaupo" was not mellifluous and chose Repauno "for reasons of euphony."[16] The new spelling also served to differentiate the company's rail spur from a nearby one "already dubbed Repaupo."[17]

The other DuPonts supported the Repauno enterprise indirectly but became more involved as the business rapidly grew. The family had already heavily invested in other high-explosives companies, most notably the California Powder Works, maker of Hercules Powder. The DuPonts also acquired Hercules Powder Company proper, plus the Atlantic Giant Powder Company in Kenvil, New Jersey. The

Left: The Atlantic Giant Powder Company was DuPont's largest early competitor in the eastern United States. DuPont bought stock in the company and then took it over. Atlantic Giant Powder, adapted from advertisement, Frederick Abel, *Mining Accidents and Their Prevention* (New York: Scientific Publishing, 1889), A-4.

Something About High Explosives

As a sample of the printers' and designers' art, and as an educational treatise on the qualities of High Explosives, the catalogue recently distributed by the Repauno Chemical Co., Wilmington, Del., deserves more than passing mention. It is of eighty-one pages, on heavy calendared paper; the cover is artistically designed, and shows Atlas with the world upon his shoulders , and the words "Atlas Powder" outlined in a blasting fuse. The illustrations, of which there are over fifty, are very instructive, besides being of real artistic merit.

The frontispiece represents an explosion with Atlas Powder in the Chicago Drainage canal. Scattered through the pages are other realistic representations: Mixing Atlas Powder at the Repauno works, group of employees at the Repauno works, machine shop, interior of shell-making house, birdseye view of the Judson Powder works, and acid works and refineries, office and laboratory, the Atlas Powder and Repauno gelatine works and the laboratory, Repauno works being the principal ones. Other views show the effect of various high explosives manufactured by the company in ice blasting, in breaking log jams, stump blasting, etc. The letter press in is in red, and everything is stated in a clear and concise manner.

The name "high explosives" is now generally applied to that class of explosives of which nitro-glycerine is the active principle. "Atlas Powder" and Repauno Gelatine, manufactured by the Repauno Chemical Co., are classed under this head. The value of an explosive depends on three things: Strength or disruptive power, safety in handling, and keeping qualities. The productions of the Repauno Co. are said to possess all of the above essential qualities.

The result of DuPont's acquisition of most of its major competitors was that Hercules, Atlas, Judson, and Giant cartridges and bags also bore the DuPont oval logo from 1907 to 1912. *Author*

Left: DuPont realized early on that echoing a product's quality with advertising of equal caliber enhanced credibility. Early DuPont calendars were adorned with scenes commissioned from noted artists. Adapted from "Something about High Explosives," *Hardware*, October 10, 1895, 22.

Atlantic Giant Powder Company had been launched in 1870 as a branch of the original Giant Powder Company of California. A factory was built on farmland purchased from Jonas Halse near Kenville (now spelled "Kenvil"), in Morris County, New Jersey, in 1871. The plant began operation the following year. The Atlantic Giant Powder Company was formally incorporated in 1876.

The DuPonts accumulated enough stock in the Atlantic Giant Powder Company to reorganize it and rename it the Atlantic Dynamite Company of New Jersey in 1882. In 1895 the Atlantic Dynamite Company was again reorganized and incorporated by DuPont in New Jersey. The Atlantic Dynamite Company also became the distributor of Judson Powder, which was also produced at Kenvil.

A new company, the Eastern Dynamite Company, was formed in 1895 as a holding company for the Atlantic Dynamite Company, Repauno Chemical Company, and Hercules Powder Company. DuPont black-powder competitor Laflin & Rand Powder Company initially held Eastern Dynamite Company stock but was bought out by DuPont in 1902. Laflin & Rand never made dynamite.

A holding company owns the stock in other companies. The purpose is to consolidate the properties of majority-owned operating companies without forming a separate corporate entity under which the companies would operate as subsidiaries. This allows for greater freedom in operating a variety of companies separately while still maintaining control through stock ownership.

Above: The very first Red Cross dynamite cartridges, produced in 1906, did not have the DuPont oval logo, which debuted the following year. Ads dated as late as 1910 show cartridges without the oval logo. The 1911 edition of DuPont's *High Explosives* pictures cartridges with the oval logo. *Author*

Right: Note that not a single brand from this 1910 chart was labeled with a numerical strength. *United States v. DuPont, Rebuttal Record* (Washington, DC: USGPO, 1910), 2955.

GRADE CHART FOR STANDARD BRANDS OF DU PONT HIGH EXPLOSIVES.

Strength	N. G. Dynamite — Atlas / Hercules Giant*	Red Cross**	"Extra" or Ammonia — Atlas Hercules Giant* / Red Cross	Repauno (Gelatin)	Forcite Giant* / Hercules (Gelatin)
80%				Gel. A+	Gel. No. 1 XXX
75%				Gel. A	Gel. No. 1 XX
70%					Gel. No. 1 X
60%	B+	No. 1	Triple Extra	Gel. B+	Gel. No. 1
50%	B	No. 2 SS	Double Extra	Gel. B	Gel. No. 2 SS
45%	C+	No. 2 S	Extra +	Gel. C+	Gel. No. 2 S
40%	C	No. 2	Extra	Gel. C	Gel. No. 2
35%	K	No. 2 B	5F	Gel. K	Gel. No. 2 B
33%	D+	No. 2 C	FFFF		
30%	D	No. 3	SP 1		
27%	E+	No. 3 B	SP 2		
25%	H	No. 4	SP 3		
20%	E	No. 4 B			
20%	S+	S+			
17%	SSS	SSS			
17%	S	S			
15%	SS	SS			

*Rights of manufacture and sale owned by the DuPont Co., only for the States east of and including North Dakota, South Dakota, Nebraska, Kansas, Oklahoma and Texas.

**Red Cross is not made in strengths less than 20%.

The Red Cross line was soon expanded to include gelatin formulations. Then original-formula Red Cross cartridges were marked "Extra," indicating that ammonium nitrate is the main ingredient, as it always was. *Author*

For all intents and purposes, Repauno Chemical Company, from its inception, and the Atlantic Dynamite Company, after 1882, were branches of the DuPont Company. Previously, the Atlantic Dynamite Company was the property of the California Giant Powder Company. Thus, Atlantic Dynamite Company was never really its own production company, instead selling Giant-, Judson-, and DuPont-branded explosives. Atlantic Dynamite Company was formally dissolved on June 3, 1904. Nevertheless, the Repauno plant, the Kenvil plant, and DuPont's Barksdale, Wisconsin, plant were sometimes referred to in newspaper articles and other sources as Atlantic Dynamite Company properties into the 1930s.

On July 31, 1907, the US government announced it was charging DuPont with violations of provisions of the Sherman Antitrust Act of 1890, which prohibited unfair competition. The DuPont family, under the guise of the Eastern Dynamite Company, managed to acquire and dissolve sixty-four competing black-powder and dynamite companies, with another sixty-nine companies in the process of acquisition or dissolution (or both) at the time of the lawsuit. One court filing claimed that non-DuPont interests represented only 5 percent of the total US explosives market. On June 12, 1911, DuPont was ruled to be in violation of the Sherman Act. On December 15, 1912, the company was divided into DuPont, Atlas Powder Company, and Hercules Powder Company. The story is enormously more complicated, and I refer you to *The Powder Trust*, written in 1912 by William Stevens, for a detailed treatment. According to Stevens, DuPont became "one huge concern at the head of the greater part of the explosives business in the United States."[18] Because of DuPont's ownership of the various companies, its original line of dynamite produced by Repauno had included Atlas, Giant, Judson, and Hercules brands (Repauno began selling Judson Powder in 1893 under an agreement with Judson and Atlantic). However, all retained their own labels until 1905, when the first DuPont-marked dynamite was produced at the Barksdale, Wisconsin, plant. The Barksdale plant was named for Hamilton Barksdale, a DuPont

executive who had established DuPont's Eastern Research Laboratory at the Repauno site in 1902. The laboratory was devoted to improving dynamite, and its innovations included production-line-style manufacture and vastly improved explosive formulations, particularly regarding low-freezing and permissible explosives.

Repauno and the original Hercules Powder Company were dissolved as subsidiaries of Eastern Dynamite Company in 1902. All assets of the Eastern Dynamite Company were then taken over by a newly organized E. I. DuPont de Nemours and Company, formed in 1903 for the purpose of this consolidation. The Eastern Dynamite Company technically endured as an empty shell until formal dissolution in 1912. Repauno Gelatine (later Gelatin) was produced from 1893 to 1905 under its own label and from 1906 until 1924 as a DuPont-branded product. The Repauno location stopped making dynamite in 1953, but the

plant continued to produce industrial chemicals.

DuPont had opened plants in Barksdale, Wisconsin, in 1905; in Louviers, Colorado, in 1908; and in DuPont, Washington, in 1909. After the antitrust suit was settled, DuPont built plants in Hopewell, Virginia, in 1913, and Ramsey, Montana, in 1917. In 1924 the General Explosives plant at Carl Junction, Missouri, was purchased and converted to produce DuPont products. A plant near Mineral Springs, Alabama, opened in 1927. The facility was 10 miles southeast of Birmingham and was known as the Birmingham Works. A plant near Seneca, Illinois, 90 miles southwest of Chicago, opened in 1928. The Potomac River plant near Martinsburg, West Virginia, opened in 1953.

Red Cross dynamite was DuPont's most popular early label. The Red Cross brand was first applied to dynamite in 1895, when the label appeared on a gelatin dynamite line sold by the

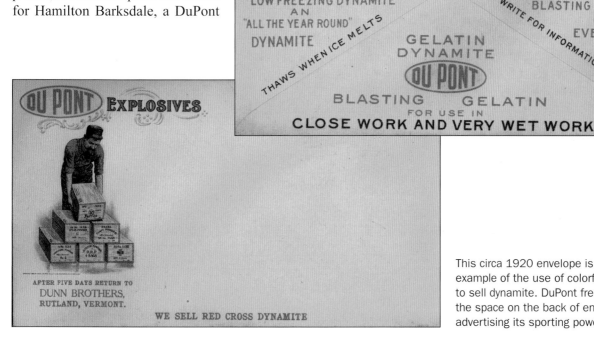

This circa 1920 envelope is a fabulous example of the use of colorful graphics to sell dynamite. DuPont frequently used the space on the back of envelopes advertising its sporting powders. *Author*

American Forcite Company of New Jersey. DuPont acquired the brand along with the Forcite brand when DuPont's Eastern Dynamite Company acquired American Forcite Company in 1899.

The American Red Cross humanitarian relief organization was founded in 1881. In 1900, Congress passed a law called a charter that protected the Red Cross symbol. Some companies, most notably Johnson and Johnson and DuPont via Forcite, had "prior-use exemptions" and were free to use a red cross on their products.

According to Van Gelder and Schlatter, DuPont established a Machinery Commission in 1903 that "was especially active between 1906 and 1911."[19] Introduced were a variety of innovations to mechanize the production of dynamite, adding improved, high-capacity machines for tasks such as shell assembly, packing, and box printing. It took the most modern (at the time) equipment to transform dynamite from a handmade product to one that was mass-produced in the hundreds of millions of units per year.

Enormous World War I defense contracts necessitated DuPont's capital expansion, with the company holding four times the assets at the end of the war

Top right: This image was originally captioned "Mixing house, where the nitroglycerin is thoroughly incorporated with an absorbent to make dynamite." E. I. DuPont de Nemours, *High Explosives, First Section* (Wilmington, DE: E. I. DuPont de Nemours, 1920), 12.

Center right: This image bore the caption "Here are made the paper shells into which dynamite is loaded. The loaded shells are known as cartridges." E. I. DuPont de Nemours, *High Explosives, First Section* (Wilmington, DE: E. I. DuPont de Nemours, 1920), 12.

Bottom right: This image was captioned "Packing the cartridges into wooden cases ready for the market." E. I. DuPont de Nemours, *High Explosives, First Section* (Wilmington, DE: E. I. DuPont de Nemours, 1920), 14.

as prior. Aetna Explosives Company had done the same thing on a much-smaller scale; yet, the debt needed to finance the new plants drove the company into insolvency. According to a DuPont-produced history of the company, "DuPont had to find peace-time outlets for its expanded physical and personnel resources. The general answer was clear enough: translate the chemistry of explosives into the chemistry of the consumer market."[20]

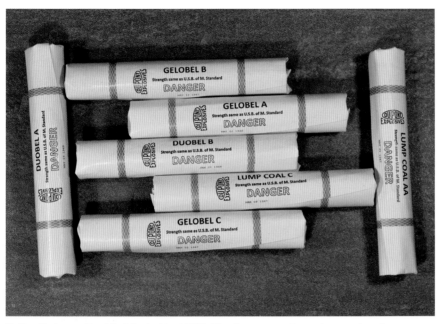

DuPont's Lump Coal A debuted in 1936. The line was not nearly as extensive as DuPont's other brands of permissibles and grew to only three varieties. *Author*

Some editions of the *Blasters' Handbook* (*top row, from left*): 1920, 1922, 1938, 1939, and 1942; (*bottom row, from left*): 1952, 1969, 1980, 1998, and 2011. *Author*

Nitrocellulose was one of DuPont's products that was already being used in nonexplosive applications such as photographic film and lacquers. DuPont had acquired the International Smokeless Powder & Chemical Company of New Jersey and its lacquer plant in 1904 and the Fabrikoid Company in 1910. Fabrikoid was a type of faux leather made from nitrocellulose. (Most editions of the *Blasters' Handbook* were bound with Fabrikoid. Many editions also noted the use of DuPont's pigments for the inks used in printing.) The Arlington Company, makers of a nitrocellulose-based plastic called pyralin, was acquired in 1915.

DuPont then began a period of rapid growth that included investing in the young General Motors Company and producing dyestuffs, which are chemicals used for coloring everything from metal to fabric. In 1921, DuPont's manufacturing business was split into four divisions: Explosives, Cellulose Products, Pyralin, and Paints and Dyestuffs. Just one year later, advertisements boasted of eight divisions: Fabrikoid, Paints and Varnishes, Lithopone, Explosives, Pyralin, Chemical Products, Acids and Heavy Chemicals, and Dyestuffs (lithopone is a white pigment).

DuPont's lightning success in transforming itself into a consumer products behemoth was unprecedented. The pantheon of iconic DuPont inventions includes nylon, Corian, rayon, Lucite, Styrofoam, Tyvek, Teflon, Lycra, and neoprene. Notice that some product names are not capitalized. DuPont decided to let a few brands such as nylon become generic industrial terms. In promotional material such as *This Is DuPont*, published in 1949, "rayon" is not capitalized. The strategy helped certain products gain acceptance as generic yet vital industrial materials.

The company's famous slogan, "Better Things for Better Living . . . through Chemistry," was introduced in 1935. By 1948, DuPont had more

Note that the artist has reconfigured the information on the cartridge to accommodate the hand of the working man. E. I. DuPont de Nemours, advertisement, *Saturday Evening Post*, September 21, 1918, 75.

than 1,200 industrial and consumer product lines.

The result of this spectacular diversification was that explosives were only 5 percent of DuPont's business by 1935 and 1 percent by 1960. Nevertheless, the DuPont Explosives Division continued to be far and away the country's most prolific manufacturer of dynamite until it ceased making the product in 1977. DuPont credits its explosives business for providing the chemical research know-how and mass-production techniques that led to its being among the biggest corporations in the United States in the 1950s.

The Repauno plant was the largest in the world until 1930 and consisted of more than four hundred buildings on 1,640 acres, producing up to 50,000,000 pounds of dynamite a year and more than 2.5 billion pounds of dynamite over the course of its operation. Add in DuPont's other plants and the figures are enormous. According to the 1952 *Blasters' Handbook*, "During a peak year DuPont plants manufactured about 500,000,000 cartridges of dynamite."[21] However, at that time the

company was also on the cutting edge of developing the other explosives that would all but replace dynamite in the coming decades.

DuPont Nitramon was introduced in 1934, the basic formula consisting of a cap-insensitive mixture of ammonium nitrate, TNT, and paraffin. The explosive came in waterproof cans that were threaded to screw together in long columns, which were then placed into boreholes. The cans could be detonated with dynamite, or with special primers that also came in cans and were screwed into sections of the columns.

DuPont made ANFO mixtures beginning in 1955, but its most-popular nondynamite products were its Tovex water-gel explosives, introduced in 1958. The formulation evolved over the years but primarily consisted of monomethylamine nitrate and ammonium nitrate suspended in water. It was sold in cartridged chubs and various pourable forms.

The success of Tovex led to DuPont's phasing out of dynamite, closing the Carl Junction plant in 1961; the

Barksdale, Birmingham, and Louviers plants in 1971; and the DuPont, Washington, plant in 1976. The Barksdale facility had stopped making dynamite in 1961. The Seneca, Illinois, plant closed in 1968. In 1977 the Potomac River plant stopped making dynamite but continued to produce DuPont's other explosive products until 1988, when DuPont sold its explosives division to Explosives Technologies International (ETI). (DuPont had closed its Hopewell plant in 1919 and its Ramsey plant in 1921. The Repauno location ceased operations in 1953.) In the 1970s, even without dynamite, DuPont still commanded one-third of the US explosives market.

DuPont produced an astonishing array of printed material, both to advertise dynamite and to instruct on its safe and effective use. Most of these were not small booklets; the *Farmers' Handbook* (also spelled *Farmer's Handbook* in some versions) was published in at least seven editions and grew to 184 pages—longer than early editions of the general *Blasters' Handbook. Increasing Orchard Profits*

•*Useful Information for Practical Men*, 1908 •*The Achievement*, 1908 •*What a Massachusetts Farmer Did to Boulders and a Minnesota Farmer Did to Stumps*, 1910 •*Talking Points on DuPont Explosives for Jobbers and Dealers and Their Salesmen*, 1910 •*New Farms for Old through Deep Plowing with Red Cross Low-Freezing Dynamite*, 1911 •*Subsoiling with Dynamite*, 1911 •*A Few Pointers on Charging and Tamping of Boreholes*, 1911 •*High Explosives*, 1911, 1915, 1920, 1921 •*Farmer's Handbook of Explosives*, 1911, 1912, 1915, 1920, 1922, 1925, 1927 •*Agricultural Blaster*, 1912 •*Tree Planting with DuPont Dynamite*, 1913 •*Agricultural Blasting: A Money-Making Profession*, 1913 •*Farmer's Handbook: How to Use Red Cross Dynamite*, 1913 •*The DuPont Farmer – Uses of DuPont and Repauno Stumping Powder*, 1914 •*Road Construction and Maintenance*, 1915 •*Blasting Pole and Post Holes*, 1915 •*Increasing Orchard Profits with Red Cross Dynamite*, 1915 •*Explosives for Quarrying*, 1915, 1920 •*Developing Logged-off Lands of the Pacific Northwest with DuPont Explosives*, 1916 •*Explosives for Shale and Clay Blasting*, 1916 •*Handbook of Explosives: Instructions for Clearing Land, Planting and Cultivating Trees, Ditching, Subsoiling and Other Purposes*, 1917 •*The Giant Laborer*, 1917 •*Fighting Fires with Dynamite*, 1918 •*Agricultural and Miscellaneous Uses for DuPont Explosives*, 1918 •*Blasting Ice with DuPont Explosives*, 1920 •*Ditching with Dynamite*, 1922 •*Clearing Logged-Off Lands*, 1923 •*Building New York's Newest Subway*, 1927 •*Explosives in the Pacific Northwest*, 1934, 1941 •*Blasting Ditches with Explosives*, 1936 •*Ditching with Dynamite for Mosquitos and Flood Control, Pipeline and Highway Construction, Agricultural Ditching and Control of Water Flow*, 1945

•*Blasters' Handbook*: 1918, 1920, 1922, 1925, 1928, 1930, 1932, 1934, 1938, 1939, 1942, 1949, 1952, 1953, 1954, 1958, 1963, 1966, 1967, 1969, 1977, 1980. The 13th edition was printed in 1952, 1953 and 1954, the 14th edition in 1958 and 1963, the 15th edition in 1966, 1967 and 1969, and the 16th edition in 1977 and 1980. When Explosives Technology International, Inc. (ETI) acquired DuPont's remaining explosive business in 1988, ETI placed its own dustjackets on the 1980 printing. The International Society of Explosives Engineers (ISEE) published a 744-page, completely revised 17th edition in 1998. A 1030-page 18th edition that debuted in 2011 is now in its fourth printing and is also available as an electronic edition.

Above: The first edition of DuPont's *Blasters' Handbook* was 122 pages long. The book would grow to more than five hundred pages. The work was used as a training textbook by the US Bureau of Mines Explosives Research Center. *Author*

Left: This list is not comprehensive but is certainly representative of the company's broad efforts to sell as much dynamite to as many different users as possible. *Author*

These 1970s-era dynamite cartridges represent some of the last produced by DuPont. *Author*

with *Red Cross Dynamite* has more than two hundred testimonials spanning twenty pages, all about dynamite and orchards! The publications bear a sequential code: *Blasting Pole and Post Holes* is number A-397, *Clearing Logged-Off Lands* is A-1418, and *DuPont Explosives in the Pacific Northwest* is A-4629. This means that there were more than 4,600 different DuPont promotional publications by 1934 (most were for DuPont's other, nonexplosive products).

There were even more DuPont publications in the form of its *Service Bulletin*, *Farmer's Bulletin*, and *DuPont Agricultural Newsletter*.

DuPont also ran advertisements in publications, from local newspapers to trade journals to general-interest magazines. The company sent posters aimed at farmers to rural movie houses and provided "Instructional Motion Pictures" such as *Farming with*

Dynamite free of charge. One two-reel film called *Green Valley* told the story of a farming town saved by irrigation facilitated with dynamite.

DuPont salesmen distributed advertising specialty products such as signs, ink blotters, playing cards, letter openers, small tools, notebooks, and calendars to hardware stores, mine offices, and individual farmers.

DuPont published a magazine called *Vertical Farming* from 1915 to 1917. "Vertical farming" meant explosively "tilling" the soil much deeper than was possible with a plow. The company's flagship *DuPont Magazine* ran from 1913 to 2003, heavy with explosives-related copy and images through the 1920s. The magazine was in fact launched to specifically promote explosives: the first five issues of *DuPont Magazine* were titled *DuPont Magazine and Agricultural Blaster*.

Dyno Nobel

The history of Dyno Nobel is quite complex, owing to 150 years of multinational mergers, acquisitions, and the repeated renaming of companies and subsidiaries. Even the published corporate accounts of the entities involved are sometimes inconsistent, but here is a rough overview.

Alfred Nobel founded Nitroglycerine AB in 1864 with a group of wealthy businessmen. "AB" stands for *aktiebolaget*, Swedish for "stock company," and is roughly equivalent to the United States' "Limited." The company became informally known as the Nitroglycerin Company. Nobel's original production facility was in Vinterviken (Winter Bay), Sweden, just south of Stockholm. The plant was managed by Nobel's older brother Robert. Beginning in 1867, the company produced #1 and #2 dynamites with 75 percent and 64 percent strength, respectively.

In 1915, Nitroglycerine AB acquired Gyttorps Sprängamnes (*sprängamnes* is old Swedish for explosives) AB and moved the production facilities of Nitroglycerine AB to Gyttorp, a village in Nora, in central Sweden. In 1921 the Vinterviken plant shut down and the buildings were used as company warehouses. Nitroglycerine AB changed its name to Nitro Nobel AB in 1965. In 1986 the company became a subsidiary of Norwegian chemical company Dyno Industrier. At that time, the company was organized into six divisions: Dyno in Norway, Dyno/IRECO in the United States, Nitro Nobel Asia Pacific, and, in Sweden, Nitro Nobel, Nitro Consult, and Dyno Nitrogen. Dyno Nobel had acquired IRECO in 1984 and made it a subsidiary. In 1988 the US operations became known as Dyno Nobel, USA, headquartered in Salt Lake City. The company's facility in Carthage, Missouri, is the only plant in the United States currently manufacturing dynamite.

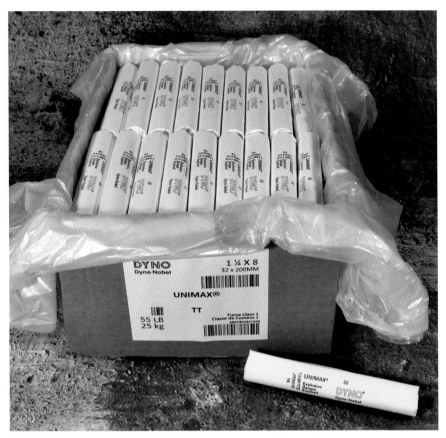

According to Dyno Nobel, "Unimax is an extra gelatin dynamite formulated to consistently deliver high detonation velocity and excellent water resistance."[22] Unimax dynamite debuted in 1986. *Author*

In 1999 the company changed its name to Dyno Nobel Sweden AB. In 2003, Dyno Nobel merged with the Ensign-Bickford Company. In 2006, Dyno Nobel acquired Explosives Technology Inc. (ETI), which had purchased DuPont's Explosives Division in 1988. Dyno Nobel was acquired by Australian explosives behemoth Orica in 2006. The company has been owned by Incitec Pivot, an enormous, multinational chemicals firm also headquartered in Australia, since 2008.

Nitro Nobel engineer Anders Persson invented Nonel nonelectric signal tubing in 1967, and the innovation became commercially available in 1972. The system was revolutionary, in that all worries about extraneous electricity, save for a direct hit by lightning, were eliminated. Other companies soon manufactured their own versions (Atlas Powder Company's was called Blastmaster), and new takes are still being developed to go with advancing computerized initiation systems.

Dyno Nobel stresses safety during all phases of production, transport, and use. The company operates a consulting division called Dynoconsult to provide cutting-edge advice and training. *Author*

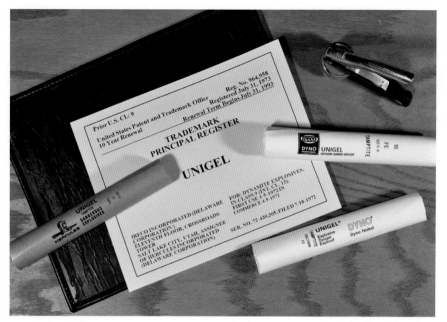

As we have seen with Aetna, Atlas, and DuPont, explosives companies kept popular inherited brands active. Hercules Powder Company's Unigel label, introduced in 1972, was acquired by IRECO and passed on to Dyno Nobel (Unigel was actually first used for a British "permitted" explosive that was introduced in 1954). *Author*

Hercules Powder's Red H was another carryover brand acquired by Dyno Nobel through IRECO. The Tamptite label is applied to modern, cartridged-emulsion explosives as well as Dyno Nobel's Unigel and Unimax nitroglycerin dynamites, and its Z-Powder permissible dynamite. *Author*

According to the company website, Dyno Nobel operates "thirty-two manufacturing plants on three continents"[23] and hundreds of distribution centers around the globe. In addition to dynamite, the company makes bulk ammonium nitrate, packaged emulsions, cast boosters, and its famous nonelectric detonating systems.

In February 2020, Dyno Nobel reported that its explosives were used in the world's largest mining blast in Queensland, Australia. The explosion consisted of "2,194 tons of bulk explosives across 3,899 holes."[24]

General Explosives Company

General Explosives Company was incorporated in Missouri in July 1916 with a capitalization of $50,000. The initial major stockholders were E. William Hawley, Walter Wallace Edwards, Joseph M. Owen, and Albert J. Rawlings. According to Van Gelder and Schlatter, the company was formed "to take over the business of the Home Powder Company, which sold but did not manufacture powder around Joplin, Missouri, and furthermore so that the new corporation could manufacture dynamite."[25] Home Powder Company was started in 1908 by explosives salesman and executive George N. Spiva.

Elmer William Hawley was a St. Louis–based railroad executive who served as vice president of the new General Explosives Company. Walter Wallace Edwards was trained as a mechanical engineer and had been superintendent of the Aetna plant in Indiana from 1904 to 1915. Edwards served as the first president of General Explosives and went on to help found the Apache Powder Company in 1920. Albert J. Rawlings was a New York businessman who became the new company's secretary. George Spiva became a wealthy banker and served as president of General Explosives

General Explosives Company's Special dynamite was its ammonium nitrate-based formulation, with "special" meaning "extra." General 1 was one of the company's thirteen brands of permissibles. *Author*

HIGH EXPLOSIVES

Brand	Purpose	Fumes	Water Resistance
L F Dynamite	Rock excavation Quarrying	Below 40 % fair Above 40 % poor	Good
L F Special Dynamite (Ammonia)	Soft stone Earth and shale, Stumping, etc.	Good	Fair
L F Gelatin	Wet quarries Tunneling Underground and	Excellent up to and including 60% grade	Excellent

APPROXIMATE RELATIVE DISRUPTIVE POWER OF DIFFERENT GRADES

One cartridge 60%	75% .86	50% 1.12	40% 1.28	30% 1.50	Cartridges of same size

GENERAL EXPLOSIVES COMPANY
7 SOUTH DEARBORN STREET
CHICAGO

Above: Multiple sales offices helped General Explosives Company grow into a sizable producer in just six years. *Keystone Mining Catalogue*, 1921, 440.

Right: General Explosives Company was one of very few producers to specify "high-freezing" in its advertisements. However, no cartridges of dynamite from any company were ever marked "high-freezing," since doing so would imply a higher freezing point than the normal freezing point of nitroglycerin. *Author*

Company from 1920 until 1924.

A factory comprising some fifty buildings was completed in under a year and began production in January 1917. The plant was on Center Creek, near Carl Junction, Missouri, with main headquarters in Chicago. The site was originally about 3 miles northwest of Joplin but is now a suburb. Peak yearly production was more than 14,000,000 pounds of dynamite. The company was sold to DuPont in August 1924, and DuPont operated the plant until 1960.

Another well-known but unrelated company called General Explosives Inc. was near Hull, Quebec. Yet another company called General Explosives Company was launched in Latrobe, Pennsylvania, in 1930 and was purchased by American Cyanamid Company in 1933, becoming a division thereof. This General Explosives Company resold dynamite but made only electric blasting caps. Still yet another company operating as General Explosives Company of New Jersey was incorporated in 1905 and sold dynamite until its plant was leveled by an explosion in 1908.

Giant Powder Company

The Giant Powder Company was incorporated on August 13, 1867, as the pioneer producer of dynamite in the United States. A plant was built at what is now Glen Canyon Park, at the south end of San Francisco. The first batch of dynamite was made during the week of March 19 to March 26, 1868, and consisted of 1,300 pounds of Giant Powder dynamite.

The first plant was destroyed in an explosion in 1869 and was rebuilt south of Golden Gate Park. This plant blew up in 1879. The operation then moved to Fleming Point, in Alameda County. This plant was leveled by an explosion in 1883. It was rebuilt, only to explode once more in 1892. In 1892, Giant Powder Company merged with the Safety Nitro Powder Company and became Giant Powder Company,

Giant Powder Company continued to sell "Giant Powder" as its own label until the mid-1920s. By that time, the company also sold dozens of other formulations, including stumping powders, quarry powders, and permissibles. *Author*

The Coalite brand was shared by Atlas Powder Company and Giant Powder Company after 1915. When referencing the dynamite, the US Bureau of Mines listed the brand under "Atlas-Giant." *Author*

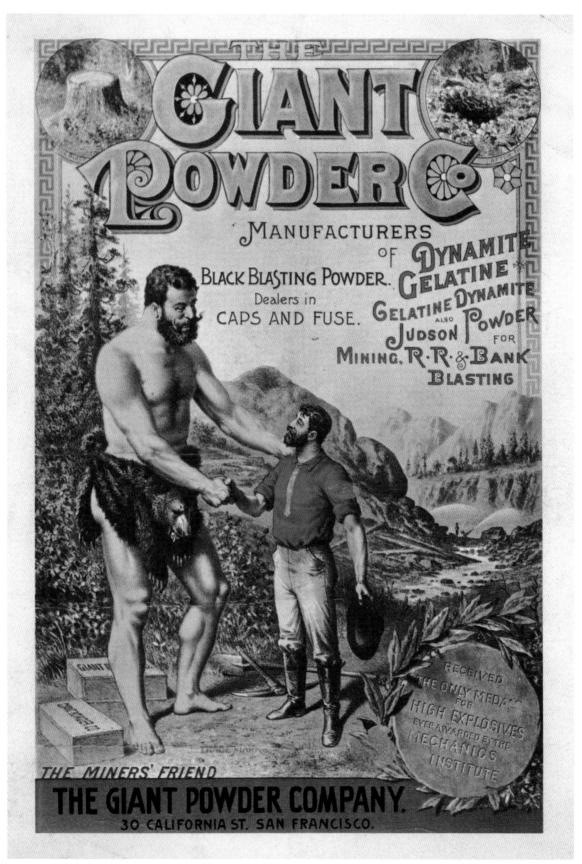

Although Giant Powder obtained trademark approval in 1873, legions of small dynamite makers would refer to their products as "giant powder." *Author*

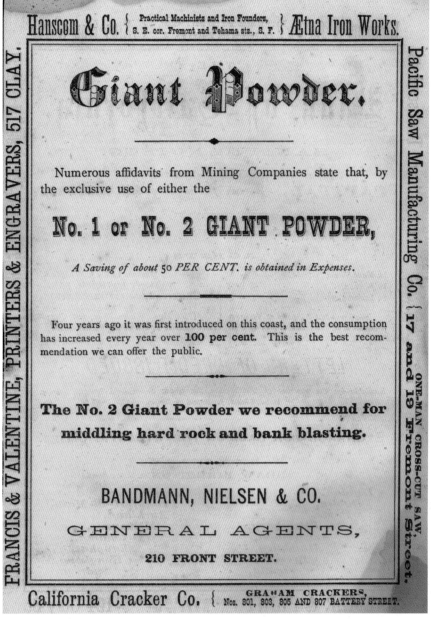

This ad dates from 1872 and touts the increase in consumption of the product. However, the ad fails to mention actual sales figures. Giant Powder, advertisement, source unknown.

Consolidated. The plant of the Safety Nitro Powder Company became Giant Powder Company's main production facility. The factory was 4 miles west of Pinole. Pinole is 10 miles north of Oakland, in Contra Costa County.

The original organizer of Giant Powder Company was German immigrant Julius Bandmann, who ran a San Francisco importing firm called Bandmann, Nielsen & Company, founded in 1849. Bandmann's brother Charles worked for Nobel in Sweden.

Julius imported nitroglycerin oil and even planned to launch an American firm for its production, but Charles informed him of Nobel's newest invention, dynamite. Julius (who went by "Juls") interested a group of prospective local investors, and Charles persuaded Nobel to send a representative to demonstrate the power of the new explosive. The agent, Theodore Winkler, made 3 pounds of dynamite, and the demonstration on August 10, 1867, sufficiently impressed Bandmann

and his associates to launch a new company to manufacture dynamite.

Egbert Putnam Judson was one of some twenty founding stockholders, and the Giant Powder Company's initial line of products consisted of Giant Powder, Judson Powder, and Nobel's Explosive Gelatine. Judson patented Giant Powder #2 in 1873 and eventually established his own Judson Powder Company in 1890 after a falling-out with Giant over competing products.

An eastern branch known as Atlantic Giant Powder Company was launched in 1870. A factory was built near Kenvil, New Jersey, in 1871 and began operation the following year. This branch became Atlantic Dynamite Company in 1882 and was acquired by the DuPonts. Egbert Judson also established an eastern plant for his Judson Powder on the same property. Arthur Pine Van Gelder served as superintendent of the Kenvil plant from 1909 until 1917.

The original Giant Powder formulations were so popular that the uncapitalized term "giant powder" immediately became a synonym for dynamite and continued to appear in legal definitions of dynamite through the twentieth century.

In 1895, Repauno Chemical Company, Atlantic Giant Powder Company, and the eastern branch of the Hercules Powder Company merged to operate under the Eastern Dynamite Company. This left the western Giant Powder Company, Consolidated, to operate independently until 1915. The location produced 14,000,000 pounds of dynamite in 1906.

Giant Powder Company, Consolidated, was acquired by Atlas Powder Company in 1915. All Giant Powder Company products made thereafter were sold "under authority" of Atlas. According to Van Gelder and Schlatter, "Since 1916 the affairs of the Giant Powder Company, Consolidated, have been managed by the Atlas Powder Company, although a separate corporate organization is maintained."[26] In practice, the arrangement resulted in some cartridges bearing both the Atlas and Giant logos. Advertising and promotion concepts were shared as well, and some product catalogs released by the two companies were nearly identical except for the brand names.

The last Giant-branded dynamite made by the Giant Powder Company was produced in 1960. Giant Gelatin

Gelodyn was another brand shared by Atlas Powder Company and Giant Powder Company. *Author*

Giant Gelatin was, in the 1920s, one of a very few twentieth-century dynamites to be packed in white shells. *Author*

was sold by Atlas Powder Company into the 1990s.

The site of the original plant was declared an official California Historical Landmark (CHL 1002) in 1991. A memorial plaque was approved in 2015 and placed at the Glen Canyon Park entrance in 2017 (the actual site of the plant is thought to be where the recreation center is currently located). Another plaque, placed in 1992, is displayed at Pinole Point, designated CHL 1002-1. This plaque is at 5550 Giant Highway. The plaque reads, "Pt. Pinole is the last site of the Giant Powder Company, the first company in America to produce dynamite. Following devastating explosions at its San Francisco and Berkeley sites, the business moved to this isolated location in 1892, incorporating the established Croatian community of Sobrante. The company town of Giant quickly grew into one of the North Bay's industrial centers. Explosives were produced here until 1960 and were essential to mining, dam, and other construction projects throughout the Western Hemisphere."

Grasselli Powder Company

The Grasselli Powder Company was incorporated in August 1917 as a subsidiary of Grasselli Chemical Company, which was founded by Eugene Ramiro Grasselli in 1839. Grasselli Chemical Company had opened a sulfuric acid plant in Cleveland, Ohio, in 1866 and gradually expanded the company to produce a variety of chemicals, including fertilizers, insecticides, and acids. Eugene died in 1882, and the firm was taken over by his son Caesar.

Grasselli Powder Company was formed by the acquisition and merger of three existing companies: the American High Explosives Company, the Burton Powder Company, and the Cameron Powder Manufacturing Company. Joseph Burton became president of the new company, having served as secretary and treasurer of his uncle Job Burton's

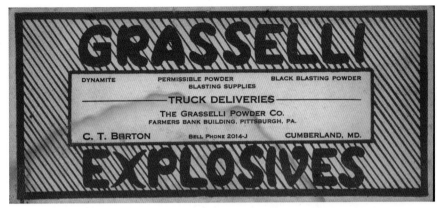

An advertising ink blotter measuring 3¾" by 8" and used to absorb excess ink from the tip of a fountain pen. A common advertising giveaway from the 1910s to the 1940s, the blotter was rendered obsolete by the invention of the ballpoint pen. *Author*

The Cameron label was inherited from Cameron Powder Company, and the Mine-ite brand was originally a Burton Powder Company product. *Author*

American High Explosives Company.

By 1922, Grasselli Powder Company had fifteen chemical plants in addition to warehouses in Pennsylvania, Ohio, West Virginia, Illinois, Wisconsin, and Michigan. Grasselli was a major supplier of acids to DuPont, and the company and its facilities were acquired by DuPont in 1928. Grasselli did business under its own name until 1938, when it became the Grasselli Chemical Division of DuPont.

The United States did not officially enter World War I until April 1917, but millions of tons of munitions and other war materiel were supplied to Britain, France, and Russia throughout the conflict. When war broke out in 1914, entire new munitions plants sprang up, along with housing and infrastructure to support their workers. DuPont made more than 1.5 billion pounds of explosives during the conflict, starting with 8,000,000 pounds of smokeless powder shipped to France in 1914. Employment at the company ballooned from 5,300 to 48,000. Hercules Powder Company had fourteen plants producing TNT and ramped up production of black powder from 155 pounds a day to 215,000 pounds a day. Atlas Powder Company, already the world's largest producer of ammonium nitrate, made 350,000,000 pounds of it for military use. Trojan Powder Company made 50,000,000 pounds of grenade- and mortar-shell powder, and Grasselli made TNT and countless tons of nonexplosive chemicals. (While an ideal explosive for war, owing to its stability and power, TNT could not be used in underground mines because of its fumes and was never widely used for quarrying because it is harder to detonate, is less powerful by weight, and is more expensive.)

Hercules Powder Company

Hercules Powder dynamite was originally a product of the California Powder Works near Santa Cruz, California. California Powder Works began selling the explosive in 1869 and formed a Hercules Powder committee to monitor the new product. After the first year of sales, the committee reported no profits for Hercules Powder but was "not willing to recommend immediate discontinuance of the product."[28] DuPont invested early on in the company, soon owning 40 percent of its stock. DuPont became complete owner in 1903. The Hercules brand was then produced by DuPont from 1903 to 1911.

Another plant producing Hercules Powder operated from 1877 to 1894 near Cleveland, Ohio, where California Powder Works already had a facility. Hercules Powder Company as such did not formally exist until it was organized in 1882 by Lammot DuPont as a Delaware corporation.

The compound was named by John W. Willard, who ran the Cleveland operation. The name was intended to be a counter to Giant Powder, since Hercules was a slayer of giants. "Powder" was used as a synonym for dynamite by early producers to more liken the product to blasting powder, which was already in wide use.

Hercules Powder #1 originally consisted of 77 percent nitroglycerin, 1 percent sodium nitrate, 2 percent

Hercules Powder was patented in 1874. In 1877 a separate complex of "eight or ten buildings" was established at the California Powder Works. The facility was dubbed the Hercules Powder Works. *Library of Congress Prints and Photographs Online Catalog*, LC-DIG-pga-00310.

wood pulp, and 20 percent magnesium carbonate. #2 contained 42 percent nitroglycerin, 43.5 percent sodium nitrate, 11 percent wood pulp, and 3.5 percent magnesium carbonate. Because of the large amount of black powder in the latter formulation, it was billed as Black Hercules.

In 1882, western production of Hercules Powder moved to north of San Francisco, and the company town of Hercules emerged. Hercules stopped producing explosives in California in 1964, outlasting the nearby Giant Powder facility by four years (Giant Gelatin was sold by Atlas Powder

Company into the 1990s).

By 1906, DuPont had managed to acquire almost every major producer of explosives in the United States. The government filed antitrust litigation in 1907, and the end result was the breakup of DuPont into Hercules Powder Company, Atlas Powder Company, and DuPont.

The new Hercules Powder Company was incorporated in October 1912 and began production January 1, 1913. The company operated until 1985, when it was purchased by IRECO. IRECO had been absorbed by Dyno Nobel in 1984 but still operated as a subsidiary.

Alliant Powder Company acquired the Hercules brand for its sporting powders in 1995, and the Hercules logo appears on Alliant products to this day.

Hercules was initially assigned the Atlantic Dynamite Company plant at Kenvil, New Jersey, and the factory in Hercules, California. Hercules also received the Lake Superior Powder Company's old dynamite plant in Michigan, but it was in too populated an area to be restarted. All the other plants initially assigned to Hercules as part of the antitrust settlement produced black powder or gunpowder only. In 1913, Hercules built a new dynamite

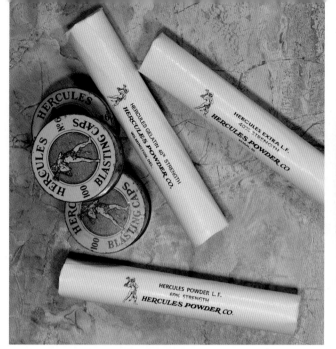

The new Hercules Powder Company's initial offering of dynamites included the usual straight, extra, gelatin, and gelatin-extra formulations. Hercules introduced its 100 percent strength blasting gelatin in 1914. *Author*

Hercules Red H was the company's first permissible formula, debuting in 1913 with seven grades. Six grades of Hercules Xpdite made the list the following year. *Author*

THE CALIFORNIA POWDER WORKS.

HERCULES SLAYING THE GIANTS

HERCULES POWDER.

HERCULES POWDER, celebrated for its strength, uniformity and safety, chemically compounded to neutralize poisonous fumes; the miners hail it as their friend. Mines run on the most economical principles use Hercules Powder.

No. 1, XX.

IS THE STRONGEST EXPLOSIVE KNOWN.

No. 2-Is superior to any of that grade.   No. 3-For Pipe, Clay and Cement.

Patented in the U. S. Patent Office.

All kinds of Blasting, Sporting and Military Powder, Fuse and Caps.

THE CALIFORNIA POWDER WORKS,

Office, 230 California Street,   SAN FRANCISCO.

plant in Coonville, Utah, 18 miles west of Salt Lake City, and started production there the following year. In 1915, Hercules bought Independent Powder Company of Carthage, Missouri. In 1921, Hercules acquired Aetna Explosives Company, inheriting dynamite plants in Birmingham, Alabama; Ishpeming, Michigan; Emporium, Pennsylvania; and Fayville, Illinois. A plant in Sinnemahoning was acquired but closed.

Russell H. Dunham became president of the new corporation, with T. W. Bacchus and J. T. Skelly serving as vice presidents. Dunham, chairman of the board and president until 1939, had worked for Loraine Steel and Bethlehem Steel Corporation before joining DuPont in 1902 as comptroller. English businessman Thomas Wally Bacchus worked as a salesman for the original Hercules Cleveland location beginning in 1893. In 1915, Coonville was renamed Bacchus and he became superintendent of the new plant. James T. Skelly, a champion target shooter,

Left: Giant Powder was Hercules Powder's largest competitor through the 1880s. California Powder Works, advertisement, *Blue and Gold* (San Francisco: Bacon, 1883), 234.

Scene in one of the Hercules Powder Company's packing houses.

The Dynamite Maker and the Food Supply

The dynamite maker's service to all of us does not consist solely of placing explosives in the hands of the miner for the production of our coal and metals. The labor of these men in the Hercules plants is also closely connected with the most fundamental of all industries—agriculture. Their work helps to provide the food that nourishes us.

With the increase in the country's population, new agricultural lands are required to sustain it, and these are being secured by reclaiming our vast areas of stump and swamp land. Hercules Dynamite is being used extensively in developing these sources of food supply that have hitherto lain dormant and unproductive.

The sixty million acres of swamp land in this country—now a menace to public health—await the product of the dynamite maker to transfer them into fertile, productive farms. It has been stated by Government authorities that one man with dynamite can dig as much ditch as six men with picks and shovels.

In many sections of the country, our Agricultural Service Men are demonstrating the use of explosives to land owners and contractors. If you desire further information, write the Agricultural Department of the Company at Wilmington, Delaware. "Progressive Cultivation", a 68-page booklet, gives full information about the use of explosives for agricultural purposes.

HERCULES POWDERS

HERCULES POWDER CO.

Chicago	St. Louis	New York
Pittsburg, Kan.	Denver	Hazleton, Pa.
San Francisco	Salt Lake City	Joplin
Chattanooga	Pittsburgh, Pa.	Wilmington, Del.

Using explosives for large-scale land improvement is largely in the past. Swamplands are now called wetlands and are considered vital to ecosystems. The US Farm Agency even operates a Farmable Wetlands Program "to restore previously farmed wetlands."[29] Hercules Powder, advertisement, *Everybody's Magazine*, June 1920, 91.

Before the construction of the Interstate Highway System in the late 1950s, waterways were a much more important means of transport. Currently 60 percent of US commercial transportation is by truck, and just 8 percent by ship. Hercules Powder, advertisement, *Literary Digest*, March 29, 1919, 74.

started working for Laflin & Rand at age fifteen and went on to head the DuPont Sporting Powders Division in 1903.

The divestiture of DuPont allowed for both Hercules and Atlas to avail themselves of DuPont expertise for a period of five years. However, the companies already had quite a bit of DuPont help, since the executives of both companies were former DuPont employees. Atlas was originally headquartered in the DuPont Building in Wilmington, with the main offices of Hercules Powder Company across the street. Moreover, the DuPont family continued to hold substantial stock both in Hercules and Atlas. The three companies were investigated several times for collusion and were found guilty of price fixing in 1945.

Like DuPont and Atlas, Hercules began diversifying into nonexplosive chemicals and in 1916 won a large contract to supply England with acetone, an important solvent and explosives ingredient. The contract stipulated that the acetone be produced from novel sources, and Hercules found a way to extract it from sea kelp.

In the 1920s, Hercules expanded into other chemicals and by 1935 consisted of five divisions: Explosives, Naval Stores, Nitrocellulose, Chemical Cotton, and Paper Products. Naval stores are turpentines, glues, and gums made from tree sap. This diversification, while significant, paled in comparison with that of DuPont, and explosives were still 20 percent of Hercules's business in the 1950s. By 1970, Hercules Inc. was one of the ten largest chemical producers in the country.

Hercules introduced its Spiralok threaded, connectable cartridges in 1944. The cartridges were specifically designed for use in seismic exploration. DuPont started making its Fast-Coupler in 1942, and Atlas began selling its Twistite connectors in 1943. The connectors are used mainly for seismic prospecting and for presplitting dimension stone. Dyno Nobel still sells versions for both applications.

All producers, including Hercules, converted to military munitions for World War II. After the war the company began producing plastics. In the 1960s, rocket fuels and military chemicals overtook commercial explosives in revenue. An aerospace division was launched in 1978.

Hercules Powder Company changed its original logo to a silhouette version in 1937. The logo became just the head of Hercules in 1963, and the company's name was changed to Hercules Inc. in 1966. A more streamlined version

Unusual Angle Yields Good Basis for Dynamite Campaign

Illustrated Trip through Plant Is Used in Hercules Powder Company's Advertising to Decrease Popular Dread of Dynamite

Dynamite is an old copy subject. It has been long and well advertised. The news columns also are frequently telling its story. It has not over many obvious, hit-you-in-the-eye advertising angles at that. Pretty well everything that could be told about it for those who need it, was told in the last Hercules campaign, which graphically showed all the major uses of this explosive in mining, engineering, quarrying, drainage, and farming. Yet for the current campaign an entirely new angle has been hit upon – a trip through a dynamite plant!

"Come along; don't be afraid!" these advertisements say in effect. "Come through the plant and see dynamite made, trucked about and handled. Take up a stick yourself and get familiar with it. No, no, it won't harm you. Why yes, it will blow up if you get rough with it; but it is so made nowadays that it has the sweet temper of a saint and you can't make it get nasty unless you set yourself deliberately to get its goat. Mind those little hand buggies. See, they have pneumatic tires. Yes there are moments during the process of manufacture when a jolt isn't altogether desirable. But our men know and take care. Watch that one over there. He is mixing nitroglycerin. Come along – nothing to be afraid of!"

The average man is frankly scared of dynamite. He knows little or nothing about it except that it is an explosive with the reputation of having an extremely nasty temper, and he prefers to keep out of its way. The Hercules Powder Co. feels that this attitude on the part of the public mind goes too far and is not beneficial.

So here we have an advertising campaign actually planned to popularize knowledge about dynamite! The Hercules company does not exactly seek to put a stick or two of dynamite in every housewife's market basket each shopping day; but it would have even the average housewife know that, while dynamite is not intended as a fire lighter or for baby to cut his teeth on, with proper care it is not by any means so dangerous as is commonly supposed. The explosives company believes that a vast deal of work of the utmost important to the welfare and progress of the United States, especially the reclamation of our stupendous areas of waste land for farming, is very largely hindered by the dread the ordinary man has of dynamite.

Hence the new campaign was planned to introduce dynamite to the average man as one of his best friends. It seeks to make the man-in-the-street so familiar with the facts about dynamite that he will lose his dread of it and also cease to spread that dread.

"Listen!" whispers the copy between its lines; "from the steel bed springs on which you sleep so comfortably at night to the diamonds you give your wife, there is hardly anything that increases your enjoyment of life, but dynamite helps to get it for you. Hundreds of things you use everyday and consider necessities, from the stove that cooks your breakfast to the telephone wires over which you send words that you will be late for dinner, would, but for dynamite be so expensive as to be impossible. Come along and get to know this little friend!"

This article ran in *Printer's Ink*. Established in 1888, the magazine was the first national publication devoted to examining the "science" of advertising. Adapted from "Unusual Angle Yields Good Basis for Dynamite Campaign," *Printer's Ink*, November 4, 1920, 117.

of the head logo is in current use by Alliant Powder Company for its sporting powders. (The history of sporting powders is material for another book.)

The Fayville location was closed after an explosion in 1923. The Emporium plant closed in 1931. The Hercules, California, plant closed in 1964. The Bacchus, Utah, plant was retooled to make solid rocket fuels in 1968 and still operates, now owned by ATK Launch Systems Inc., a subsidiary of Northrop Grumman Innovation Systems Inc. The Birmingham plant closed in 1986. The Kenvil, New Jersey, plant closed in 1996.

The Ishpeming location still operates, now as Pepin-IRECO, but does not make dynamite. The Carthage location is home to Dyno Nobel and is the only plant in the country still producing dynamite.

Like DuPont, and to a lesser extent Atlas, Hercules produced an enormous amount of printed material related to explosives. Most publications and articles were instructional, outlining current best practices regarding use. However, Hercules, more than any other producer, employed goodwill advertising directed toward the public at large. The article (see p. 169) describes

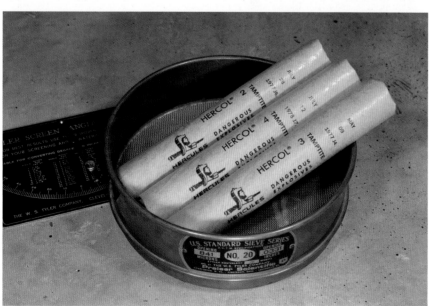

Top left: Hercules Hercomite was a very low-density ammonium nitrate formulation that was first offered in 1920. The original 25 percent strength Hercomite was replaced by Hercomite 1 through Hercomite 7, with strengths ranging from 20 to 60 percent. *Author*

Center left: Hercules Powder Company began selling "King-Size" small-diameter cartridges in 1950. The sticks came in 12", 16", 20", and 24" lengths in diameters of 1¼", 1½", 1¾", and 2". The obvious advantage was that deep holes took less time to load. *Author*

Bottom left: Hercules Hercol was an ammonium nitrate formulation designed for bulk loading for quarrying. Introduced in 1958, Hercol was ideal for "the economy-minded quarry and pit operator,"[30] according to advertisements. *Author*

Hercules Unigel semigelatin dynamite was introduced in 1972 as "a replacement for many of the more expensive specialty-grade dynamites that offer varying energy value with each grade,"[31] according to Hercules. This was a stark reversal of the early trend toward more varieties and grades. *Author*

The lightest mixtures in the Hercomite line had cartridge counts of 175 sticks. *Author*

one such ad campaign launched in the late 1910s. Two ads from this campaign are shown (see pp. 17, 20).

Illinois Powder Manufacturing Company

Illinois Powder Manufacturing Company was launched in 1907 by a group of executives from other explosives firms. J. Lowe White was a representative for Austin Powder Company and Keystone Powder Manufacturing Company. A. C. Blum worked for Keystone Powder as well, and Almont Lent was former president of Austin Powder. White became secretary and treasurer of the new firm, and Adolphus C. Blum was president until 1917, when White became president.

Note the term "splendid explosive." DuPont once described its blasting powder as having a "bright, attractive appearance."[32] Illinois Powder Manufacturing, advertisements, *MK & T Employee's Magazine*, May 1917, 41, and June 1916, 55.

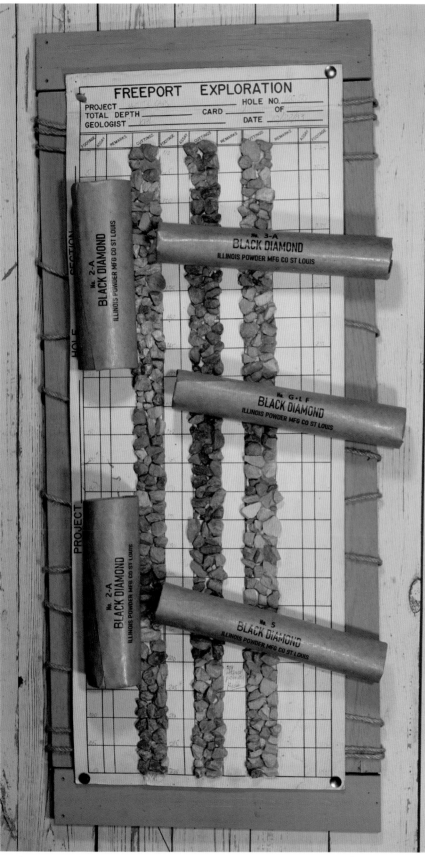

According to Van Gelder and Schlatter, "A suitable location for the factory was found on Sherman Hollow,"[33] a narrow limestone valley off the Mississippi River 1 mile east of Grafton, Jersey County, Illinois. Construction began in May 1907 and production started in December 1908, with 2,400,000 pounds of dynamite produced the first year and 10,000,000 pounds in 1926. Another plant was established in 1940 near Spanish Fork, Utah, now a suburb of Provo. An article in 1957 related that the sales of Illinois Powder Manufacturing

The Black Diamond line of permissible dynamites was introduced in 1909, making the second official permissibles list issued by the US Bureau of Mines. The number would grow to seven varieties and was produced until the 1950s. *Author*

The folding matchbook was patented in 1892 and immediately became another way for advertisers to reach prospective consumers. Collectors of matchbook covers are called phillumenists. *Author*

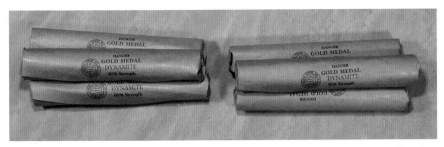

The original Gold Medal dynamite was a straight nitroglycerin formula. Illinois Powder Company never made a broad range of dynamites like DuPont, Atlas, and Hercules, instead focusing on gelatin and permissible formulas. *Author*

Illinois Powder Manufacturing Company hired an advertising agency specifically to promote its Gold Medal explosives. *Author*

Company were "concentrated west of the Mississippi."[34] The company was purchased by American Cyanamid in 1957. An ad in November 1957 declared that "in a move to extend and broaden the range of our services to explosives users, American Cyanamid Company has purchased the plants, magazines and sales facilities of Illinois Powder Manufacturing Company."[35]

One early product that Illinois Powder offered was a 6" long, 1¾" diameter cartridge, which it described as convenient due to its larger diameter and invariable weight of 0.5 pound. While this was a decade before DuPont started offering its wide choice of lengths and diameters, Atlas Powder had been sold in 6-inch lengths and diameters of up to 2 inches since the 1880s. In all early cases the somewhat

larger-diameter cartridges could be used only in relatively soft rock, or to fill coyote tunnels or other cavities in harder rock.

J. Lowe White had received a gold medal for his display of explosives at the 1898 Trans-Mississippi Exposition. At that time, White was a salesman for both the Oriental Powder Company and the Austin Powder Company. Both Illinois Powder Manufacturing Company and Austin Powder Company carried Gold Medal dynamite beginning in the 1910s. All the Gold Medal brand of dynamite for both companies was produced by Illinois Powder Manufacturing Company until 1931, when Austin Powder Company opened its first dynamite plant. Likewise, Illinois Powder Manufacturing Company sold Austin Blasting Powder. Illinois

Powder also made the dynamite sold by King Powder Company from 1920 to 1957. King Powder Company ceased business the following year.

It was surely J. Lowe White and Almont Lent's association with both Illinois Powder and Austin Powder that resulted in the shared Gold Medal label. Illinois Powder also made the Black Diamond brand of permissibles.

Van Gelder and Schlatter reported that Illinois Powder Manufacturing Company supplied explosives for "many of the active coal, lead, zinc and other mines, many large quarries, as well as numerous drainage and other improvement projects of the southwestern section of the United States."[36] The Grafton plant was torn down in 1964, and the site was sold by American Cyanamid to Trojan Explosives Company in 1973. Trojan used the site for storage.

Independent Explosives Company

Independent Explosives Company started as a family business founded in 1936 by Leonard E. Weitz. Weitz was a prominent Ohio banker and businessman and onetime president of the Cuyahoga County League of Banks and Loan Associations. Weitz had served as mayor of his hometown of Rocky River Village (now Rocky River City), on the shore of Lake Erie. He worked for Burton Explosives Company from 1930 to 1934. Weitz attended Case Western University in Cleveland and has a scholarship named in his honor.

The firm established headquarters in Scranton, Pennsylvania, and built a plant in Suscon, near Pittston, in Luzerne County. Weitz served as president until 1961, when his son John took over. John was president until 1969, having also served as the company's geologist and secretary from 1952 to 1961. Son Joseph also worked for

By 1986, Independent Explosives Company offered twenty-eight different labels of dynamite, including Tamite, Ditchrite, and Index. The company's Unitgel brand predated Hercules's Unigel label, introduced in 1972. *Author*

This box dates from 1981 or earlier because of the "Explosive A" label. The designation was replaced with "Class 1.1 Explosive" the following year. *Author*

the firm as a geologist and "Assistant to the President" from 1955 to 1958. Local businessman Frank T. Dainese was president from 1969 until 1988.

Independent Explosives Company began by making dynamite and then expanded into ammonium nitrate explosives, and finally bulk emulsion explosives. The firm received a patent for a water-gel explosive in 1986.

An iteration of the company still operates as Independent Explosives Company Inc. (IEX). According to the company's website, the original company became four separate firms in the mid-1950s: Independent Explosives Company, Independent Explosives Company of Pennsylvania,

Independent Explosives Company of West Virginia, and Chemex Supply Corporation.

Independent Explosives Company of Pennsylvania became a wholly owned subsidiary of IRECO in 1988, after IRECO was absorbed by Dyno Nobel. According to the IEX website, the firm is an "independent retail distributor of commercial explosives" and "offers quality Dyno Nobel products."[37] The company is currently headquartered in Bloomfield, Connecticut, and concentrates on serving the Northeast, with sales offices in New Hampshire, Pennsylvania, and New York.

IRECO

IRECO Chemicals Company was formed in Salt Lake City in 1962 by a merger of Intermountain Research and Engineering Company and Mesabi Blasting Agents Inc. The Intermountain Research and Engineering Company was founded in 1958 by Melvin Alonzo Cook, Douglas Pack, Robert Keyes, Robert Clay, Wayne Ursenbad, Kirkwood Collins, and Earl Pound. Cook was a chemist and professor who had invented slurry explosives two years earlier. The other six were graduate students of his. Cook started working for DuPont during graduate school in 1934. He was employed for DuPont as a research chemist for thirteen years and then taught at the University of Utah. His most important invention was the water slurry explosive in 1956. The mixture of aluminum powder, ammonium nitrate, and water was the first explosive to incorporate water as a means of suspension. He remained president of IRECO until 1972, when son Melvin Garfield Cook took over.

Mesabi Blasting Agents was launched in 1960 by a group of Intermountain Research and Engineering Company scientists including Cook, who served as president and was a major shareholder. The company was named for the Mesabi Iron Range in northeastern Minnesota. Mesabi

Blasting Agents continued to operate as a subsidiary of IRECO Chemicals Company.

Cook's slurry mixture of ammonium nitrate, aluminum powder, and water proved so safe, effective, and easy to use that it and similar compounds all but replaced dynamite in less than two decades, reducing dynamite's share of the explosives market to 15 percent by 1971. Water gels and slurries, emulsion explosives, and ANFO now make up 97 percent of the explosives used in the United States.

IRECO was purchased in 1975 by Gulf Resources and Chemical Corporation and continued to operate as IRECO Chemicals Inc. In 1984, IRECO was acquired by Norwegian chemical company Dyno Industrier and became IRECO Inc., a subsidiary. IRECO Inc. was dissolved in 1985 and was phased out as a subsidiary of Dyno Nobel in July 1993.

IRECO the company obtained more than a hundred slurry-related patents. The product line of IRECO did not include dynamite until 1985, when the company, now operating as a subsidiary of Dyno Nobel, bought the Hercules suite of explosives.

Cook himself was granted more than a hundred patents and copatents while working for DuPont and IRECO. He improved DuPont's vaunted Nitramon formulation and wrote hundreds of articles and six books, including the seminal *The Science of Industrial Explosives* in 1974. He was awarded one of only three ever Nitro Nobel Gold Medals for his work.

Dynamite with the IRECO logo was produced for only about nine years. Dyno Nobel picked up and continued all three of the brands shown. *Author*

Early 1990s IRECO cartridges bore a combination IRECO/Dyno logo. *Author*

Jefferson Powder Company

According to Van Gelder and Schlatter, "Jefferson Powder Company was organized in 1905 in cooperation with the Pratt Coal Company and the Republic Iron and Steel Company to manufacture black powder and high explosives for the mines in the neighborhood of Birmingham, Alabama."[38] This rationale was similar to that for establishing the Apache Powder Company in Arizona two decades later. Jefferson Powder Company was incorporated in March 1905 in Birmingham by a group of prominent local businessmen, including Culpepper Exum, Erskine Ramsey, and Moses V. Joseph. Exum had earlier founded the Birmingham Fertilizer Company and served as Birmingham mayor and chamber council president. He was a major shareholder and president of the Jefferson Powder Company for its entire existence.

Erskine Ramsey was a financier and coal-mining engineer with patents for more than forty coal-mining-equipment innovations. Moses V. Joseph was one of the founders of the firm Loveman, Joseph, and Lowes, at that time described as "one of the greatest mercantile establishments of the South."[39] The stores, which became known as Lowes, were large department stores with locations throughout the southern states and as far north as Philadelphia. The firm went bankrupt in 1979.

All three men were civic leaders and philanthropists who were heavily involved in the development of the region. The Jefferson Powder Company even provided trophies for the local high school.

The balance of the board of directors of the new company comprised southern coal and steel company executives,

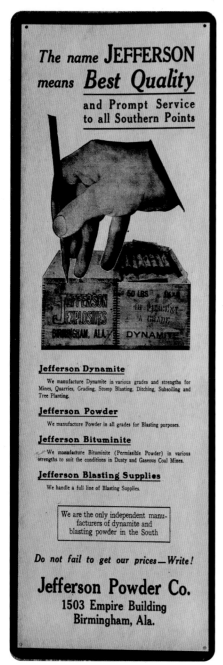

Jefferson Powder Company advertised more than most smaller producers and even some larger ones. Apache Powder Company did less advertising than Jefferson Powder Company, and when they did, they sometimes ran the same advertising copy verbatim for decades. *Author*

Note the British spelling on the cartridges: nitroglycerine. This was common in the early days of US dynamite. *Author*

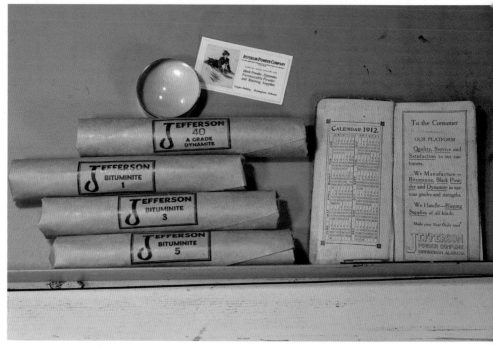

Jefferson Powder Company's Bituminite 1 label made the third list of permissible dynamites in 1910. By the time Jefferson Powder Company was sold to Aetna Powder Company, the Bituminite line had grown to four grades. "A Grade" was short for ammonium nitrate. *Author*

and the main stated purpose of the company was to provide explosives for mining interests in the region that were owned by the shareholders.

In 1908 the company made 3,000,000 pounds of dynamite, triple its 1907 output. By 1914 the company was producing 15,000 pounds of dynamite and 15,000 pounds of black powder a day and boasted one hundred employees.

While the company was headquartered in Birmingham proper, the production facility was about 6 miles northwest of the city. The executive offices were in the newly built Empire Building, at sixteen stories then the tallest building in Alabama. The Empire Building still stands, recently renovated as the Elyton Hotel.

Jefferson Powder Company was acquired by Aetna Explosives Company in 1915. Aetna in turn was purchased by Hercules Powder Company in 1921. The plant in Birmingham was operated by Hercules until it was acquired by IRECO, along with Hercules, in 1985. The plant was closed the following year.

Van Gelder and Schlatter wrote that "the Jefferson plant had been established at an opportune time. Mining had just been placed on a paying basis in the district, and the basic open-hearth steel plants, for which Birmingham is now famous, had but recently been firmly established. The South was beginning to enjoy a rapid increase in general prosperity, all of which involved the use of explosives."[40]

Keystone National Powder Company

Keystone National Powder Company was founded as Keystone Powder Manufacturing Company in Emporium, Pennsylvania, in 1900. In 1910, Cameron County, Pennsylvania, was a contender for top producer of dynamite in the United States, with three large plants, including Keystone in Emporium and two plants in Sinnemahoning.

Keystone was launched by George Jones and U. A. Palmer, who had worked for Climax Powder Manufacturing Company. In 1908 Keystone reported sales of 6,100,000 pounds of dynamite. The company served mostly western Pennsylvania until winning a huge contract in 1909 for explosives to be used during the blasting of the Panama Canal. Then, in 1910, the Keystone Powder Manufacturing Company, Emporium Powder Company, and Sinnamahoning Powder Manufacturing Company merged to meet the demand for the Panama Canal contract. The resultant company, Keystone National Powder Company, was absorbed by Aetna Powder Company in 1914.

Keystone National Powder Company provided all 10,000,000 pounds of dynamite used to blast the Panama Canal in 1910, bidding $1,017,232 to beat out DuPont's bid of $1,051,850. Trojan Powder Company also supplied $375,037 worth of Trojan Powder that

Lois F. Holberg Jr. owned a hardware store in Macon and likely sold Jefferson Powder Company dynamite. Adapted from *The Tradesman*, February 16, 1912, 5.

Keystone National Powder Company offered its Farm Right dynamite in 25 percent, 30 percent, 35 percent, and 40 percent strengths. Its Gelatine was a conventional gelatin formula. *Author*

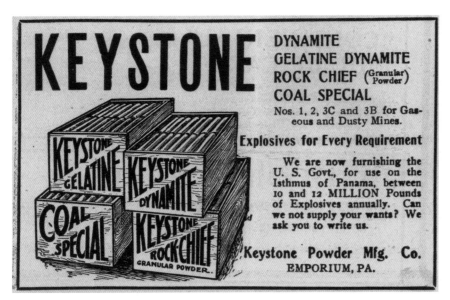

Rock Chief was Keystone's take on nitroglycerin-coated black powder, also known as railroad or contractors' powder. Keystone Powder Manufacturing, advertisement, *Engineering Record*, March 5, 1910, 97.

year. Keystone split the contracts for dynamite with DuPont for the next three years.

Were it not for the government contracts awarded to Keystone for explosives for the Panama Canal, the company would not have acquired two of its local competitors. The expansion resulted in an attractive company for Aetna to acquire.

History of the Panama Canal, published in 1918, related that "prominent among the companies furnishing explosives for the excavating and quarrying work on the canal was the Keystone National Powder Company, with factories at Emporium and Sinnamahoning, Pa. In fulfillment of successive contracts, begun in 1909, this company made dynamite shipments aggregating nearly 30,000,000 pounds in the course of the five years following. The largest single shipment reached 1,225,000 pounds, or far more than was required in the removal of the once-famous Hell Gate as an obstruction to the water traffic of New York City."[41]

Keystone released a nineteen-page booklet titled *Farming with Dynamite* in 1912. In it, the company noted that its dynamite "can be carted in a wagon without any hazard to one's self, as it will not explode by an ordinary shock."[42] The booklet also reported that, as of that time, Keystone had supplied 20,000,000 pounds of dynamite for the construction of the Panama Canal.

Keystone had two entries on the first permissibles list. In 1914 the number of approved Keystone dynamites had risen to fourteen.

G. R. McAbee Powder & Oil Company

G. R. McAbee Powder & Oil Company was incorporated in 1904 "for the manufacture of powder, high explosives, fuzes, lubricants, oil and greases."[43] Fuzes, pronounced "fuzees," were electric blasting caps. The company did not produce blasting caps, instead reselling them.

George Robert McAbee also had served as president of Standard Powder Company and was general manager of the Crescent Powder Company, which was destroyed by fire in 1904. McAbee was one of fourteen children, two of whom were brothers who also ran explosives businesses, plus two more who worked in the industry. He died in 1938 at age seventy-four.

George McAbee built a factory consisting of twelve buildings in Pequea, in Lancaster County, Pennsylvania. The plant produced more than 3,000,000 pounds of dynamite in 1905 but was

---

PERMISSIBLE EXPLOSIVES

Capacity,
8 Cars Per Day

Passed by the U.S. Geological Survey, Pittsburgh, Pa.

COLLIER POWDERS
By the Sinnamahoning Factory.

COAL SPECIAL POWDERS
By the Keystone Factory

Made in all grades to meet any requirements in gaseous and dusty mines and for any kind of coal.

All of Our Products Are Guaranteed.

Keystone National Powder Company

Emporium, Pa.

BRANCH OFFICES:

New York City
Pittsburgh, Pa.
Buffalo, N. Y.
Clifton Forge, Va.
Huntington, W. Va.

FACTORIES:

Keystone Works,
Emporium, pa.
Emporium Works,
Emporium, Pa.
Sinnamahoning Works,
Sinnamahoning, Pa.

HIGH EXPLOSIVES

Ammonia and Nitroglycerine Dynamites for Earth, Rock or Submarine Work, and Gelatine Dynamite of unsurpassed quality for Tunnel Work.

Write us and we will send you the official record made by over eleven million pounds of Dynamite furnished by us to the United States Government for the Isthmus of Panama las year, surpassing all previous records in efficiency. Present contract with the Isthmus of Panama, Four Million Six Hundred Thousand Pounds.

"Carloads" ranged from 20,000 to 60,000 pounds net weight of dynamite per railcar. Keystone National Powder, advertisement adapted from *Engineering News*, June 29, 1911, 74.

destroyed by an explosion in 1907. A new factory was built in Saltsburg, Pennsylvania, near Tunnelton, in Indiana County, Pennsylvania.

Brother Eugene B. McAbee founded the Acme Powder Company in 1887 in Black's Run in Allegheny County, Pennsylvania, and had launched E. B. McAbee Oil & Powder Company in the 1890s. Acme Powder Company was purchased by the Eastern Dynamite Company in 1897. Eugene McAbee was killed in an explosion when cleaning a furnace with explosives in 1919. Dynamite is still used this way to blast away "salamander," which is excess iron and slag.

Charles Wesley McAbee worked at DuPont, served as president of Acme Powder Company, and then cofounded Independent Powder Company of Missouri in 1902 (no relation to Independent Explosives Company of Scranton, Pennsylvania). Yet another brother, Albertus, worked for Eugene and was killed in an explosion in 1893. Still another brother, James Gerry McAbee, served as the secretary of the Standard Powder Company.

G. R. McAbee Powder & Oil was acquired by Atlas Powder Company in 1922. Atlas operated the plant until 1924.

National Powder Company

National Powder Company was founded in 1935 by William F. Grow in McKean County, Pennsylvania. Grow inherited his father's Duke Center, Pennsylvania, grocery store in 1901 and ran it until 1940. Grow was also vice president of Bankers Trust of Bradford, Pennsylvania. He was president of National Powder until his death in 1953.

The original production facility was at Duke Center, about 5 miles west of Eldred, and began production in 1937. This plant, which employed about fifty of the town's thousand residents, burned to the ground after an explosion in 1939. A new plant was built and became Eldred's largest

G. R. MCABEE POWDER & OIL CO.

General Office   Eastern Sales Office
PITTSBURGH, PA.   PHILADELPHIA, PA.

Distributing Points
TUNNELTON, PA.   WILSONBURG, W. VA.
FOXCOURT, PA.
HORRELL, PA.   WELCH, W. VA.   CATTLETSBURG, KY

G. R. McAbee Powder & Oil Company kept its sales local, focusing on Pennsylvania, West Virginia, and Kentucky. G. R. McAbee Powder & Oil, advertisement, *1918 Mining Catalogue* (Pittsburgh, PA: Keystone Consolidated Publishing, 1918), 820.

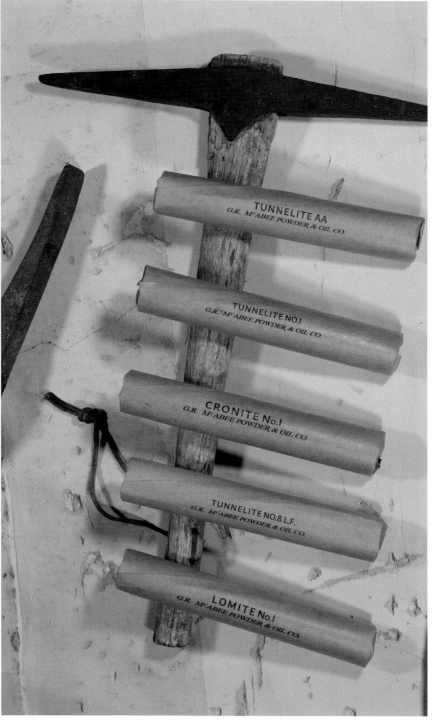

By the time Atlas Powder Company acquired G. R. McAbee Powder and Oil, Tunnelite came in eleven grades. Lomite and Cronite permissible dynamites were offered in only a single grade each. After the acquisition of G. R. McAbee by Aetna, these two brands were joined by McAbee X, McAbee Y, and McAbee Z. *Author*

National Powder Company sold a full range of dynamites, including gelatins from 20 to 100 percent strength, ammonia (extra) dynamites from 15 to 60 percent strength, and ammonia gelatin (gelatin extra) from 30 to 80 percent strength. *Author*

Stamping dynamite shell paper by hand, while common in the late 1800s, was done only occasionally during the twentieth century. The stamps were likely used to fill out short runs of shell paper, or to function as a stopgap during press repair or maintenance. *Author*

business, occupying 200 acres and at one time employing about 120 people. In 1941, National Munitions Company was established as a subsidiary to fill government contracts during World War II. An enormous military explosives plant was built and operated until 1946.

The company reported sales of 1,000,000 pounds of dynamite the first year; 38,000,000 pounds of explosives, including military munitions, were made in 1944. In 1966 the company produced eleven grades of dynamite in addition to other explosives such as Napotex, a free-flowing, bagged ammonium nitrate mixture. Dolomite was National Powder Company's line of "high stick count" ammonium nitrate dynamite, according to advertisements. The brand echoed DuPont Dumorite's promise of more cartridges for the money.

National Powder Company closed in 1970. In 1972 the site was purchased by a group of investors led by New York businessman James Joyce. A plant manufacturing Joyce's patented ammonium nitrate explosive operated as Joyce National Powder Company until a 1979 fire forced the company to close in 1980. National Powder Company and Independent Explosives Company were the last dynamite companies founded in Pennsylvania and two of the last in the country. Over the course of its history, the state has hosted at least fifty black-powder companies and more than thirty dynamite companies.

Digging Deeper

Aetna Powder Company

Early-twentieth-century advertisements for Aetna were peppered with the quaint hyperbole common to the era. Some of my favorite claims: "Every prosperous and up-to-date farmer in America is using Aetna Dynamite."[44] "Progressive farmers all over America are writing us every day that Aetna Dynamite is the greatest agent for economy and efficiency on the farm that the world has ever created."[45] "Every farmer in the world is deeply interested in the many uses of Aetna Dynamite."[46] "A few pounds of Aetna Dynamite will turn a wilderness into a productive field."[47]

In the spirit of "waste not, want not," Aetna implored farmers to "blow up your boulders with Aetna Dynamite and use the pieces of stone for road and path building. Blast your stumps with Aetna Dynamite and use the splinters for fuel."[48]

While most demonstrations of stumping dynamite were conducted on farms, in 1913 Aetna held an exhibition in Main Street Park in what would become Houston, Texas, announcing in the local newspaper that "there are a large number of big stumps in the park[,] and Aetna Dynamite will be used to blow these out,"[49] to the delight of an anticipated large crowd.

An article in 1915 pointed out that Aetna's acquisition of Jefferson Powder Company "was negotiated by the New York concern in order to gain additional capacity for the output of munitions for the warring powers of Europe."[50] Another article the same year revealed that "the Aetna Company has signed contracts for smokeless powder, etc., for delivery by January 1916, aggregating in value more than $22,000,000, the estimated profits from which are expected to be large."[51] Yet another article related that the contracts "would take substantially the full production of its plants for 1916."[52] The profits, it turned out, were not large. The war ended much too soon for Aetna to recoup its massive capital outlays.

Apache Powder Company

Just prior to opening the plant, a group of representatives of the mining interests who would soon be served by the new business took a tour of the Apache Powder Company facilities. "For the benefit of the visitors, 'dummy'

dynamite and gelatine were manufactured and packed,"[53] according to a newspaper account ("gelatine" was often used as a synonym for any gelatin dynamite). The stated purpose of the plant was reiterated in the same article: "Through operation it is expected a large saving will be made in production costs, for the powder freight rate into Arizona is very high."[54] Another article quantified the savings to "the mining industry of Arizona" at "no less than one-half million dollars per year,"[55] about $7,000,000 today.

Apache Powder Company had its own company town, but on a much-smaller scale than those of Aetna, Atlas, DuPont, and Hercules. Apache's Benson neighborhood initially consisted of 4 acres of tiny homes.

An article in 1922 reported that at the Apache plant, "There is not an electric light or wire in any of the buildings where any explosive ingredient is manufactured; instead, the wires are on the outside of the building and the electric globes are placed outside, but directly against the window panes, with reflectors throwing the light inside."[56] Another safety precaution was that "the only dangerous part of the plant, the mixing house, has been placed by itself, behind a hill from the rest of the works."[57] "The most dangerous" might be more accurate than "the only dangerous," since accidents occurred in all parts of dynamite plants.

Atlas Powder Company

In 1882, Atlas Powder dynamite was offered in the following ten grades, at that time strictly reflecting nitroglycerin content: A, 75 percent; B+, 60 percent; B, 50 percent; C+, 45 percent; C, 40 percent; D+, 35 percent; D, 30 percent; E+, 25 percent; E, 20 percent; and F+, 15 percent. The product was sold in 6" or 8" lengths in diameters from ⅞" to 2". Assuming ¼" diameter increments (there are six of them in the range specified: ⅞", 1", 1¼", 1½", 1¾", and 2"), 120 choices are possible.

And this was forty years before G. M. Norman complained about too many dynamite options!

Austin Powder Company

Austin Powder's McArthur dynamite facility employed one hundred workers when it opened in 1931. The figure had risen to 250 in 1963, and 350 in 1972. In 1980, five hundred of the company's 1,400 total employees worked at the McArthur location, which was referred to as the "Red Diamond Plant," named after Austin's popular line of dynamites. In 1963 an article reported that "the Red Diamond plant is strategically located within 500 miles of important explosives[-]consuming industries in the eastern United States. This assures [*sic*] delivery of standard-size explosives from production line to any Austin magazine within twenty-six hours or less."[58]

The article gave this description of the production of dynamite at Austin Powder:

Step 1. Preparation of the explosive ingredients begins with receipt at Red Diamond of tank car shipments of mixed acid, glycerin, and glycol. Mixed acid (approximately 50 percent nitric, 50 percent sulphuric) is then pumped to storage tanks. Next it is pumped to a weighing station and, by gravity, released to the nitrator. In the nitrator, mixed acid is cooled by coils filled with 0° to 6° F. brine supplied by a seventy-foot refrigeration plant. A glycerin glycol mix is added to the mixed acid. Finally, nitroglycerin is separated from the spent acid. After washing, it is transferred to storage tanks. Step 2. Preparation of the non-explosive ingredients takes place in the dope house. Here, nitrate of soda, nitrate of ammonia, bagasse, wood flour, rice hulls, starch, chalk, kaolin, sulphur, etc. are weighed, mixed, and screened. Step 3. Mixing the explosive and non-explosive ingredients takes place in the dynamite and gelatin mixing houses. From there the finished high explosive is sent to the machine pack and case houses.[59]

(Bagasse is plant pulp remaining after the extraction of juices, and kaolin is a type of clay.)

DuPont

In November 1924 the *Wall Street Journal* wrote of DuPont, "The manufacture of explosives is a highly developed chemical industry, and the scope of the company's operation has gradually and logically grown to include allied lines of chemical manufacture."[60] As we have learned, "gradual" is quite an understatement.

The *Encyclopedia of Explosives and Related Items* states that "the Repauno Works became by 1910 the largest in the world and remained so until the middle of 1930[,] when its production was exceeded by a plant near Modderfontein, South Africa."[61] In 1927, DuPont announced that in the prior five years it had produced more than 1 billion pounds of explosives of all kinds.

Dyno Nobel

A 2005 advertisement related that "Dyno Nobel is the world's leading commercial explosives company with over 5,200 employees in thirty-six countries, research and technology facilities on four continents, and sales of over 1.2 billion dollars annually."[62]

In 1948 the US Bureau of Mines published an extensive overview of the dangers of lightning to electric blasting operations. The study summarized dozens of instances where lightning strikes initiated blasting circuits, including one instance where a bolt of electricity from the sky detonated dynamite nearly a mile underground via the mine's cabling. The study called for lightning arrestors and capacitors to be built into a mine's electrical system, and mainly not to conduct blasting during storms. Toward this end, Nitro Nobel patented the world's first lightning detector in 1970. The device, which senses electricity in storms up to 10 miles away, saw use

beyond blasting. In 1976, one of the devices was installed at the ninth green of the main course at the Congressional Country Club in Bethesda, Maryland, for the 1976 PGA tournament. Many similar gadgets have appeared since then, and some jurisdictions mandate their use whenever electric blasting is conducted.

The only other winner of the Nitro Nobel gold medal besides IRECO founder Melvin Cook in 1990 and Dyno Nobel engineer Anders Persson in 1968 was Robert Van Dolah, for research in liquid explosives, in 1967. The medal is made of solid gold but, unlike the Nobel Prize, is not accompanied by a cash award.

General Explosives Company

Before Home Powder Company in effect became General Explosives Company, it sold dynamite made by and branded as Aetna Dynamite. Contrast with King Powder Company, which had its own branded dynamites such as Detonite produced by Aetna and then Illinois Powder Company. Advertisements run by distributors often still claimed to be the manufacturer of the products they sold. This was also the case with blasting accessories, which were made mostly by companies other than those that sold them.

Giant Powder Company

Crooks adapted to the use of dynamite. In 1900, Giant Powder dynamite was used to blow open a safe in Ashley, Connecticut, and steal $800.00.

A common way to attract investors into becoming involved with bogus mining schemes was to "salt" the mine with valuable ore. In 1899 an article told how a man fell victim to the ploy, even though he had used sticks of Giant Powder to blast down several feet into the rock himself. The con man selling the mine had removed the dynamite wrappers, painted the sticks gold, and replaced the wrappers. The blast embedded gold flecks into the rock

fragments, giving the appearance of bonanza-grade ore. As with any good, dishonest scheme, the mark thought that the crook was supplying the means to prove the deal was honest.

Finally, in 1900 a New Jersey woman divorced her husband over inappropriate behavior, including "calling her vile names and threatening to blow her up with Giant Powder."[63]

Giant Powder Company was fully aware of the appropriation of its name as a synonym for dynamite, writing in ads, "Because the superiority of Giant Powders is so generally acknowledged, other explosives are frequently offered as 'giant powders.' Insist upon having the genuine—always bearing the Giant brand."[64]

When Atlas took over the Giant plant in California, a safety makeover included replacing all metal tools with wooden counterparts, forbidding workers to wear jewelry, and using different brooms for sweeping indoors and outdoors, lest pebbles be carried inside. In a similar vein, all walkways between buildings were made of elevated wooden planks, so that dirt was not tracked indoors. The goal was to reduce the potential for sparks. While not likely to detonate a fully completed stick of dynamite, a spark could spell havoc elsewhere in the operation, and a fire of any kind could be catastrophic.

Grasselli Powder Company

During World War I, Grasselli specifically recruited military draft rejectees, declaring that "this opportunity should appeal especially to men who have been rejected for military service because of some minor disability. By going to the powder plant and securing a job they will be serving 'Uncle Sam' possibly in an even more valuable way than if they were actually on the firing line."[65] In 1922, Grasselli was still soliciting workers, running an ad seeking "dynamite men, experience preferred; steady work, wages forty-five to seventy-five cents an hour, ten-hour day. Good board, eight dollars per week."[66]

Old dynamite and empty dynamite boxes were not all that was best burned for disposal. In 1930 a wrecking company "razed by fire" the buildings, "one by one,"[67] of the then-abandoned Grasselli plant in Cameron County, according to a newspaper article. The old buildings of the Repauno works were dispatched in similar fashion in 1954. The rationale was to avoid having workers come into contact with explosive residues remaining in the structures.

For a time, Grasselli dynamite cartridges were marked with a horseshoe and the words "GOOD LUCK." The slogan was inherited from the Burton Powder Company. Whether or not this blessing engendered confidence in the product is debatable.

Hercules Powder Company

An article in 1922 about Hercules Powder's Birmingham location related that "the concern manufactures over fifty different grades and kinds of powders and explosives, and can make explosives to fit any particular work or condition. Frequently heavy orders are given on short notice, and the completed special quality explosive is made and delivered in forty-eight hours."[68]

Hercules retained its smokeless-powders division until 1995, when it was sold to Virginia-based Alliant Powder Company. Once a division of Honeywell Corporation, Alliant became an independent firm in 1990. Alliant brand powder with the Hercules logo is still sold by Alliant Ammunition and Powder Company, founded in 1999.

Illinois Powder Manufacturing Company

Not all accidents at dynamite plants were horrific explosions. In 1915 the *Alton Evening Telegraph* reported that "William Sikes, while working at the Illinois Powder Co., Friday, had the misfortune of mashing the index finger of his left hand. Mr. Sikes said his finger is very painful."[69] Snakes were also an issue that year. The *Telegraph* wrote that a night watchman was recovering after being bitten by a copperhead snake that had "popped up from beneath a platform"[70] in the Illinois Powder factory. One week later the *Telegraph* let the community know that another snake had been captured at the plant. The timber rattlesnake was described as "a very beautiful snake, four feet long and has ten rattles."[71]

Finally, in May of that same year, newspapers reported that "a tornado at Grafton, Illinois, near here last night, threaded a winding course through a large area dotted by the twelve buildings of the Illinois Powder Co., containing fifty carloads of dynamite. The wind uprooted trees, snapped others off uniformly ten feet from the ground and entirely missed the buildings of the plant."[72]

Independent Explosives Company

Independent Explosives Company was not to be outdone by Illinois Powder: in 1961 a newspaper reported that a worker "suffered a compound fracture of the left little finger when he caught it in a roll of paper while working at the Independent Explosives Co."[73] One year later, another man was treated "for a laceration of the left middle finger. He was hurt while at work as a shell house operator."[74]

In 1987 the plant at Suscon, 3 miles from Pittston, consisted of forty buildings on 400 acres and employed seventy workers. A story that year about the decline of dynamite opened, "Frank Dainese, self-proclaimed 'nitroglycerin man,' ran his fingers along a 2" × 16" tube of brown waxed paper. He handled it as he might a fine piece of jewelry. 'That's a damn good stick of dynamite,' he said. 'Feel the density. It's well-filled. It's a beautiful stick. I hate to see it replaced.'"[75] Dainese

served as president of Independent Explosives from 1969 until 1988.

In 1951 it was announced that plant workers were to receive an eight-cent-per-hour raise, up from $1.00. In 1973, Independent Explosives paid its starting plant production workers $3.00 an hour "plus all fringe benefits."[76]

A company can be incorporated at any time during its existence. Independent Explosives Company was incorporated in 1940.

IRECO

In 1985, IRECO said it ran thirty-five production facilities around the globe, employed eight hundred, and specialized "in the manufacture of explosives for the mining industry and defense, and licenses patents to about thirty other explosives manufacturers."[77]

IRECO operated its own fleet of explosives transportation trucks as a subsidiary called ECONEXPRESS. Austin Powder, Hercules, and Atlas also operated trucking divisions, with DuPont's private fleet of two hundred tractors and four hundred semitrailers hauling 2 billion pounds of explosives a year in the 1980s (none of it dynamite by that time).

Jefferson Powder Company

In 1912, Jefferson Powder Company embarked upon an "education campaign,"[78] as one newspaper put it, giving dozens of demonstrations on farms in Alabama, North and South Carolina, Mississippi, and Tennessee. Boulders and stumps were blasted and copies of the company's booklet, *What Dynamite Will Do for the Farmer*, were distributed to all comers. Illinois Powder Manufacturing Company had published its own booklet, *What Dynamite Will Do*, in 1911.

In 1913, Jefferson Powder bragged that "the Jefferson Powder Company, manufacturers of high explosives, can ensure Prompt Shipment of Fresh Goods at all times because of the constant and unlimited amount of Central Station

Power furnished by the Birmingham Railway, Light & Power Company[,] over their high[-]tension long[-]distance transmission line."[79] Electricity was still a novelty and was provided by innumerable private entities in patchwork fashion. There was no "grid," and half the homes in America did not have electrical service until the late 1920s.

Jefferson Powder Company purchased its boxes from the Acme Box Company in Chattanooga, Tennessee.

Keystone National Powder Company

Keystone commenced operation the morning of September 6, 1900, according to newspaper articles, which also related that the company employed "regularly eleven laborers—much of the time more. None of them receive less than two dollars a day—some of them more."[80]

Another, much-earlier and much less successful Keystone Powder Company existed near Chambersburg, Pennsylvania, from 1871 to 1874. The company published a notice in July 1871 declaring that "having erected and completed our Powder Mills, we are now prepared to furnish Blasting Powder of a superior quality to the trade, and also to Rail Road Companies and Contractors."[81]

G. R. McAbee Powder & Oil Company

George McAbee continued to work for Atlas Powder until his death in 1938 at age seventy-three. Businessmen such as McAbee networked through a variety of business and service organizations, as well as private clubs. McAbee was a member of the Union League of Philadelphia, Philadelphia Country Club, Masonic Order and the Shriners, Knights Templar, Duquesne Club of Pittsburgh, and Hannastown Country Club in Greensburg, Pennsylvania. Memberships in exclusive country clubs were particularly vital for developing business contacts.

National Powder Company

The company logo is modeled after the Union Explosives Company logo, which in turn was patterned after the Printer's Union trade union emblem. I could not find a direct link between Union Explosives Company and National Powder Company.

I have seen three different National Powder Company handstamps, all of the correct size for a standard shell. Even if the plant was running a printing press to print most of its shells, the stamps may have been used for very small orders, or as a fallback if the press run was short or defective shells were discovered. If a rotary press using metal dies damaged a die with no spare on hand, handstamps would have kept the operation running. Alternately, it is possible that such stamps were used for stamping precut callback cards, for stamping the tops of the cardboard interior cases that were often inserted into wooden cases, or only for stamping oversize cartridges, which were hand-packed through the 1940s.

CHAPTER 5

Dynamite Look-Alikes

Enormous numbers of cartridged explosive formulations appeared in the United States between 1870 and 1900. Nitroglycerin dynamites that came and went included Gotham's Powder, Jupiter Powder, Blackjack, Dualin, Rendrock, Vulcan Powder, Vigorite, Maximite, Forcite, Neptune Powder, Virite, and Americite. Nonnitroglycerin formulations, often still billed as dynamite, included Cerberite, Joveite, Masurite, Big Chief Powder, Bellite, Kinotite, Oxonite, Tonite, Witherite, Carboazotine, Triplastite, and Panclastite. Of the true dynamites, only a few, such as Vigorite and Forcite, survived as brand names for other companies.

During dynamite's reign, a handful of other cartridged explosives were widely used. These included Trojan Powder, various brands of pelletized black powder, and government-issued Sodatol and Pyrotol.

Military "dynamite" became available in the 1950s but was not used in combat. The US armed forces also employs commercial explosives for peacetime use and for stateside construction projects during wartime.

Although dynamite is no longer the explosive of choice for most applications, the cartridge is still vital as a means for safely handling and loading modern explosives in small-diameter and nonvertical boreholes.

Trojan Powder

Trojan Powder was born when ex-DuPont chemists Fletcher B. Holmes and Jesse B. Bronstein Sr. devised a process to stabilize nitrostarch in 1904. Nitrostarch, despite its name, is not a type of nitroglycerin, with the "nitro" referring to the nitrates that are the essential oxidizing constituent of most explosives such as trinitrotoluene (TNT) and pentaerythritol tetranitrate (PETN), which is the explosive used in detonating cord. Nitrostarch was first made in 1833 but was initially too unstable for military or commercial explosives.

Trojan Powder contained from 30 to 40 percent nitrostarch, plus ammonium nitrate and sodium nitrate. Because nitrostarch is not a liquid, it cannot freeze and, in the correct formulation, is also more stable than dynamite. And, like the Trojan Powder motto proclaimed, handling the explosive would not cause the "nitro headache" that accompanied contact with dynamite. Combustion fumes were also minimal. Of all the early nonnitroglycerin brands, only Trojan Powder was truly successful.

The primary disadvantages of nitrostarch explosives were low water resistance, low plasticity, which made

Explosives were carried by mule to places that were impossible to reach by other means. Mules were also used to pull railcars of dynamite from the main rail line to the mine. Trojan Powder, advertisement, *Engineering Record*, March 5, 1910, 97.

for more-difficult tamping, and lower shattering strength relative to nitroglycerin dynamite. Trojan Powder topped out at 75 percent strength.

The company making Trojan Powder was first called the Independent Non-Freezing Powder Company and was in Paulsboro, New Jersey, for a short time in 1905. The operation moved to Allentown, Pennsylvania, becoming the Allentown Non-Freezing Powder Company. The company became Pennsylvania Trojan Powder Company in early 1906, Trojan Safety Powder Company later in 1906, and Trojan Powder Company in 1907. Production for 1907 was 570,000 pounds. Trojan launched the Pacific High Explosives Company in Roberts, California, in 1906 and renamed it California Trojan Powder Company in 1912. Trojan opened a plant in Overton, 8 miles north of Pueblo, in 1907. Trojan Chemical Company was formed in 1915, and in 1919 all parts of the enterprise were consolidated into one new Trojan Powder Company.

Trojan Powder was first offered in #1 and #2 strengths, and then in standard strengths of 20 percent, 30 percent, 40 percent, 50 percent, 60 percent, and 75 percent. The explosive was sold in 1", 1¼", and 1½" diameters, as well as 12.5-pound bags. Large-diameter cartridges were made beginning in the 1920s. *Author*

Trojan Powder's "WR" was introduced in 1962. "WR" stands for "water resistant," and the formulation addressed a main objection to nitrostarch explosives. Nitrostarch explosives are hygroscopic, meaning that they absorb ambient humidity. *Author*

Trojan made all the explosives used in World War I hand grenades. During World War II, Trojan began producing TNT and PETN, which became the primary constituents of its commercial boosters.

According to Van Gelder and Schlatter, "The company met with numerous troubles before it finally reached success, for its capital was limited and its organization largely untrained. Among the consumers, prejudice had to be overcome against a new and untried explosive, and plant accidents were not an infrequent occurrence."[1]

Trojan Powder's Coal Powder brand debuted on the third permissibles list in 1910 and soon expanded to six grades, but Trojan stopped making permissible formulas in the late 1920s.

While Trojan Powder is not dynamite, it was eventually so successful that in 1980, DuPont bent its definition of dynamite to include nitrostarch: "Modern dynamites may be defined as cap-sensitive mixtures which contain nitroglycerin (or nitrostarch) either as a sensitizer or the principal ingredient."[2]

Trojan Powder Company was acquired by the Commercial Solvents Corporation of New York in 1967 and was retained as a division called Trojan-U.S. Powder. In 1963, Commercial Solvents had also acquired black-powder maker U.S. Powder Company, founded in Terre Haute, Indiana, in 1907.

Also in 1963, U.S. Powder Company reopened an old factory near Marion,

Illinois, to produce dynamite. This situation led to some nitroglycerin-based products bearing the Trojan name. Included were Trojan Ditching, Trojan Special Gelatin, and Prima-mite. Trojan Powder Company began producing nitrostarch slurry explosives in the late 1950s. Trojel was introduced in 1960 and came both in cartridged and pourable forms. Another of Trojan's slurry explosives called Trojatol was sold in very long cartridges ranging from 8' to 18' in length.

A plant in Spanish Fork, Utah, was acquired in 1964 and operated until 2002. Trojan was acquired by Ensign-Bickford in 1996 and, through Ensign-Bickford, by Dyno Nobel in 2003. The plant in Spanish Fork, Utah, was closed in 2006 after an enormous explosion the prior year. Dyno Nobel continues to carry Trojan-branded products, most notably Trojan cast boosters.

DuPont experimented with nitrostarch explosives, selling Nyalite from 1905 to 1909 and Arctic Powder from 1906 to 1920. No other producers ever achieved success with nitrostarch "dynamite."

Sodatol and Pyrotol

Pyrotol *was* technically dynamite but had an unusual composition and story. The US military was left with large amounts of surplus explosives after World War I. This had also happened after the War of 1812, the Mexican War, and the Spanish-American War. After the Civil War, so much extra black powder entered the market that prices were depressed for almost a decade.

In 1919 the government distributed surplus cartridged TNT and another military explosive, picric acid, free of charge to farmers for use in clearing land. From 1919 to 1923, 18,000,000 pounds of Sodatol, a cartridged mixture of excess military sodium nitrate and TNT, was sold at cost, at a price of about one-third that of dynamite. Sodatol contained no nitroglycerin

Pyrotol was used by farmers for stumping and boulder removal. Some was also used by governmental entities such as state highway departments and the US Bureau of Public Roads. *Author*

and was rated by various sources as equivalent in strength to 20 to 40 percent dynamite. Sodatol sticks weighed 7 ounces each.

In 1923, researchers at the University of Wisconsin College of Agriculture formulated Pyrotol, a mixture of 60 percent smokeless powder, 34 percent sodium nitrate, and 6 percent nitroglycerin dynamite. Smokeless powder is a nitrocellulose-based propellant for ammunition. John Swene of the college's Department of Agricultural Engineering led the team and was also responsible for the development of commercial Sodatol.

An article in the *1923 Virginia Dept. of Agriculture Yearbook* introduced Pyrotol:

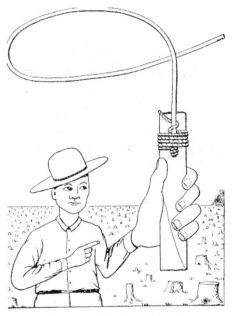

This illustration from a Pyrotol brochure shows a side-primed cartridge with string securing the fuse. US Department of Agriculture, *Prime Pyrotol This Way* (Washington, DC: USGPO, 1925), 1.

For the past four years the United States Department of Agriculture has been re-working the explosives remaining at the close of the War and putting them into shape for commercial use. By virtue of a special Act of Congress these explosives have been made available for agricultural

use. The department is now engaged in distributing an explosive called Pyrotol which has been giving excellent satisfaction for stump and boulder blasting. Already more than 12,000,000 pounds have been used by farmers with good results.

Pyrotol is put up in cartridges, the same size and shape as the ordinary cartridge of commercial dynamite.

The Pyrotol cartridge will do the same work, under ordinary conditions, as a cartridge of commercial dynamite, but as they weigh only six ounces each, while a dynamite cartridge weighs eight ounces, there are 150 cartridges in a fifty-pound case of Pyrotol but 100 cartridges in a box of dynamite of the same weight. So a case of Pyrotol is about one and one-half times as effective as a case of commercial agricultural dynamite. It is a non-freezing explosive and can be used without causing headaches or other ill effects. If stored in a dry place it will keep for at least several months without deterioration.

The government is not distributing this explosive to farmers because it is a better explosive than commercial dynamites, as this is not the case, nor is it doing so because the Department of Agriculture wishes to go into the business of making and selling explosives. It is being done because it is the only feasible way in which an economical use can be made of the surplus war explosives. On this account, the Government makes no charge for the ingredients of Pyrotol, but the farmer is required to pay the cost of preparing and cartridging it, and the freight charges. The average cost of Pyrotol delivered in carload lots at any shipping point in Virginia is 8⅔ cents per pound. When it is considered that the retail prices charged for commercial dynamite range from twenty-two to twenty-four cents per pound, and that it takes one and one-half pounds of dynamite to do the same work as can be done with one pound of Pyrotol, it will be seen that the farmer saves from twenty-four to thirty-one cents per pound when he uses Pyrotol rather than commercial dynamites. Since July 1, 1924, 150,000 pounds of Pyrotol have been distributed to Virginia farmers which saved them more than $36,000.00 at the very lowest possible estimate. This year, Virginia has been allotted 200,000 pounds which still means a saving of about $48,000.00 if there is sufficient demand to use up all of the allotment. Distribution is made only in carload lots through the Extension Division, V.P.I. All persons wishing to obtain some of the Pyrotol should confer with the county agent or, in case there is no extension agent in the county, information can be had by writing to the Department of Agricultural Engineering, V.P.I., Blacksburg, Va.[3]

EXPLOSIVE HERE

Get your Pyrotol at................................. Spencer, Wis. ....

....If your neglect

getting your order before four o'clock it will be sold to other parties.

We have made arrangements with .... Du pont ....

to sell fuse at reduced rates on the day of unloading. Caps will be given with every 200 pounds of Pyrotol.

Come prepared to buy fuse.

W. J. ROGAN, County Agent

All Sodatol and Pyrotol was sold only in standard-size cartridges. *Author*

Neither Sodatol nor Pyrotol was sensitive enough to detonate other cartridges for ditching by propagation; each column had to have a primer with its own cap. Toward this end, free #6 blasting caps were given out with every order of Pyrotol until 1926, when supplies of the caps ran out. *Author*

A contract for production was awarded to DuPont, and Pyrotol was manufactured at DuPont plants in Gibbstown, New Jersey; Barksdale, Wisconsin; and DuPont, Washington. Cases of the explosive were marked "CARTRIDGED BY E. I. DUPONT DE NEMOURS & CO." The prices charged for Sodatol and Pyrotol varied slightly because of varying distribution costs to different regions of the country.

About 60,000,000 pounds of Pyrotol were sold through noncommercial, authorized government distributors from August 1924 to March 1928. Resale at a

The Wisconsin College of Agriculture was established in 1889 and is now known as the Wisconsin College of Agricultural and Life Sciences at the University of Wisconsin–Madison. Wisconsin College of Agriculture, advertisement adapted from *College of Agriculture, University of Wisconsin Circular*, February 1925, 31.

profit was strictly prohibited. According to government-issued bulletins, Sodatol and Pyrotol cartridges were "double dipped" in paraffin, meaning that the paper for the shell was sprayed with wax prior to assembly and then the entire cartridge was then sprayed or briefly immersed.

Pyrotol was so popular with farmers that when supplies were exhausted in 1928, DuPont introduced its own commercial brand called Agritol, which it advertised as "similar to Pyrotol."[4] DuPont reminded users that "with its knowledge of making Pyrotol and an extensive variety of explosives for various purposes, the du Pont Company is in a position to make an explosive to replace Pyrotol and so enable land-clearing and farm improvements to go on without interruption."[5]

Hercules sold its version under the name Hercotol beginning in 1931. Both these explosives were conventional ammonium nitrate mixtures and were appealing to farmers because of their high cartridge count. DuPont had already marketed its Dumorite, which originally incorporated

government-surplus black powder, as an economical dynamite offering "1/3 More Per Dollar" than money spent on regular dynamite. DuPont's Dumorite was originally offered in one grade but grew to a line of four.

Some users judged Pyrotol and Sodatol to be as high in strength as 40 percent dynamite. In all likelihood the true figure was closer to 20 percent, which was the stated strength of DuPont's Agritol and Hercules Powder Company's Hercotol.

Pellet Powder

Pellet powder was not dynamite but was cartridged in similar form. Each 8" by 1¼" cartridge contained four 2-inch-long pellets of compressed sodium nitrate ("B") black blasting powder. The pellets had 0.375" diameter center holes designed to accept safety fuse. Pellet powder was quite popular and much safer and convenient than loose black powder, especially for loading horizontal and upward holes.

Compressed pellets of potassium nitrate ("A") black powder were used for firing guns starting in the 1860s, and in mines in Europe shortly thereafter. An article in 1880 relates that Nobel sold "cylinders of sizes suitable for the miner's use, with a hole drilled right through its center."[8] Canadians began using commercial mining products such as Acadia Pellet and Sampson Pellet Powder in the early 1900s. According to the US Bureau of Mines, cartridged pellet powder for blasting was "first manufactured for general sale in 1925"[9] in the United States. In fact, pellet powder was not made in the United States until 1926, when DuPont began production. By 1930, pelletized blasting powder had captured almost 40 percent of the domestic blasting-powder market. By 1936, more pellet powder was sold than loose powder. US pellet powder consumption peaked that same year but continued to outsell loose powder. In the 1930s, DuPont sold five grades in seven different cartridge sizes ranging

According to DuPont, the company's pellet powder was "for use in non-gaseous and non-dusty mines." The powder was "safer than granular blasting powder, more convenient to handle and has more water resistance. Makes much less smoke, gives better execution and lower cost." *Author*

REG. U.S. PAT. OFF.

Pellet Powder

A Deflagrating Explosive Safer Than Black Powder and Produces More Lump Coal and Causes Less Smoke

The pellets are made in several diameters and one length. Each pellet has a hole through its center. All the pellets are of the same density and pellets of the same size weigh the same.

Four pellets are packed into a cartridge. When a fractional part of a cartridge is needed in order to charge accurately, the pellets may be unwrapped and loaded singly.

Where several cartridges are to be loaded in one bore hole which is dry, some blasters think it advisable to tear the paper off the ends. Others think this unnecessary. Do not remove the remainder of the paper.

The best results are secured by firing with electric squibs.

The electric squib is placed in the borehole and the wires hitched around the cartridge as shown.

Pellet powder may also be fired with safety fuse or a miner's squib.

The fuse should run clear through the cartridge and the end is doubled back and put into the hole. The fuse is kept from being pulled out in loading and tamping.

How to Use a Miner's Squib

If a miner's squib is to be used for firing pellet powder, place the needle and tamp the hole so that when the needle is withdrawn it will leave

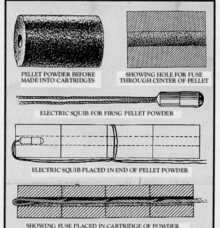

PELLET POWDER BEFORE MADE INTO CARTRIDGES

SHOWING HOLE FOR FUSE THROUGH CENTER OF PELLET

ELECTRIC SQUIB FOR FIRNG PELLET POWDER

ELECTRIC SQUIB PLACED IN END OF PELLET POWDER

SHOWING FUSE PLACED IN CARTRIDGE OF POWDER

a clear path for the squib through the stemming into the central perforation of the pellets.

Do not insert the needle between the paper wrapping and the side of the pellets. This leaves more room for the tamping bar but it gives the squib very little chance to ignite the powder and a misfire is likely to result.

Do not break up the last pellet with the tamping bar and merely insert the needle into the crushed powder. This may lead to a misfire or hang-fire.

Use a deep grooved tamping bar. A deeper groove permits the bar to ride the needle back to the charge and thus makes it much easier to load pellet powder for firing with miner's squibs, especially in 1¼-inch holes.

Pinch off the wax plug on the powder end of the squib and run the squib into the hole about 3 inches, leaving one-half inch of the body of the squib and the entire paper end extending from the hole, sticking straight out, and light the paper end of the tip. One of the principal causes of the failure of miner's squibs to burn through is pushing them too far into the hole.

Pellet powder does not fire nearly as easily as black powder, does not flash into a flame at once when lighted in the open, but it is an explosive and does explode by catching fire.

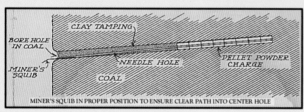

CLAY TAMPING

BORE HOLE IN COAL

MINER'S SQUIB

NEEDLE HOLE

PELLET POWDER CHARGE

COAL

MINER'S SQUIB IN PROPER POSITION TO ENSURE CLEAR PATH INTO CENTER HOLE

SOLID CLAY TAMPING AROUND NEEDLE

BORE HOLE IN COAL

POSITION OF NEEDLE

COAL

PELLET POWDER CHARGE

PROPER POSITION FOR MINER'S NEEDLE IN HOLE LOADED WITH PELLET POWDER.

1   Do not take more than one day's supply into the mine at any one time. Pellet powder is susceptible to moisture and will deteriorate when exposed to a damp atmosphere.

2   Where powder is stored in the mine it should be kept in non-conductive waterproof bags or containers.

3   Care should be taken in handling pellet powder where open lights are used to avoid any possibility of the light coming into contact with the powder.

4   Exercise judgement in gauging the amount of powder necessary to pull the cut. Overcharging is not only dangerous but costly.

5   When using more than one cartridge in a bore hole, be sure that the hole is thoroughly cleaned and push the entire charge back at one time. This prevents the possibility of any slack or dust getting between the cartridges.

6   Be sure that the bore hole is of sufficient size to allow the free entry of the powder. Never force it back with the tamping bar.

7   Do not crush the pellets with the tamping bar.

8   Where electric squibs or fuse is used for ignition, have it placed so that ignition will begin at the center of the charge.

9   Where ordinary miner's squibs are used, insert the needle in the hole in the pellet. This is necessary to insure the uniformity in burning.

10   Always use non-abrasive material for tamping, preferably clay. To insure best results and also to meet requirements of the mining law, it is necessary to tamp the bore hole solidly its full length.

11   When excessive flame and smoke are given off with pellet powder, cut down the size of the charges. Excessive charges make excessive smoke and flame.

12   In cases where little or no smoke is given off, but where the fumes give the miners a headache, it is best not to enter the place until a period of five or ten minutes has elapsed.

13   In case of misfires where electric squibs are used, do not always blame the squibs or powder. Damaged leading wires and batteries of insufficient strength are the cause of more misfires than defective squibs. Provide good leading wires or cables and be sure that the battery is of sufficient strength to fire the squib.

14   In case results obtained with this powder are not satisfactory, do not immediately blame the powder. Any of the following reasons may be the cause of unsatisfactory results: Improper placing of holes, overcharging, or insufficient or improper tamping. When used properly, pellet powder will give excellent results producing a very good grade of lump coal with much less smoke and at a lower cost to the miner than any other powder used in coal mining.

E. I. DU PONT DE NEMOURS & COMPANY, Inc.

Explosives Department   WILMINGTON, DELAWARE

Posters such as this graced the walls of explosives magazines and mining offices. This one measures 13½" × 17". *Author*

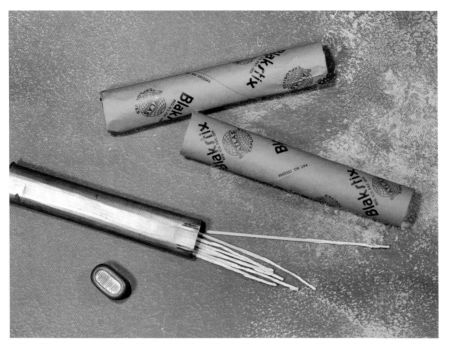

Atlas Blakstix was originally packed in a "spirally wound, glued shell," per Atlas. Also pictured is a container of miners' squibs. *Author*

from 1⅛" × 8" to 2½" × 8". By 1958, DuPont sold only its #4 and #5 Pellet Powders and discontinued the product, along with loose powder, in 1974.

One 1932 ad by DuPont proclaimed that its #4 Pellet Powder was "the first low-density pellet powder stick made with four uniform pellets alike in size, solidity, weight, and execution. Each pellet produces the same velocity, force, and effect."[11] Gauging charges with pellets was more accurate than with loose powder.

Another advantage of pellet powder was minimizing "the hazards of transporting or handling powder in electrified mines and the careless usage of metal tools for opening containers, all recognized hazards associated with granular powder in metal kegs."[12] Even though granular blasting powder kegs came with bungs for opening, many miners punched a hole in the top of the keg with a pick. Despite this and other dangers associated with loose black powder, powder continued to be used into the 1970s. Pellet powders were never deemed permissible, but their slow, heaving action made them popular with miners in less dusty

western coal mines, as well as with miners of other soft materials such as clay and shale. Single pellets were exploded to produce relatively mild explosions ideal for the production of lump coal.

Pellet powder was sold by dynamite makers Aetna, Atlas, Austin Powder, Hercules, DuPont, and the short-lived Burton Explosives Company. Companies charged an 8 to 10 percent premium over uncartridged powder. Some black powder companies such as King Powder Company and Equitable Powder Company also offered pellet powder. Many sole producers of black powder carried precartridged, nonpelletized blasting powder. And DuPont and Hercules Powder Company made cartridged nitroglycerin-coated blasting powder in a range of grain sizes (see p. 24).

Atlas sold a cartridged, "aerated" pellet powder under the name Blakstix beginning in 1932. In 1935, Blakstix was given a special new wrapper featuring cardboard disc ends and designed so a stick could be easily divided into quarters without unwrapping. In the 1940s, Atlas replaced Blakstix with

Atlas Pellet A, B, C, D, and E varieties, which roughly corresponded to DuPont's numbered variants. The formulations offered a range of blasting characteristics suited to coal and soft-rock mining.

Military Dynamite

Military dynamite was developed by Hercules Powder chemist W. R. Baldwin Jr. in 1952, with cartridges marked "M1," "M2," and "M3." "M" stands for medium velocity, rated at 20,000 feet per second, with a strength equivalent to 60 percent strength commercial dynamite. M1 cartridges weighed 8 ounces and came one hundred to a case. A low-velocity version was introduced in 1957.

US military dynamite was not dynamite, despite the label. Only compounds containing nitroglycerin meet the strict technical definition. The most recent mixture was 75 percent RDX, 15 percent TNT, 5 percent oil, and 5 percent cornstarch. RDX stands for Research Department Explosive and is a strong but stable compound developed by the British during World War II. TNT is trinitrotoluene, another relatively stable compound synthesized in the 1890s.

Military dynamite cartridges were water resistant for up to twenty-four hours and were nonfreezing. Recommended uses included culvert blasting, quarrying, and demolition. Military dynamite was also used to clear boulders and stumps for road building and other construction. I could not find any accounts of US military dynamite being used as a weapon. Commercial dynamite is much too sensitive to shock and heat for standard use on the battlefield.

Procurement of military dynamite was discontinued in 1987, and supplies were allowed to run out. In 1990 the US Department of the Army stated, "Now, construction and quarry blasting requires procurement of local commercial explosives."[13] Commercial

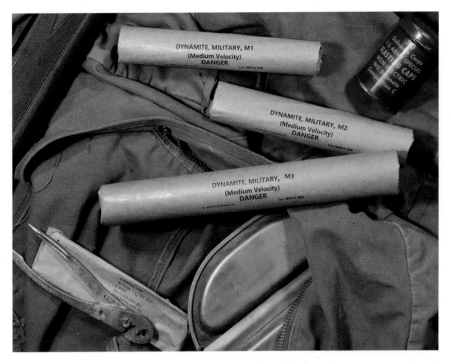

US military dynamite was produced in three sizes. M1 cartridges measured 1¼" × 8", M2 cartridges measured 1½" × 8", and M3 cartridges measured 1½" × 12". *Author*

dynamite was already in wide use by the military for peacetime applications. (It appears the Army may have had a change of heart, since the use of military dynamite is still addressed in the latest field manuals.)

Because of the lower sensitivity of military dynamite and other military explosives such as TNT and C-4, stronger blasting caps were used to guarantee detonation. Military blasting caps are much longer than commercial caps because they contain more explosives.

The US Army uses many other explosives, including the vaunted plastic explosive C-4, which, like military dynamite, is mostly RDX. Blocks of nearly pure TNT are the most widely known form of military explosive, and explosive power of any kind, even nuclear, is still rated relative to the explosive power of a given number of tons of TNT. Military dynamite was produced both by Hercules and DuPont.

Cartridged TNT was rarely used for mining. Nonetheless, in popular culture "TNT" and "dynamite" are practically synonymous. The reasons for this are likely rooted in World War I, when TNT was first used as a military explosive. In 2018 an article by the Bradbury Science Museum reported that entirely new explosive formulations were being tested by the US Army, with the goal of replacing TNT.

Although US military dynamite as such did not exist in World War II, for some reason replica military dynamite is often dated and presented as if it did.

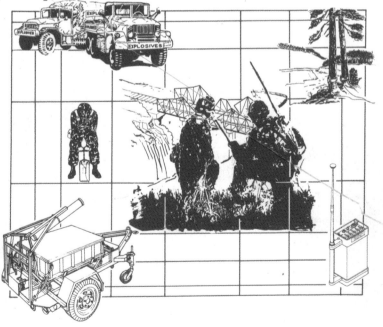

FM 5-25
explosives and demolitions
MARCH 1986
HEADQUARTERS, DEPARTMENT OF THE ARMY

DISTRIBUTION RESTRICTION. This publication contains technical or operational information that is for official government use only. Distribution is limited to US government agencies. Requests from outside the US government for release of this publication under the Freedom of Information Act or the Foreign Military Sales Program must be made to Commander, TRADOC, Fort Monroe, VA 23651-5000.

This *Army Field Manual* was first issued in 1940 as *FM-25* and has been updated ever since. Note the 1980s use of lowercase letters. The latest issue is designated FM 3-34.214. US Department of the Army, *FM 5-25 Explosives and Demolitions* (Washington, DC: USGPO, 1986).

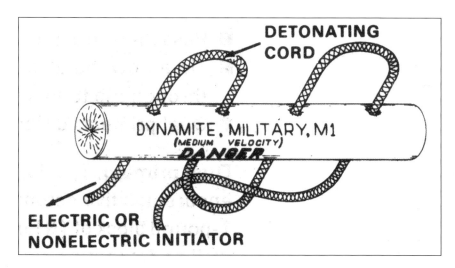

Left: The need to lace detonating cord through a cartridge of military dynamite speaks to the explosive's relative insensitivity. With commercial dynamite, a single hole is punched into the end or side of the cartridge. US Department of the Army, *FM 5-25 Explosives and Demolitions* (Washington, DC: USGPO, 1986), 2-29.

Other Cartridged Explosives

The current convention in the industry is to refer to explosives in stick form as "cartridge (rather than cartridged) explosives" or "packaged explosives." As with dynamite, modern explosives are categorized as to water resistance, fume production, speed, and strength.

It took until the 1970s to perfect paper-cartridged emulsion explosives. Austin Powder Company's Emulex brand debuted in 1986. Emulex was one of the first cartridged emulsions to have the same plasticity and loading ease of gelatin dynamites.

Many modern cartridged explosives are packaged in "chubs," which are

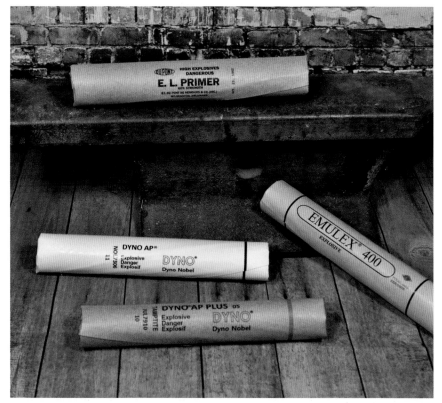

Left: A few cartridged emulsion explosives look or looked just like dynamite. This endurance is a testament to the simplicity of the concept of shaping a charge to fill a hole. *Author*

Below: Hard-hat stickers promoting cartridged explosives. Austin Powder's Emuline is a detonator-sensitive, cartridged explosive formulated for presplitting and other delicate work. *Author*

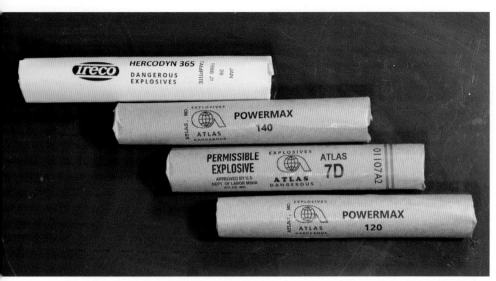

Atlas Powder introduced its Powermax cap-sensitive, cartridged emulsion explosives in 1977. Upon its introduction, Atlas emphasized the explosive's similarity in handling to dynamite and even touted that it was cartridged in plain brown wrappers. *Author*

plastic wrappers twisted at the ends like sausage casings. According to the US Department of Transportation, "Packaged in a thin, tough, plastic film, emulsion cartridges have a good degree of rigidity and resistance to rupturing during normal handling but maintain the ability to rupture and spread when tamped."[14] Emulsions and water gels can come in cartridges and tubes up to 20' in length.

One drawback of emulsion explosives is that they sometimes fail to completely detonate in very wet ground. Another is that the shock from one blast can affect the explosives in nearby boreholes, causing incomplete detonation to occur in certain soil and rock conditions.

Emulsion cartridges over 3" in diameter are a relatively recent development with limited applications. Since most large-diameter holes are vertical, it is much easier to park a slurry truck next to an array of boreholes and pump explosives into the holes.

Digging Deeper

Trojan Powder

In 1909, specifications for government contracts for explosives for the building of the Panama Canal sought bids for dynamite and Trojan Powder. DuPont complained that since only one company manufactured "Trojan Powder," soliciting bids for that product by name obviously discriminated against all other companies. In all likelihood the bureaucrats tasked with writing the bid requests made the mistake that many before them had, which is assuming that trade names such as Giant Powder, Vulcan Powder, or Trojan Powder were synonyms for dynamite. At that time, DuPont happened to be the only other company to offer nitrostarch explosives, specifically its Arctic brand. So if the government had nitrostarch explosives in mind, it should have said so. Regardless, both companies supplied explosives for the project, and DuPont did just fine, providing between 1905 and 1912 "33,000,000 pounds of dynamite, 3,570,000 electric blasting caps, 2,800,000 regular blasting caps, 800 miles of fuse, 20,000 pounds of leading and connecting wire, and 230 blasting machines."[15] All DuPont dynamite sent to Panama was produced at the Repauno facility.

In 1911 the world was notified that "the Penna. Trojan Powder Co. put a punch time clock in their office, which is something new in this vicinity. The employees did not like it at first, but now they are satisfied."[16]

Sodatol and Pyrotol

In 1924 the *Wall Street Journal* reported that "the du Pont Co. has recently received what is believed to be the largest contract for commercial explosives ever awarded in the history of the explosives industry. It calls for the preparation, cartridging, packing and shipping of approximately 100,000,000 pounds of Pyrotol, the ingredients of which will be ground smokeless powder, sodium nitrate and a nitroglycerin sensitizer."[17] The same year a United States Department of Agriculture press release claimed that the 100,000,000 pounds of Pyrotol intended for distribution would be "enough to load a freight train forty miles long."[18] In fact, only about 60,000,000 pounds were eventually sold, reducing the length of the train to a mere 24 miles.

Pyrotol was available in amounts of as little as 100 pounds, and a limit of 1,000 pounds was placed on any individual farmer. Railcar loads were delivered to distribution points and then broken up into smaller lots. The explosive could not be sold by dealers and was prohibited for use in quarries, except that "a farmer can use it to get out limestone for use on his own farm."[19] All this was to "prevent speculation and give everybody a chance,"[20] per a 1926 bulletin from the Virginia Polytechnic Institute.

The claim that Pyrotol was "one and one-half times as effective" as commercial dynamite, owing to its lower density, depends on how you define "effective." If you could achieve your objective with half a stick of regular dynamite versus a full stick of Pyrotol, then Pyrotol was not more effective, pound for pound, which is a more honest assessment of strength. Farmers seemed to prefer more sticks over stronger sticks.

A healthcare products firm called Pyrotol Chemical Company was headquartered in Beaumont, Texas, and advertised heavily in the early twentieth century, but the purveyor of dandruff remedies and toothpastes ceased advertising in 1924. This makes me speculate that the government bought the name to avoid confusion with the explosive. Adding fuel to this theory is the fact that the Texas company had a cartoon mascot named "Pyrotol Paul," while the government deployed "Pyrotol Pete," also known as University of Wisconsin College of Agriculture's Walter Rowlands. Rowlands conducted almost a thousand demonstrations of the explosive for farmers.

In 1928, DuPont wrote, "The Federal Government has stopped supplying Pyrotol. But that doesn't mean that there isn't any more low-priced, efficient farm explosive. The DuPont Company which cartridged Pyrotol has made a new land-clearing explosive similar to Pyrotol."[21] This was Agritol, sold through the 1950s.

In 1923 the Washington State Grange, a large community service organization that still operates, launched Grange Powder Company of Seattle. The company produced two products: Grange Stumping Powder and Grange Special Stumping, rated at 20 percent and 25 percent strength, respectively. The explosives had similar formulations and performance relative to Pyrotol. Competition from Pyrotol through 1928 led the company to cease production. The company then operated on a very small scale until at least the 1940s and reported employment of five workers in 1939. Grange Powder was used primarily in northwestern Washington State.

Pellet Powder

DuPont experimented with various pellet powder formulations in 1924 and 1925, conducting tests and providing samples to miners (DuPont made pellet powder for guns beginning in the 1860s). According to the 1942 *Blasters' Handbook*, "The convenience of using pellet powder is obvious—the necessity of making up, filling, and waterproofing cartridges is eliminated and the charges may be more accurately gauged."[22]

The fact that pellet powder was never deemed permissible by the Bureau of Mines was always an issue of contention. One writer in 1952 noted that Alabama law allowed pellet powder in coal mines, provided that the mine was nongassy and very wet, which reduced dust. In his view, permissible dynamites presented their own hazard: "Permissible powders require a blasting cap. Sometimes they fail to go off and are loaded out in the fine coal and explode in grates and stokers. Recently a man in Birmingham lost his sight this way. There are a large number of people who have lost their fingers and sight on account of this."[23] While the last claim is an overstatement, such incidents did occur. In 1924 a man found an entire tin of blasting caps in a shovelful of coal, but the caps did not make it into the stove. In 1935 a woman in Baltimore was killed when a cap fragment "no larger than a pea"[24] was launched from her coal-fired furnace. And in 1942 an Ashland, Pennsylvania, woman was severely burned when a single cap exploded in her coal-burning stove.

Military Dynamite

In 1943 an article related that "commercial explosives should not be confused with military explosives. Dynamite cannot be shot from a gun or used as a bursting charge in shells. On the other hand, smokeless powder, and trinitrotoluene (TNT), widely used in war, are not suitable for commercial work."[25] According to the *Encyclopedia of Explosives and Related Items*, "After WWII, when large quantities of scrap, cast TNT were available as surplus material, some blasting cartridges for use in strip mining were made by filling cylindrical cardboard containers (3" to 4" in diameter and about 3' long) with lumps of TNT"[26] ("cast" in this context means formed into blocks). The cartridges were detonated with a dynamite primer acting as a large blasting cap. And while *Army Technical Manuals* state

that "military dynamite M1, M2, and M3 has been standardized for use in military construction, quarrying, and service demolition work,"[27] the same manuals recommend commercial dynamite for stateside peacetime blasting. The manuals also note that "military dynamite is safer to transport, store, and handle than 60 percent straight nitroglycerin commercial dynamite and is relatively insensitive to friction, drop impact, and rifle bullet impact."[28] Modern TNT and RDX compositions are also nearly impossible to freeze and are packaged to be extremely water resistant. Friction and drop impact safety comparisons to straight dynamite are a little unfair, since gelatin mixtures would serve and are quite rugged. Bullet resistance is another matter, and any nitroglycerin-content explosive will almost certainly detonate if shot.

Unlike in the United States, the British did employ nitroglycerin dynamite in combat, mainly for antitank tactics. Again, the main obvious disadvantage of transporting dynamite on to an active battlefield is its "rifle-bullet sensitivity." Military explosives such as RDX, C-4, and TNT are much less likely to explode if impacted by ordinary-caliber bullets.

In 1897 a New Jersey inventor named W. A. Eddy, who used giant kites to carry cameras for early aerial photography, suggested that the kites could also be used to deliver dynamite bombs into enemy territory. The kite-borne cameras were actuated via very long cables and worked. The dynamite scheme relied upon the natural descent of a released kite, carried by the wind, into enemy territory. There would be no reliable way to detonate dynamite launched this way, a fact not mentioned by the inventor.

Other Cartridged Explosives

A typical emulsion formula is 60 to 80 percent ammonium nitrate, 10 to 20 percent sodium nitrate, up to 10 percent aluminum powder, and up to 3 percent mineral oil. A typical ANFO formula is about 95 percent ammonium nitrate and 5 percent fuel oil. ANFO prices in 1971 averaged five cents a pound versus twenty-five cents a pound for dynamite. Dynamite composed only 30 percent of the explosives market in 1970 and only 5 percent in 1980. Dynamite's market share now is well below 1 percent.

An article in 1974 portending DuPont's phasing out of dynamite quoted the company as stating that its improved Tovex "has made dynamite obsolete."[29] According to the story, "Water gels were first developed in the late 1950's but have been difficult to detonate reliably in drill holes less than five inches in diameter. This difficulty has been overcome with Tovex through the use of its patented amine nitrate sensitizer to provide the needed reliability in small[-]diameter applications. The safety advantages of Tovex are demonstrated by its resistance to detonation by bullets, fire, and impact."[30] The same article noted that, at that time, "of the 2.55-billion-pound explosives market in the United States, cartridged explosives such as dynamite and Tovex represent some 325 million pounds. The remainder includes fertilizer[-]grade ammonium nitrate blasting agents (about two billion pounds) and pumped or bagged water gels and slurries (225 million pounds)."[31] The difficulty in detonating smaller charges of most water gels lies in their insensitivity; enough must be used such that a booster, at a minimum consisting of a pound of TNT or stick of dynamite, can be inserted into the mix. Tovex was sensitive enough for ordinary blasting caps and could be used in small-diameter boreholes. The Tovex brand, inherited by Explosives Technologies International, was licensed to other companies. Tovex explosives are still produced by Biafo Industries of Pakistan.

CHAPTER 6

More Dynamite Companies

When digesting these shorter histories, please keep in mind that some of the geography has changed since the 1890s. A few county names and boundaries are different now, and some small towns no longer exist.

I certainly must have overlooked a couple of small producers of dynamite, but remember that companies such as Laflin & Rand, King Powder Company, Fort Pitt Powder Company, and many smaller entities did not manufacture their own dynamite, despite their claims to the contrary in ads. For example, the Shenandoah Dynamite Company was merely a reseller of dynamite made by Keystone Powder Manufacturing Company. Even the US Bureau of Mines was sometimes confused. In the 1913 roster of permissibles, Fort Pitt's Mine Powder No. 1 is listed as "manufactured by the Fort Pitt Powder Co." In fact, according to Van Gelder and Schlatter, Cameron Powder Company made the dynamite. Fort Pitt Powder Company produced only blasting powder and blasting caps.

There were many other "dynamite companies" that were distributors of the products of others, as well as "powder companies" that sold black powder or dynamite (or both) but made neither. Firms with tantalizing names in unusual parts of the country, such as Tennessee Dynamite Company, Southern Dynamite Company, and Dixie Dynamite Company, all were resellers.

Acme Powder Company

Acme Powder Company "was one of a number of small dynamite concerns established in Pennsylvania during the late eighties and early nineties, due to the demand of the local coal mines,"[1] according to Van Gelder and Schlatter. Eugene B. McAbee, one of George's explosive brothers, founded the firm with L. D. Stickney, building a plant at Black's Run, 14 miles north of Pittsburgh, in Allegheny County, Pennsylvania, in 1887. The Eastern Dynamite Company acquired the business in 1897 and dissolved it in 1904.

THE ACME POWDER CO.
MANUFACTURERS OF
DYNAMITE
AND
Blasting Powder
806 DUQUESNE WAY, PITTSBURGH, PA.

Mining Directory (Chicago: Poole Brothers, 1892), 548.

Ajax Dynamite Works

The Ajax Dynamite Works was founded in 1878 by Civil War veteran and oilman Henry H. Thomas near Kawkawlin, in Bay County, Michigan. The plant was destroyed five

AJAX DYNAMITE WORKS.

A manufacturing concern of Bay City, Mich., the Ajax Dynamite Company, which, independent of the trust, manufactures all the standard and competing grades of dynamite suitable for all kinds of blasting, also caps, safety fuses, electric fuses, batteries, leading wire, stump augers, combination cap crimpers and fuse cutters, and all blasting supplies. It also manufactures nitro-glycerine for use in increasing flow in oil, gas and water wells.

Michigan Federation of Labor Official Yearbook, 1907, 163.

times but operated until about 1913. The company reported sales of 425,000 pounds of dynamite in 1908, with annual production topping out at around 500,000 pounds. The company was large enough to run advertisements in regional magazines and even offered a pamphlet on stumping. Given the economies of scale involved, it is likely that the various blasting supplies listed in advertisements were Ajax-branded products made by other manufacturers.

American Cyanamid Corporation

American Cyanamid Corporation was founded by engineer Frank Washburn in 1907 to manufacture cyanamide, which is a mixture of lime, nitrogen, and carbide that was initially used to make fertilizers and then a host of other products. Cyanamid was the company's trade name for its formulation, first produced at a plant near Niagara Falls in Ontario, using the waterway for power. The firm diversified into pharmaceuticals and industrial chemicals, renamed itself American Cyanamid and Chemical Corporation, and bought the American Powder Company in 1929; General Explosives Company of Latrobe, Pennsylvania, in 1933; Burton Explosives Company in 1934; and Illinois Powder Company in 1957 (Joseph Burton had founded a new company and built a dynamite factory in New Castle, Pennsylvania, in 1930). American Cyanamid Company sold dynamite as one of its five thousand products from 1933 until 1969. Headquarters were moved from Wayne, New Jersey, to Rockefeller Plaza in New York in 1934. The corporation experienced too many ownership changes to outline here.

American Forcite Manufacturing Company

American Forcite Manufacturing Company was launched in 1883 by Manuel Eissler, with the Forcite name licensed from the Belgian patent owner. Eissler was a mining engineer, chemist, and prolific author who is said to have "designed, erected, and managed"[2] the plant. On April 2, 1884, Forcite became the first gelatin dynamite produced in the United States. Arthur Pine Van Gelder served for a time as supervisor of the plant near Landy, New Jersey. The company operated two main offices: Forcite Powder of New Jersey and Forcite Powder of New York. The company was acquired by DuPont and dissolved in 1905, but DuPont carried the label until 1927. (Atlas Powder Company received the Forcite plant in New Jersey as part of the dissolution of DuPont but retooled the factory and never made Forcite-labeled dynamite.) Canadian Explosives Company, a firm with heavy DuPont ties, also began using

the brand in 1919. Canadian Explosives Company became Canadian Industries limited (C-I-L) in 1929, producing its Forcite until the 1960s.

> FORCITE.—A variety of gelatine dynamite or gelignite made in Belgium. It contains blasting gelatine 86 to 64 per cent., sodium or ammonium nitrate, wood meal, magnesia and sometimes bran.
>
> An American explosive of the same name is a dynamite containing wood tar—
>
> | | |
> |---|---|
> | Nitroglycerine | 49 |
> | Collodion cotton | 1 |
> | Sodium nitrate | 88 |
> | Sulphur | 1·5 |
> | Wood tar | 10 |
> | Wood pulp | 0·5 |

Arthur Marshall, *Dictionary of Explosives* (Philadelphia: P. Blakiston's Son, 1920), 41. Forcite stick, *author*.

American High Explosives Company

American High Explosives Company was founded in 1907 by Job Burton of the Burton Powder Company, which never made dynamite. Dynamite sold by Burton Powder Company was made by the American High Explosives Company. The American High Explosives Company plant was near New Castle, in Lawrence County, Pennsylvania, near the original Burton powder mills. Job Burton was president, and his nephew Joseph Burton became secretary and treasurer in 1913. The company was acquired by Grasselli in 1917, and the plant was kept operating until being heavily damaged by an explosion in 1926. When DuPont acquired Grasselli, the remainder of the plant was dismantled. Just prior to that, the plant had a capacity of 12,000,000 pounds of dynamite a year, up from 3,000,000 pounds in 1908. Burton Explosives Inc. was founded by Joseph Burton in 1930 and built a dynamite factory on the old Burton Powder Company site.

Anthony Powder Company

Anthony Powder Company "seems to have done a good business among the mines of the northern peninsula of Michigan,"[3] per Van Gelder and Schlatter. The firm was launched on January 31, 1890, near Ishpeming, Michigan, by Edward C. Anthony, a timber salesman and mayor of Negaunee, Michigan, who had also founded a powder mill in 1879. The new dynamite plant was run by David McVichie, a quarry owner who had served as Ishpeming's postmaster. The Eastern Dynamite Company purchased almost half of the Anthony Powder Company's stock in 1901 but could not quite take over the company. The plant produced 943,000 pounds of dynamite in 1905. In September 1906 the plant was lost to an explosion, and the company was dissolved by stockholders on July 30, 1907.

Blue Ridge Powder Company

Blue Ridge Powder Company was founded in 1891 and headquartered in Allentown, Pennsylvania. A plant was built in Bowman, in Carbon County, in 1892. The firm's president was quarry owner Jonathan H. Pasco, and the other founders were W. H. Deshler, James Webb, A. P. Berlin, Sylvester Hower, and M. O. Bryan, all local businessmen. In 1898, sales for the company were 171,680 pounds. The company manufactured black powder as well and was purchased by the Eastern Dynamite Company in 1899 and dissolved in April 1904.

Burton Explosives Inc.

Burton Explosives Inc. was launched in 1930 by Joseph Burton, who erected a factory near New Castle, Pennsylvania, on 400 acres where the American Powder Mills had once stood. The plant had a capacity of 18,000,000 pounds

Author

of dynamite a year. The company was backed and staffed by former associates of Burton through his various business concerns, which most recently had included president of Grasselli Powder Company. The firm was purchased by the American Cyanamid Company in 1934 and became the Burton Explosives Division of American Cyanamid and Chemical Corporation. American Cyanamid produced Burton-branded explosives until the early 1940s, but dynamite cases continued to be marked "Cleveland, Ohio," where Burton Explosives offices were headquartered.

California Powder Works

California Powder Works was incorporated in 1861 by John H. Baird, a steamboat captain and one-time California state senator. A black-powder mill was erected 4 miles from Santa Cruz but did not begin production until May 1864. In 1869 the company transitioned to dynamite, feeling competition from the newly established Giant Powder Company. A nitroglycerin plant was erected on Baird's own land, and the company began selling Hercules Powder and Black Hercules, which was nitroglycerin and black powder. The location became known as the Hercules Powder Works and then the Hercules Powder Company even before the establishment of the first actual Hercules Powder Company as a branch location in Cleveland, Ohio, in 1877. The Santa Cruz location was closed and a new California plant was built on 22 acres near Pinole in 1880. This plant gradually expanded to cover 3,000 acres and leased part of the land for grazing cattle. The formulation of Hercules Powder dynamite was refined over the years and in 1882 consisted of nitroglycerin absorbed in wood pulp. The price of Hercules Powder then was twenty-five cents a pound for its "#1" dynamite, but the price would be cut in half ten years later under pressure from the innumerable small producers that crowded the market. Many would use low prices or exaggerated claims to gain customers for a short time and then go out of business. The DuPonts gradually bought nearly half the stock in the California Powder Works and finally acquired the rest in 1903. The DuPonts acquired the Cleveland location in 1881 and renamed it the Hercules Powder Company of Cleveland, which was incorporated by Lammot DuPont in 1882. In 1898 the California Hercules plant was deemed the largest in the world, producing 15,000,000 pounds of dynamite a year. The Cleveland plant closed in 1894, and the Pinole location stopped producing explosives in 1964. California Powder Works the company was formally dissolved in 1907. William Russell Quinan was superintendent of the Hercules plant in Pinole from 1883 to 1899. In 1884 his hand-packing machine went online at the plant. He then established the American Powder Packing Company to lease the machines to other producers.

Cameron Powder Company

Cameron Powder Company was founded by a group of Cameron County, Pennsylvania, businessmen, including George P. Jones, John Rice, and George W. Huntley Sr. The company was incorporated in 1909 and built a factory in Wayside, just southeast of Emporium, with headquarters in Sinnemahoning. The plant produced ammonium nitrate dynamites from June 1910 until 1917, when it became the property of Grasselli Powder Company. Grasselli continued to operate the plant, with an annual capacity of 12,000,000 pounds of dynamite, and carried the Cameron label until Grasselli was acquired by DuPont in 1927.

Climax Powder Manufacturing Company

Climax Powder Manufacturing Company was incorporated in 1889 by Fred Julian, E. G. Coleman, J. D. Billard, and the firm of Arthur Kirk & Son. Julian had worked for Giant Powder Company and managed the Eldred Powder Company. Coleman was Julian's bookkeeper, and Billard was a prolific New York and Pennsylvania businessman and investor. A plant was built near Emporium, Pennsylvania, and reached a peak yearly production of 10,000,000 pounds of dynamite. Like many others, Julian "was impressed" with Emporium "as a suitable central location for a dynamite plant,"[4] per Van Gelder and Schlatter. Dynamite was assembled entirely by hand for the whole of the company's

existence; this was true of most of the small producers listed in this chapter. Climax was purchased by Eastern Dynamite Company in 1903 and dissolved in 1905. DuPont operated the plant until 1912, with Arthur Pine Van Gelder as superintendent from 1906 to 1908.

Clinton Dynamite Company

Clinton Dynamite Company was founded in 1884 by William P. Foss and Smith M. Weed. Foss operated a two-man nitroglycerin factory in 1881 and sold nitroglycerin in New York. Weed was a wealthy Pittsburgh businessman. Another factory that Foss had established in Plattsburg, New York, operated for two years and then was moved to near Haverstraw, New York. The Eastern Dynamite Company purchased the company in 1895 and used the plant until 1897 before dismantling it, "and the trade of that district was thereafter supplied from the plants of the Eastern Company, where more modern and economical methods of manufacture and large-scale production made possible sufficiently low costs to more than compensate for the difference in freight,"[5] according to Van Gelder and Schlatter. Like many of the acquisitions by Eastern Dynamite, Clinton Dynamite Company technically existed as a "shell" until formal dissolution, in this case by DuPont in 1904.

Columbia Powder Company of New Jersey

Columbia Powder Company was founded in Farmingdale, New Jersey, in 1891 as a subsidiary of the Phoenix Powder Company. Fordyce Laflin Kellogg, who had worked for the Laflin & Rand Powder Company, launched the Phoenix Powder Company in 1888 and then attempted to branch out from black powder to dynamite, but the firm failed in 1894. The equipment was bought by the Dittmar Powder Works.

Columbia Powder Company of Washington

Another Columbia Powder Company was founded in Tacoma, Washington, in 1935 by Franklin W. Olin, who built a plant built near Frederickson, 10 miles southeast of Tacoma, in Pierce County. Olin had established the Equitable Powder Manufacturing Company in 1892 and the Western Cartridge Company in 1899, both in East Alton, Illinois. In 1908, Olin purchased the Buckeye Powder Company near Peoria, Illinois, and renamed it the Western Powder Company. In 1910, Olin bought the Egyptian Powder Company in Herrin, Illinois, which had opened in 1904. In 1931, Olin acquired the Winchester Repeating Arms Company and the one-year-old Liberty

Missouri State Gazetteer, 1893–94, 472.

Explosives Company of Mount Braddock, Pennsylvania, changing the name to Liberty Powder Company, a subsidiary of Olin Corporation. In 1944, Columbia Powder became a subsidiary of Olin Industries, which grew into a multinational behemoth. Son John took over as president of Olin Industries that year, serving until 1953. In 1953, Olin Industries acquired the United States Powder Company, founded in 1904 in Terre Haute, Indiana. Olin Industries became Olin Mathieson in 1954, after merging with Virginia company Mathieson Chemical Corporation. In 1961, Olin Mathieson ceased producing dynamite, according to the Institute of Makers of Explosives. In 1963, Commercial Solvents Company of New York purchased the remaining explosives business of Olin and renamed it U.S. Powder Company, a division of CSC (CSC also bought Trojan Powder Company in 1967).

Both Equitable Powder Company and Egyptian Powder Company sold Alton Dynamite, which was produced by Illinois Powder Company until 1931, when Austin Powder Company took over.

Author

Columbian Powder Company

Columbian Powder Company was started in 1892 by L. D. Stickney, who had also helped organize the George R. McAbee Powder and Oil Company. Other founders included George E. Hunter and Charles Dickson. The firm was headquartered in Pittsburgh, with a plant in Shannopin, then 10 miles northwest of Pittsburgh. The facility was making only about 150,000 pounds of dynamite a year when it was bought in 1898 by Eastern Dynamite Company, which kept the plant running until 1902.

Crescent Powder Company

Crescent Powder Company was founded in 1895 as Crescent Powder Company, Limited. The company did not begin production until 1899 and reincorporated as Crescent Powder Company in 1902. The plant near Franklin Forge, Pennsylvania, produced 7,000,000 pounds of dynamite a year but was leveled by an explosion in 1904. George McAbee then reorganized as the George R. McAbee Powder & Oil Company, with a factory in Pequea, Lancaster County, Pennsylvania, but that factory was destroyed by an explosion in 1907. Yet another factory was erected near Tunnelton, in Indiana County, and operated until being taken over by Atlas in 1922. Atlas sold the plant to local blasting firm Amos L. Dolby Company in 1992.

Dittmar Powder Works

Dittmar Powder Works was founded by chemist and Prussian army captain Carl Frederick Wilhelm Ernst Dittmar in 1869. Dittmar held a patent for smokeless powder and had worked for Nobel. He established his first powder mill in Neponset, Massachusetts, and began selling Dualin in May 1870. Dualin was a mixture of wood pulp, sodium nitrate, and from 25 to 67 percent nitroglycerin. It was originally sold in #1 and #2 formulations. A different formulation called Dualin had been patented in 1868 in Britain by Dittmar's erstwhile business partner J. F. E. Schultze, but Schultze lacked the capital to produce it. Dittmar obtained his own US patent in 1870.

The plant moved to Binghamton, New York, in 1877 and was reorganized as the Dittmar Powder Manufacturing Company in 1879. The factory was moved yet again to New Jersey, then back to New York, near Baychester. Dittmar died in 1883, and his wife, Maria, took over. In 1895 the firm moved again to Farmingdale, New Jersey.

The company phased out Dualin, replacing it with Dittmar's Powder dynamite, which lacked the sodium nitrate. Maria Dittmar sold her interests in the company to Eastern Dynamite in 1899, retaining the right to continue selling dynamite to her core clientele and to act as an independent distributor of explosives from other companies. Maria died in 1912, and the firm operated as a distributor until 1936, run by Dittmar's sons Carl Jr. and Arthur Carl and grandson Arthur Carl Jr.

Carl Dittmar Sr. was successfully sued by Giant Powder over Glukodine, which he falsely claimed was a nonnitroglycerin formula. DuPont trademarked "Dittmar" in 1907, renewing in 1947, but never made use of the mark.

We
desire to
call your attention to
a new explosive compound,

"GLUKODINE,"

the recent invention of
MR. CARL DITTMAR   Patent applied for.
Its DISTINCTIVE qualities are:

1st. Greater power than Nitro-Glycerine.
2d. Less sensitiveness to shock or accidental concussion. 3d. Greater Safety in Transit and Handling.

GLUKODINE!

We are now prepared to furnish it in Cartridges of any desired size and strength, and at prices which will compare favorably with those of any "HIGH EXPLOSIVE" on the market.
Yours truly,

The Dittmar Powder Manufacturing Co.

21 Park Place,
NEW YORK.

Forest and Stream, August 24, 1876, 45.

Eldred Powder Company

Eldred Powder Company began as Alford & Dean Company in 1883. Lumberman Byron Alford and investor J. W. Dean built a plant near Eldred, in McKean County, Pennsylvania.

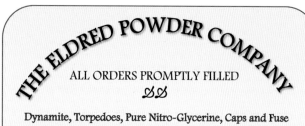

THE ELDRED POWDER COMPANY
ALL ORDERS PROMPTLY FILLED
☙☙
Dynamite, Torpedoes, Pure Nitro-Glycerine, Caps and Fuse

PRESENTED BY
W.Z. GEORGIA, Pres't

P.O. ADDRESS:
OLEAN, NEW YORK

Author

The firm was renamed the Eldred Powder Company when Alford bought Dean out in 1888. The company was run by Alford's wife, Mary, when Byron died in 1898. The firm's stock was acquired by Willis Z. Georgia of Olean, New York, who also founded the Buckhannon Chemical Company in Olean. The company sold 81,000 pounds of dynamite in 1907. Georgia became president of the company and sold it to DuPont in 1909. H. P. Hall began tinkering with shell-packing machines while employed at Eldred Powder, work he would continue as an employee of Aetna. He perfected the device while working at Repauno and patented it in 1898.

Empire Powder Company

Empire Powder Company was founded in 1916 by Emporium, Pennsylvania, engineer and investor George P. Jones, with a plant near LeRoy, in Genesee County, New York. Jones had worked for the Climax Powder Company and had cofounded Keystone Powder Manufacturing Company in 1900 and Cameron Powder Manufacturing Company in 1909. Empire Powder Company had a capacity of 3,500,000 pounds of dynamite a year. Its plant became part of the Union Explosives Company in 1920 and was upgraded to an annual capacity of 5,000,000 pounds.

6 Men Blown to Atoms

-

Emporium Powder Co's Mixing House Again Goes Up.

-

At 2:10 o'clock, yesterday afternoon the mixing house of Emporium Dynamite plant, situated on West Creek, about two miles west of Emporium, went up in smoke killing six men and injuring many others.

Cameron County Press, March 21, 1907, 1.

Emporium Powder Manufacturing Company

Emporium Powder Manufacturing Company "was another small company which was organized with local capital in the Emporium District in 1904,"[6] according to Van Gelder and Schlatter. Nevertheless, the plant averaged 5,000,000 pounds of dynamite a year. The plant, about 2 miles west of Emporium, became part of Keystone National Powder Company in 1910, Aetna Powder Company in 1914, and Hercules Powder Company in 1921. The plant was closed in 1931. The company's Stratite dynamite made the first list of permissible explosives in 1909. The 1907 article exemplifies the era's matter-of-fact reportage on the frequent early accidents at dynamite plants and later casually relates that normal operations were to be resumed in two weeks. Newspapers engaged in a macabre contest to compose the most-lurid descriptions of the most complete disintegration imaginable. "Reduction to atoms" trumped mere vaporization. Containers deemed suitable in size for the storage of remains ranged from cigar boxes to thimbles. Sometimes, dynamite boxes were actually used as temporary makeshift coffins, much to the delight of irony-loving headline composers. Occasionally, reporters were given an unexpected bonus, such as when an 1881 Aetna accident yielded a plant worker's intact upper lip complete with mustache.

Enterprise High Explosives Company

Enterprise High Explosives Company was founded in 1895 by Elmer Brode, a butcher who had become involved with explosives through his investment in a short-lived National Dynamite Company that existed near his hometown of Tamaqua, in Schuylkill County, Pennsylvania, from 1890 to 1894. Brode bought the plant and moved it to near Nesquehoning, in adjacent Carbon County. The plant made 3,500,000 pounds of dynamite a year until Brode sold it to Eastern Dynamite Company in 1897 and began working for DuPont. The plant was kept in operation until 1907, and Brode remained with DuPont until 1922.

An Enterprise Powder Manufacturing Company operated in Luzerne County, Pennsylvania, from 1892 to 1914 but made only black powder.

Gogebic Powder Company

Gogebic Powder Company was founded in 1889 by miner and merchant Joseph Sellwood, with William Ira Prince serving as director and C. S. Bundy acting as secretary and treasurer. Prince was a banker and prolific executive who served as director of the Brotherton Iron Mining Company, in which Sellwood was a partner. Prince also was

mayor of Duluth, Minnesota, from 1913 to 1917. Bundy was a businessman, rancher, and former superintendent of Bessemer, Michigan. A plant was built near Bessemer, in then newly formed Gogebic County. The plant served the miners of the Gogebic Iron Range and was moderately successful largely due to its location, producing more than 1,000,000 pounds of dynamite a year. Eastern Dynamite Company acquired the company and dissolved it in 1902.

Gogebic Iron Spirit, July 23, 1891, 4.

Greenwood Powder Company

Greenwood Powder Company was "the first dynamite company to operate in the Joplin district"[7] of Missouri, per Van Gelder and Schlatter. The company was founded in 1892 with a small plant 3 miles northwest of Webb City. The firm was owned by Clarence Gaston and Walter Koontz, who ran a coal-mining company called Gaston & Koontz in Webb City. Former Aetna Powder Company manager George Perry ran the new company, employing three men and producing only about 750 pounds of dynamite a day. Greenwood Powder Company was purchased by Eastern Dynamite Company in 1895 and dismantled.

Hancock Chemical Company

Hancock Chemical Company was established in 1884 by Robert Warren, who built a plant just across the portage from Houghton, Michigan. Warren was a "prominent pioneer in the dynamite industry,"[8] according to Van Gelder and Schlatter, who noted that Vulcan Powder was also called "Warren's Powder." Warren had founded the Vulcan Powder Company and established several plants to make Vulcan Powder in Pennsylvania and New Jersey. He was president of Hancock Chemical for only a short time, handing over the reins to Henry deShon, who was already president of Lodi Chemical Company. The Eastern Dynamite Company became exclusive distributor for Hancock Chemical Company dynamite in 1898. Hancock Chemical Company was acquired by the Calumet and Hecla Mining Company in 1909, and the plant closed in 1911. Hancock Chemical Company reported sales of 1,800,000 pounds of dynamite in 1907.

Copper Handbook Advertiser, 1902, 55.

Hecla Powder Company

Hecla Powder Company was founded in 1880 by Charles A. Morse, his brother George Franklin Morse, and their cousin Jerome Edward Morse. George became president,

Engineering and Mining Journal, August 4, 1883, xiii.

Jerome was treasurer, and Charles was superintendent. Yet another brother, Albert G. Morse, became secretary. Charles Morse was a Civil War veteran who became a merchant and then got involved in mining, serving as president of the South Springs Gold Mining Company in California. Hecla Powder Company's plant was located north of Rahway, New Jersey, with headquarters in New York City. Hecla sold granular dynamite labeled as Miner's Friend before being sued by Giant Powder Company for trademark infringement in 1882. Hecla's Miner's Friend dynamite formulation was similar to Judson Powder. Although all Hecla Powder dynamite came cartridged, the company recommended that cartridges be broken open and poured into boreholes, and even offered 3" diameter cartridges—impractical for most borehole widths of the era—designed for that very purpose. Repauno Chemical Company acquired the company in 1885 and William DuPont became president. The plant was closed in 1886. The DuPonts operated Hecla Powder Company as a distributor, changing the name to Hecla Dynamite Company in 1896. The company was finally dissolved in 1905.

Hudson River Powder Company

Hudson River Powder Company "operated in the late nineties, largely to supply the local quarries near Rondout, New York, where its plant was located,"[9] per Van Gelder and Schlatter. Founder Jacob Rice served as president. Rice had been a New York state senator and had many business ventures, including furniture manufacturing, boatbuilding, and shipping. He was a quarry owner at the time of the launching of Hudson River Powder Company. The company was sold to Eastern Dynamite Company in 1901 and dissolved in 1904.

Independent Powder Company of Missouri

Independent Powder Company of Missouri was launched in 1902 by Norman P. Rood and Alfred G. Cummings, who built a plant near Carthage, Missouri, with headquarters in Joplin, about 10 miles southwest. Cummings was president, Rood was secretary, and Charles W. McAbee was vice president. Rood was also superintendent of his father George Rood's Independent Powder Company of Indiana, a black-powder company in Coal Bluff, Indiana. Cummings was manager of the Chicago and Indiana Coal Company and president of the United States Powder Company. McAbee was George R. McAbee's brother. The company had peak production of 15,000,000 pounds of dynamite a year and billed itself as the only independent (non-DuPont-controlled) entity west of the Mississippi River at the time (Giant Powder, Consolidated, would also qualify). The location was sold to Hercules in 1915, passed down to IRECO and Dyno Nobel, and operates to this day. Herbert Talley, who was Cummings's brother-in-law, worked for Independent Powder from its inception and perfected his famous mixing and packing machines there. Both Rood and Talley continued to work for Hercules. Independent Powder Company hosted an annual hand-drilling contest, and in 1914 a team of triple jackers won the event by drilling a 37½" deep hole in fifteen minutes. A report of the event does not specify what kind of rock was being drilled.

Clay Worker, November 1912, 574.

Independent Powder Company of New Jersey

Independent Powder Company of New Jersey was established by ex-Repauno clerk John E. Alexander and oilman and investor William H. Curtis. According to Van Gelder and Schlatter, Alexander had "secured financial backing" from Curtis "sometime between 1885 and 1888."[10] We do know for certain that the company was incorporated by Alexander and Curtis in 1890. The factory at Whiting, in Ocean County, produced about 1,000,000 pounds of dynamite a year, employing "five or six hand-packers."[11] The company's signature dynamite was labeled Dart Powder and was a 40 percent strength formula containing 28 percent nitroglycerin. The firm's charter was revoked in 1897 for tax issues, and the plant was acquired by Eastern Dynamite in 1901. The plant was dismantled, but the site was used by DuPont for explosives testing during World War I.

International Smokeless Powder and Dynamite Company

International Smokeless Powder and Dynamite Company was organized in 1899 by Lewis Nixon, a New Jersey businessman and owner of the Crescent Shipyards at Elizabethport, New Jersey. A plant was constructed near South Amboy, in Middlesex County, New Jersey, and operated until being taken over by DuPont in 1904. The firm was renamed the International Smokeless Powder and Chemical Company in 1903 and diversified into lacquers. DuPont continued to operate the plant, at that time producing black powder in addition to nitrocellulose and chemicals for lacquers, until 1915.

International Smokeless Powder and Dynamite Co.

850 DREXEL BLDG., PHILADELPHIA

MANUFACTURERS OF

SMOKELESS POWDERS

Accepted and Used by the United States Government

ALSO SPORTING POWDERS OF EVERY DESCRIPTION

Safety Powders, Dynamite for Mines, Torpedoes and Blasting. Manufacturers of all Varieties of Nitrated Cellulose for Use in Making Varnish, Artificial Silks, Celluloids, etc.

FACTORY: PARLIN, NEAR SOUTH AMBOY, NEW JERSEY

The Lucky Bag of the U.S. Navy (Annapolis, MD: US Naval Academy, 1901), ii.

Joplin Powder Company

Joplin Powder Company was founded in 1903 by James Lawrence, who built a plant 5 miles southwest of Joplin, Missouri. Lawrence was a Scottish immigrant who had worked at Nobel's Ardeer plant in Scotland for ten years. Lawrence had started the Standard Explosives Company near Toms River, New Jersey, before becoming assistant superintendent at Repauno Chemical. The plant produced 15,000 pounds of dynamite a day and, according to Van Gelder and Schlatter, "was locally known as the 'Jack-cracker' plant because its product was supposed to be a good cracker for 'jack,' the zinc blend ore of the locality."[12] ("Jack" is miner's slang for the mineral sphalerite, which is the primary ore of zinc.) The plant was purchased by DuPont in 1905 and closed in 1908.

Judson Dynamite & Powder Company

Judson Dynamite & Powder Company was founded in 1890 by seventy-eight-year-old Egbert Putnam Judson, a civil engineer who had helped organize both the Giant Powder Company and the Atlantic Giant Powder Company. Judson patented Giant Powder No. 2 in 1873 and Judson Powder in 1876. Judson had begun manufacturing both products at the Giant Powder facilities, operating informally as Judson Powder Company at the New Jersey location. Giant Powder No. 2 consisted of 40 percent nitroglycerin, 40 percent sodium nitrate, and absorbents. Judson Powder was nitroglycerin-coated black powder and became known as Railroad Powder (R.R.P.) because it was used extensively for shaping land for the construction of railways. Because of its heaving action, Canadians called

JUDSON POWDER – A low grade high-explosive, invented by Egbert Judson, of San Francisco, California, and patented in 1876, since which time it has been manufactured in large quantities at Berkeley, California, and has grown rapidly in favor with all who have used it, taking the place of black powder in heavy work. It is not a high explosive and cannot be used for such work as is intended for Giant, Atlas or Hercules powder, but wherever black powder is in use Judson Powder can be substituted to great advantage. As this powder contains nitro-glycerine it becomes hard in cold weather (at about 45° F). When in this state it readily breaks up into grains by a little pressure and can then be poured like sand into the smallest crevice.

When using frozen powder, it is necessary to use a priming cartridge of Giant, and to always have this cartridge soft. For blasting or quarry work, Judson Powder is put up in water-proof paper-bags, containing 6 1/4, 12 1/2, and 25 pounds each, and 8, 4 and 2 bags respectively are put into wooden boxes holding 50 pounds. It is also put up in special water-proof cartridges of any size desired for special purposes.

Farrow's Military Encyclopedia (New York: Military Press, 1895), 147.

this type of formulation "lifting powder," and the explosive was rated in strength between ordinary black powder and low-strength dynamite.

Judson Powder was sold in bags and cartridges in several grades and underwent minor changes in formulation over the course of its existence. The powder was sold by DuPont into the 1920s, although the Judson name was eventually dropped. Egbert Judson erected his own plant south of El Cerrito, California, in 1890 but died in 1893. Nephews Charles G. Judson and Henry Clay Judson then took over. DuPont formed a special company called the California Investment Company to gain ownership of Judson Dynamite and Powder in 1904. DuPont operated the plant until a 1905 explosion rendered it inoperable, but the Judson Dynamite and Powder Company existed as an empty shell until formal dissolution in 1912. Judson Powder is dynamite because of its nitroglycerin content, which ranged from 5 to 10 percent by volume.

Arthur Kirk & Son

Arthur Kirk & Son was founded in Pittsburgh in 1867 by Scottish immigrant Arthur Kirk and his son David. In

JUDSON POWDER.
A SUBSTITUTE FOR BLACK POWDER
Guaranteed STRONGER than the Best BLACK BLASTING POWDER, and far safer to Use.
Adapted for all kinds of Blasting where Black Powder has heretofore been used.
TO RAILROAD CONTRACTORS:
GENTLEMEN. We beg that you make no contracts for Black Powder before you have tested the Judson Powder. Remember it will break your rock so that you have no sledging to do. It will do excellent work in loose and seamy ground where Black Powder will only burn up. For Blowing out Stumps and Trees it is the best and cheapest powder on the market. We will prove what we say. ADDRESS:
JUDSON POWDER CO.
Works, DRAKESVILLE, N.J.   Office, RUSTIC, N.J.

Frederick Abel, *Mining Accidents and Their Prevention* (New York: Scientific Publishing, 1889), Ad 4.

USE KIRK'S FUMELESS GELATINE DYNAMITE
FOR ALL UNDERGROUND WORK
CAPITAL $125,000.00
ARTHUR KIRK & SON CO.
POWDER   SAMPSON POWDER   DYNAMITE   GELATIN DYNAMITE
ELECTRIC EXPLODERS, SQUIBS, FUSE, ELECTRIC BATTERIES, LEADING WIRE

Railroad Gazette, March 13, 1903, 139.

addition to dynamite and black powder, the firm sold a variety of mining equipment, including blasting machines, rock crushers, and drills. An accomplished engineer, Arthur Kirk wrote *The Quarryman and Contractor's Guide* in 1891. The company was acquired by Eastern Dynamite Company and dissolved in 1904. Arthur Kirk & Son also cofounded the Climax Powder Company. Job Burton's first explosives job was with Arthur Kirk & Son in 1885.

Lake Superior Powder Company

Lake Superior Powder Company was founded in 1868 to provide black powder to the mines of northern Michigan. The DuPonts acquired control of the company in 1876 but operated it under its own name until 1903. In 1881 the company built a dynamite factory near Marquette, Michigan, that produced 2,000,000 pounds a year. In 1903 the E. I. DuPont de Nemours Powder Company was organized, and the location became known as DuPont's Marquette plant. Lake Superior Powder Company was formally dissolved in April 1905. The plant was dismantled in 1910, and the site was assigned to Hercules Powder Company as part of the DuPont antitrust settlement.

Mount Wolf Dynamite Company

Mount Wolf Dynamite Company "was a small concern operating from about 1896 to 1899 in eastern Pennsylvania,"[13] per Van Gelder and Schlatter. The company made 100,000 pounds of dynamite in 1897 and was incorporated in 1898. Mount Wolf Dynamite Company was purchased by Eastern Dynamite Company in 1899 and closed. Like many companies acquired in this way, the firm existed as a shell until formal dissolution in 1904.

Neptune Powder Company

Neptune Powder Company was founded in 1877 by Andrew J. Parker, who built a plant in Pelham, New York. Neptune Powder contained 33 percent nitroglycerin, 45 percent sodium nitrate, 17 percent charcoal, and 5 percent sulfur. The firm manufactured its signature product for only two years before Giant Powder Company obtained an injunction for patent infringement in May 1879. Neptune Powder Company was dissolved in 1880. Parker then sold dynamite for Dittmar Powder Works. Like with Vulcan Powder, the Neptune Powder name lived on for a hundred years after the demise of the company. The name continued to appear in encyclopedias, dictionaries, state and local legislation, and numerous books on the use of explosives. Without context, one could get the impression that Neptune Powder

was made for many decades. "Giant powder" was the clear king of this kind of linguistic appropriation, known as a "proprietary eponym." "Giant powder" was used as a catchall term for dynamite in countless references, from newspaper articles, to books on blasting, to ads for other company's dynamite, as shown by the Gogebic Powder ad (see p. 203). In a double twist of irony, lists of types of "giant powder" often included Neptune Powder!

New York Powder Company of New Jersey

New York Powder Company of New Jersey was founded in 1881 by Carl Walter Volney. German immigrant Volney was a chemist and early nitroglycerin producer, having operated a business out of Boston in 1872 after working for Carl Dittmar. Volney built a dynamite plant near Toms River, New Jersey, that began production in 1882. The company failed after one year. Henry deShon, who would also become president of the Hancock Chemical Company, then reorganized the New York Powder Company as the United States Dynamite and Chemical Company.

DYNAMITE.

RED STAR BRAND.

FOR MINING, QUARRYING and RAILROAD WORK.

Safety Fuse, Caps, Platinum Fuse, Leading and Connecting Wire, with every other requisite for Blasting Purposes.

New York Powder Co., Works at Mauch Chunk, Pa. On Liberty Street, New York

Colliery Engineer and Metal Miner, July 1896, xxi.

New York Powder Company of Pennsylvania

New York Powder Company of Pennsylvania was founded in 1891 by S. M. Keiper, who built a factory near Nesquehoning, in Carbon County, to serve "the anthracite coal mines and the limestone quarries of eastern Pennsylvania,"[14] per Van Gelder and Schlatter. Keiper owned the De Hart Iron Mine in Morris County, New Jersey. The company produced Red Star dynamite and had a capacity of 2,500,000 pounds a year. Eastern Dynamite acquired the company in 1897 and operated it until 1908.

Nitro Powder Company

Nitro Powder Company began in 1890 as J. B. Pennell & Company and was reorganized as the O.K. Dynamite Company in 1895. The company was reorganized once more as the Nitro Powder Company in 1896, headquartered in Kingston, New York. The plant, near Esopus, in Ulster County, New York, manufactured gelatin dynamite and two permissible formulations: Nitro Low-Flame Number 1 and Number 2. These made the official permissible lists until the plant was destroyed by an explosion in 1916. The company built TNT and picric acid plants at Esopus in 1918 to supply explosives for World War I. The dynamite plant was rebuilt in 1921 and operated until 1925.

Engineering New Record, January 25, 1912, Selling Section, 27.

Oliver Powder Company

Oliver Powder Company was founded in Laurel Run, Luzerne County, Pennsylvania, by Civil War general Paul A. Oliver. Oliver built a dynamite factory next to his existing powder mill, launched in 1873. The location, on

OLIVER'S POWDER. Superior Strength. Freedom From Smoke.

METEOR DYNAMITE,

EXTRA STRENGTH FOR HEAVY ROCK WORK and ORE MINING.

OLIVER'S FLAMELESS DYNAMITE,

Will Not Ignite Gas or Coal Dust. Does Not Shatter Coal.

ALL GRADES OF MINING POWDERS A SPECIALTY.

THE OLIVER POWDER CO.   OLIVER'S MILLS, LUZ. CO., PA.

Journal of the Central Mining Institute of Western Pennsylvania, 1900, 238.

600 acres, became known as Oliver's Mills. The company started making early permissible dynamites—then called safety explosives—in 1895. According to Van Gelder and Schlatter, Oliver was "undoubtedly the first to make such explosives"[15] in the United States. The company also offered five grades of dynamite with 30 to 60 percent nitroglycerin content, in addition to 13 percent strength Black Jack and 10 percent strength Victor Coal Powder. DuPont acquired the company in 1903 and dissolved it in 1907. DuPont continued to manufacture Oliver's Meteor Dynamite until 1909, and the brand was included in the first list of permissible explosives. Oliver's Flameless Dynamite used sodium chloride to reduce (but not eliminate) flames produced during an explosion.

Peerless Explosives Company

Peerless Explosives Company was founded by former Grasselli Powder executives Edward R. Day, Raymond I. Bradford, and Eugene Frost. The men served as president, vice president, and treasurer, respectively. A plant was built at White Haven, in Luzerne County, Pennsylvania, in 1922 and began production the following year. The facility had an annual capacity of 4,500,000 pounds of dynamite and produced "the usual grades of dynamite,"[16] per Van Gelder and Schlatter. Peerless merged with black-powder maker Black Diamond Powder Company and the Union Explosive Company and became Peerless-Union Explosives Company in 1930. The merger was organized behind the scenes by Atlas Powder Company, which then gained control of the new company in 1931. Atlas operated the plant until 1958.

PEERLESS EXPLOSIVES COMPANY

Incorporated

Offices

Wilkes-Barre, Pa.   Pittsburgh, Pa.   Huntington, W, Va.

Manufacturers of

PEERLESS EXPLOSIVES

Permissible Explosives, Dynamite, Blasting Powder and Blasting Supplies.

Mining Catalog, Coal Edition (Pittsburgh, PA: Keystone Consolidated Publishing, 1923), 1056.

Pennsylvania Powder Company

Pennsylvania Powder Company was launched in 1921 by George W. Huntley Jr., John Schwab, and Robert Schless. George Huntley Sr. had died earlier in the year, having just completed construction of a dynamite plant near Emporium under the auspices of his Modern High Explosives Company. Both Schwab and Schless had worked for the Cameron Powder Company, and Schless held a patent for an explosives compound. Almost immediately the Union Explosives Company acquired Pennsylvania Powder Company. Union Explosives had itself been recently created through the merger of the Long Powder and Supply Company and the Empire Powder Company. The plant eventually reached a capacity of 3,000,000 pounds of dynamite a year. The company originally sold dynamite "under the 'Union' trade-mark,"[17] per Van Gelder and Schlatter, then operated under its own name until a 1957 explosion destroyed the plant.

Pluto Powder Company

Pluto Powder Company was established in 1905 by Richard J. Waters, who worked for the Climax Powder Manufacturing Company. John Rice, president of the General Crushed Stone Company of Pennsylvania, became president. The company resold dynamite made by the Emporium Powder Manufacturing Company only until 1908, when a dynamite plant was erected 4 miles south of Ishpeming, Michigan. The plant began production in 1908, with 2,000,000 pounds of dynamite made the first year. The plant eventually reached a capacity of 10,000,000 pounds a year. Aetna Powder Company absorbed the company in 1914, and Hercules Powder Company inherited the plant when it acquired Aetna in 1921. The location was in turn acquired by IRECO in 1985 and became Pepin-Ireco in 1988. The facility operates to this day, although the plant no longer produces dynamite.

Potts Powder Company

Potts Powder Company was founded in 1906 by former Forcite plant superintendent George Eustis Potts, who served as vice president. Augustus J. Paine Jr. was president. Paine was a prolific businessman best known for his founding of the Castanea Paper Company in Lockhaven, Pennsylvania. The plant was built 7 miles south of Tamaqua, Pennsylvania, and began production in July 1907. The facility was modern for its time, with machine packing into machine-made shells. The plant had a capacity of 18,000,000 pounds of dynamite per year. Atlas purchased the location, known as the Reynolds Works, in 1913 and

kept it running until it was sold in 1997. In the interim the plant was expanded to make blasting caps and electric blasting caps. According to Van Gelder and Schlatter, "The permissible powder made (by Potts) under the name of 'Coalite' obtained a high reputation, and in consequence this trade name has been retained"[18] by Atlas.

Puget Sound & Alaska Powder Company

The Puget Sound & Alaska Powder Company was founded in 1906 by Peter David, who built a plant in Mukilteo, near Everett, Washington, mostly to serve local logging interests. David was a lawyer and businessman who also owned and operated a theater chain and had served as president of the Manufacturers Association of Washington. The company was originally named the Randanite Powder Company, which is odd, since "randanite" is French for kieselguhr. The new name was adopted in 1908, one year after the plant was completed. Output for the first year of production was 175,000 pounds. Atlas Powder acquired the plant in 1930 and closed it. The company's Vulcan Dynamite is unrelated to that sold by the Vulcan Powder and Vulcan Dynamite Companies. Puget Sound & Alaska Powder Company trademarked the "Kick in Every Stick" slogan in 1920.

Timberman, November 1924, 202.

Rendrock Powder Company

Rendrock Powder Company was founded in 1873 by Jasper R. and Addison C. Rand. The Rands were brothers of Albert T. Rand, who was president of the Laflin & Rand Powder Company, which also carried Rendrock. Treat S. Beach had acquired a patent for the Rendrock formula—a 40 percent nitroglycerin mixture also containing 40 percent potassium nitrate—earlier in the year. Beach then acted as salesman for the company. A factory was built in Pequannock, New Jersey, and started production in 1874, eventually making 10,000 pounds of Rendrock dynamite a month. Rendrock was successfully sued for patent infringement by Giant Powder Company, which obtained an injunction in 1880. The company then started producing Rackarock, an unusual nonnitroglycerin explosive that "consisted of a paper or cloth cartridge of potassium chlorate which was dipped into nitro benzol immediately before use,"[19] according to Van Gelder and Schlatter (see p. 214). Initially an inexpensive alternative to dynamite, the product enjoyed limited success and was even used by the US military. The company was dissolved in 1942.

J. R. RAND, President. N.W. HORTON, Sec'y. A.C. RAND, Treas.

RENDROCK POWDER COMPANY,

MANUFACTURERS OF
The Ingredients of the Patent
Rackarock Blasting Powder,

And dealers in Electric Blasting Machines, Electric Fuzes, Leading Wire, Blasting Caps, Safety Fuse, etc. These Ingredients are not explosive until combined. Can be forwarded by express or fast freight if desirable. Approximate Nitro-Glycerine in strength.

New York Traveler's Guide, 1883, 36.

Rockdale Powder Company

Rockdale Powder Company was founded in September 1901 and did business until October 1909. The York Powder Mills (founded in 1874), near Yorkville, Pennsylvania, was acquired by a group of local businessmen, including Henry Washers (listed incorrectly as "Harry Washers" in Van Gelder and Schlatter), William Koehler, and Spencer Gilbert. The production facilities moved to Hoffmanville, Maryland, and began making dynamite in addition to black powder. The company resisted buyout pressure from DuPont, and Rockdale Powder Company's former shareholders sued DuPont in 1915 for unfair competition, later dropping the suit. In 1905, Rockdale Powder Company's annual production peaked at 1,000,000 pounds of dynamite.

Safety Nitro Powder Company

Safety Nitro Powder Company was founded in 1880, according to Van Gelder and Schlatter, "to exploit the patents of Dr. Gilbert S. Dean, a San Francisco dentist, who had become interested in explosives and had carried on experiments in his dental laboratory."[20] Van Gelder and Schlatter add, "He became insane late in 1881 or early 1882,"[21] but the dentistry/explosives combo may be evidence of a head start in that department. The plant was 4 miles west of Pinole, in Contra Costa County, California, and produced seven grades of its low-nitroglycerin, sodium-nitrate formulations. The plant had a capacity of 1,500,000 pounds a year. Giant Powder Company merged with Safety Nitro

SAFETY NITRO POWDER
The Best High Explosive in the Market

 E. S. B.

None Superior for Bank and Stump Blasting. To Parties Clearing Land we call their Especial Attention

FUSE AND CAPS
SEND FOR CATALOGUE

SAFETY NITRO POWDER CO.
430 California Street San Francisco, Cal.

Blue and Gold (Berkeley: University of California, 1891), 25.

Company in 1892 to become Giant Powder Company, Consolidated. Giant continued to carry the Safety Nitro Label until the early 1910s. Giant was acquired by Atlas in 1915. ("E.S.B." in the ad was a brand of blasting powder and stood for "Extra-Strength Blasting.")

Sinnamahoning Powder Manufacturing Company

Sinnamahoning Powder Manufacturing Company was founded in 1905 by Henry Auchu, who served as president until the company became part of Keystone National Powder Company in 1910. Auchu also was an executive at the Keystone Powder Manufacturing Company and the C. B. Howard Company, a lumber supplier. Output for 1908 was 4,500,000 pounds of dynamite. The company's Collier brand made the first permissible explosives list in 1909. The brand was continued by Keystone, then by Aetna. Sinnamahoning Powder tried to trademark the Collier brand by cartridging it in a red wrapper, but the US Patent Office ruled that "the color of the package in which goods are placed cannot constitute a trade-mark for such goods."[22] (Sinnamahoning is now spelled Sinnemahoning.)

Standard Explosives Company

Standard Explosives Company was founded by S. T. Appollonio, Christian F. Schramme, and Christian Kessler, who built a plant near Tom's River, New Jersey, in 1887. The actual construction of the plant was supervised by James Lawrence, who would go on to found Joplin Powder Company in 1903. The plant had a capacity of 4,000,000 pounds of dynamite a year and produced straight and ammonia dynamites and even dabbled in nitrostarch explosives. Appollonio had been a general manager for Repauno. Schramme is described by Van Gelder and Schlatter as a "German importer and exporter,"[23] and he is listed as a merchant in business directories. "Staunton's copper mines in Michigan were the largest customers, taking mostly straight 40 percent nitroglycerine dynamite,"[24] according to Van Gelder and Schlatter, who also describe a large order that required a thirty-six-hour-straight shift, reporting that

 14,159 March 8, 1886. Blasting Apparatus, High Explosives and Powders. Standard Explosives Co., New York, N.Y.

Digest of Trademarks for Machines, Metals, Jewelry and the Hardware and Allied Trades (Washington, DC: Gibson Brothers, 1893, 44.

"light for the night work was furnished by kerosene lamps outside of the windows."[25] The plant was closed in 1889, and the company dissolved in 1890.

Sterling Dynamite Company

Sterling Dynamite Company was founded as the Standard Dynamite Company in 1889, but the name was changed in 1890. The firm was launched by a group of Tennessee businessmen including L. E. Montague, who served as president, and L. B. Searle, the company's first secretary/treasurer. Searle was a locally famous Prohibitionist, and Montague was a prolific Chattanooga businessman. A factory was erected in Bessemer, in Jefferson County, Alabama, with headquarters in Chattanooga. Annual output was 500,000 pounds of dynamite. The company was acquired by Eastern Dynamite in 1893 and kept open until 1903. Sterling Dynamite Company was formally dissolved in 1904.

Texas Dynamite Company

Texas Dynamite Company was incorporated in September 1905 and built a plant near Beaumont, Texas. The company is listed in Van Gelder and Schlatter as "The Beaumont Powder Company" and "the first and only dynamite factory to be established in Texas."[26] While the latter part is true, there was never a Beaumont Powder Company per se. (Oddly, Van Gelder and Schlatter do give the sales of the Texas Dynamite Company for 1908 as 1,350,000 pounds but do not relate the company to the "Beaumont Powder Company.") Texas Dynamite Company reported sales of only 345,000 pounds in 1907. Major stockholders were Charles W. Nelson, H. B. Smith, H. A. O'Neil, M. L. Porter,

Lumber Trade Journal, January 1910, 100.

and J. M. Carr. H. B. Smith was president; Nelson, vice president; and Porter, secretary. The firm was large enough to advertise and even submitted a failed bid to supply dynamite for the Panama Canal but went bankrupt in 1914.

Union Explosives Company

Union Explosives Company was formed in 1920 via the merger of the Long Powder and Supply Company and the Empire Powder Company. The Long Powder and Supply Company was an explosives distributor founded by J. Edgar Long and H. B. Cooper of Clarksburg, West Virginia. Union Explosives acquired Pennsylvania Powder Company in 1921. Union Explosives Company merged with black-powder maker Black Diamond Powder Company and Peerless Explosives Company and became Peerless-Union Explosives Company in 1930.

United States Dynamite and Chemical Company

United States Dynamite and Chemical Company was established in 1885 by Henry S. deShon, who reorganized the New York Powder Company of New Jersey, which he had acquired two years earlier. DeShon was president of the Lodi Chemical Works near Passaic, New Jersey. Eastern Dynamite acquired the business in 1897 and closed it in 1901. The plant produced an average of 1,800,000 pounds of dynamite a year, all "mixed, sieved, and packed by hand,"[27] according to Van Gelder and Schlatter. For perspective, the DuPont Louviers plant in Colorado alone produced 16,000,000 pounds a year.

Vigorite Powder Company

Vigorite Powder Company was incorporated in September 1877 with a consortium of California businessmen as stockholders and corporate officers. The original Vigorite dynamite was based on a formula concocted by a Norwegian chemist and was refined into two grades with about 35 percent and 45 percent nitroglycerin content. The production facilities were just southwest of present-day El Cerrito, at Point Isabel in Marin County, California, and the firm was renamed the California Vigorite Company in 1879. That same year, Vigorite was unsuccessfully sued by Giant Powder Company for patent infringement, with Giant's patent being declared invalid. Suits like this opened the way for the flood of producers inhabiting this last chapter of the book.

The company moved its plant to Point Isabel in 1900, and in 1903 DuPont gained control through stock acquisition. In 1906 the Vigorite Company was sold to the California Powder Company, which DuPont also owned. The Vigorite Powder Company ceased to be, but the facilities operated until 1912, when the site became part of the new Atlas Powder Company. Atlas closed the plant but carried over the brand name as a permissible explosive until 1925, offering six grades.

References to Vigorite in local and state laws pertaining to explosives continued into the 1960s, relics from the almost hundred-year-old lawsuit that granted Vigorite a patent separate from Giant Powder. Other formulations with individual patents, such as Hercules Powder and even Dualin, were also listed on rosters of items covered by a particular statute. The thinking was that without this level of specificity, it might be illegal to transport or use dynamite in a certain manner, but technically legal to do so with Dualin or Vigorite.

Vulcan Powder Company

Vulcan Powder was a nitroglycerin dynamite formula developed in the 1870s by Robert W. Warren, whom Van Gelder and Schlatter describe as "one of the early nitroglycerine manufacturers of the United States."[28] His Boston-area plant opened in 1873 but was shut down in 1877 over a lawsuit by Giant Powder. Warren moved west,

---

Punishment for unlawful possession.

§7. No person may knowingly keep or have in his or her possession any dynamite, vigorite, nitro-glycerin, giant powder, hercules powder, or other high explosive, except in the regular course of business carried on by such person, either as a manufacture thereof or merchant dealing in the same, or for use in legitimate blasting operations, or in the arts, or while engaged in transporting the same for others, or as the agent or employee of others engaged in the course of such business or operations. Any other possession of any such explosive substances as are named in this act is unlawful; and the person so unlawfully possessing it shall be punished by imprisonment in the state prison not exceeding five years, or by fine not exceeding five thousand dollars, or by both such fine and imprisonment.

US Department of the Interior, Bureau of Mines, *California Mining Statutes* (Washington, DC: USGPO, 1918), 236.

VULCAN BLASTING POWDER.

The Strongest, Safest, Most Uniform and Reliable "HIGH EXPLOSIVE" Manufactured on the Coast.

MINERS TESTIFY THAT IT IS FREE FROM OBJECTIONABLE FUMES. We call the attention of all desiring such a Powder to our various grades, which we are prepared to sell at LOWEST RATES.

No. 1.-- Equaling Liquid Nitro-Glycerine in Strength. We recommend this Grade in extremely hard rock, boulders, iron, etc.

No. 2.-- Will do the work thoroughly in all but the hardest kinds of rock.

No. 3.-- For bench work, pipe-clay, soft and shelly rock, outside work and quarrying.

Single and Triple Force Caps, Fuse of all Grades, Vulcan Powder Thawing Boxes, Batteries and Exploders. For Sale at the Lowest Rates.

VULCAN POWDER COMPANY,

Office, 218 California Street   SAN FRANCISCO, CAL

Mining and Scientific Press, November 1, 1879, 288.

building a plant in San Pablo, in Contra Costa County, California, in 1879. Julius Baum was owner and president of the new operation. Baum was an extremely wealthy San Francisco businessman who also owned nonnitroglycerin Tonite Powder Company. At one time, the plants of Vulcan Powder and Tonite were separated by only 100 yards. Vulcan Powder Company was purchased by California Vigorite Powder Company in the late 1890s.

Warren established his own company called Vulcan Dynamite Company in 1883 near Allentown, Pennsylvania, but sold it to Repauno Chemical Company after only two years. Repauno operated the plant for a few more years before dismantling it. Vulcan Blasting Powder was in fact ordinary extra dynamite, composed of 33 percent nitroglycerin, 40 percent sodium nitrate, and the rest sulfur and charcoal. Puget Sound & Alaska Powder Company sold an unrelated Vulcan Dynamite.

West Coast Powder Company

West Coast Powder Company was a small producer founded in 1923 in Everett, Washington, by Ralph W. Hammitt and his wife, Winnifred. A plant was built about 8 miles south of Everett and made dynamite, blasting powder, and mercury fulminate. The company employed nine men in 1938 and was incorporated in 1940. West Coast Powder was a member of the Institute of Makers of Explosives from 1939 to 1962. The firm ceased production of explosives in 1964 but maintained warehouses until 1966. The company's Wesco Coal Powder No. 1 made the Bureau of Mines' list of permissible explosives from 1930 until 1962. In 1932, West Coast Powder commissioned Norman Rockwell to paint images for a calendar.

Tonite, Masurite, and Rackarock

Three of the more popular, true nonnitroglycerin explosives besides Trojan Powder were Tonite, Masurite, and Rackarock. Tonite was a British explosive introduced in 1875 that consisted of nitrocellulose and barium nitrate. It was manufactured and sold in the United States from 1880 to 1885 by the Tonite Powder Company at Stege, near San Francisco. Tonite Powder Company was founded by William Letts Oliver, who also founded the California Cap Company.

Masurite was an ammonium nitrate explosive produced from 1902 to 1911 by John W. Masury in Ohio.

Rackarock was a Sprengel-type explosive named for German inventor and chemist Hermann Sprengel. It consisted of cartridges of potassium chlorate powder that the consumer dipped in benzene. Rackarock gained fame in

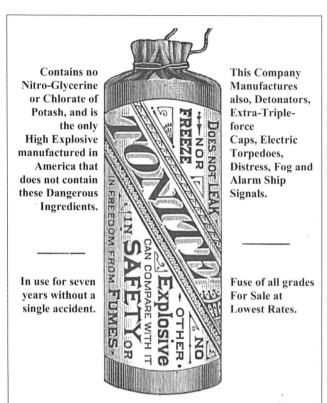

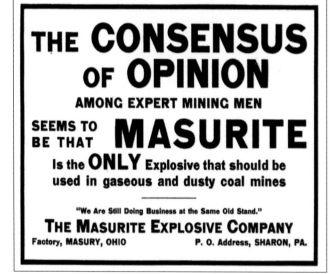

Tonite Company, advertisement. Chas. Yale, *Working Mines on the Pacific Coast* (San Francisco: Spaulding, 1882), 110. Masurite Company, advertisement, *Mines and Minerals*, July 1909, 45.

1885, when more than 100 tons were used to remove an enormous rocky obstruction known as Hell Gate in New York Harbor. The event, billed as the largest man-made explosion ever up to that time, was witnessed by 50,000 spectators and recorded by more than a hundred cameras.

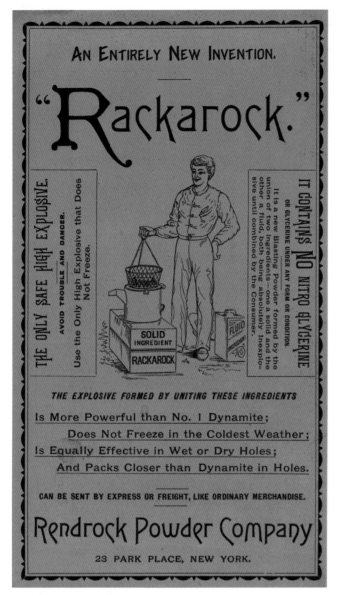

AN ENTIRELY NEW INVENTION.

"Rackarock."

THE ONLY SAFE HIGH EXPLOSIVE.

AVOID TROUBLE AND DANGER.

Use the Only High Explosive that Does Not Freeze.

It is a new Blasting Powder formed by the union of two ingredients—one a solid and the other a fluid, both being absolutely inexplosive until combined by the Consumer.

IT CONTAINS NO NITRO GLYCERINE

OR GLYCERINE UNDER ANY FORM OR CONDITION.

SOLID INGREDIENT
RACKAROCK

FLUID INGREDIENT

THE EXPLOSIVE FORMED BY UNITING THESE INGREDIENTS

Is More Powerful than No. 1 Dynamite;
Does Not Freeze in the Coldest Weather;
Is Equally Effective in Wet or Dry Holes;
And Packs Closer than Dynamite in Holes.

CAN BE SENT BY EXPRESS OR FREIGHT, LIKE ORDINARY MERCHANDISE.

Rendrock Powder Company

23 PARK PLACE, NEW YORK.

Rackarock was used as a US military explosive until being replaced by TNT in World War I. Rendrock Powder, advertisement, *Chicago Business Directory*, n.d., 1551.

Needless to say, dual-ingredient explosives were not nearly as easy to use as dynamite. The safety claims for such explosives, while true, were far outweighed by the inconvenience of the cumbersome preparatory procedures.

Rackarock was made in ever-decreasing quantities from 1881 to 1942 by the Rendrock Powder Company, described on (see p. 210).

Digging Deeper

A map in *The History of the Explosives Industry in America* showing the original locations of the Giant Powder Company, Hercules Powder Company, and California Powder Works is captioned "San Francisco was the cradle of the

dynamite industry in America and over fifteen companies established dynamite plants around the Bay."[29] Yet, only Judson Powder Company, Safety Nitro Powder Company, Vigorite Powder Company, Vulcan Powder Company, and the three shown on the map were significant enough to warrant mention elsewhere in the book.

For perspective on the production figures, in 1897 the US national output of dynamite was 30,000,000 pounds. Just two years later the number had almost tripled to 86,000,000 pounds. By 1905 the figure had risen to 130,000,000 pounds. In 1925 the total was approximately 250,000,000 pounds. And according to DuPont, in the 1940s the amount had grown to 350,000,000 pounds of dynamite per year.

Any dynamite could be called by the name of another. An article in 1884 describing the workings of the Aetna Powder Company's Indiana plant was headlined "Giant Powder—Facts about the Manufacture of Dynamite."[30] The *Fort Scott Daily Tribune* (Kansas) received this slightly snide letter in January 1887: "We notice in to-day's times, of this city, an account of an explosion of 200 cases of 'giant powder' near your city yesterday morning. This could not have been giant powder, as we have made no shipment to your section recently, but we should, however, like to see more detailed accounts of the explosive and would thank you to mail us copies of the TRIBUNE bearing upon that matter."[31] The sender, who knew that no such newspapers would be forthcoming, was the firm of Small & Schrader, "general agents of the Atlantic Dynamite Co., successors to Atlantic, Giant and Judson Powder Co."[32] It didn't help that companies such as Gogebic Powder Company brazenly used the Giant name in ads for its own products.

In 1885, Congress contemplated legislation targeting terrorists who used explosives. The law would ban anyone from manufacturing, buying, or selling, for nefarious purposes, "any metalline nitroleum, Neptune powder, Oriental powder, giant powder, Hercules powder, selenitic powder, thunderbolt powder, dynamite, or other nitro-explosive compound,"[33] as well as "ballastite, burnio powder, colonial powder, detonite, fulgurite, or any other chlorate explosive compound."[34] The futility of trying to list every explosive with a name was not recognized until well into the twentieth century. Colorado law as late as 1974 still prohibited misuse of metalline nitroleum, a type of blasting oil patented in 1869 that was never produced or used. Even some textbooks on blasting practice dutifully listed very obscure explosive names gleaned from optimistic patent applications.

In 1906, former DuPont employee Robert Waddell led the fight to have DuPont investigated for collusion, claiming, "The DuPont Trust is owned by men who daily, continuously, and openly break the laws of the state and nation."[35] Waddell argued that especially in wartime, it was a national security risk to have the country's military munitions produced by one entity, but then he contradictorily wanted the US government to be just such an entity. Earlier non-DuPont controlled interests had formed a consortium called the Independent Powder and Dynamite Manufacturers of the United States of America. The pressure worked and led to the breakup of DuPont.

More than a quarter of the smaller early dynamite firms were in Pennsylvania. The high concentration of dynamite and black-powder companies in the state was due to the proximity to nineteenth-century eastern coal, iron, and copper mines. Gold mining in California required far less in the way of blasting, and the state has no major coal reserves. In terms of conventional metals, to this day the state produces only gold and silver, and then in very small amounts compared to the state's boom years. California harvests prodigious quantities of other minerals, including rare-earth elements and a host of construction materials such as gypsum, gravel, aggregate, portland cement, and dimension stone. But mining started in the East, and it was largely unprofitable to produce explosives elsewhere until the twentieth century, and then only in a few other regions with significant mining industries. Except for satellite plants operated by the largest makers, thirty-five of the fifty states have never hosted a dynamite-producing company.

In the 1950s, Hawaii was suffering a dynamite shortage. Concerns over safety had caused shipments to be limited to 500 pounds at a time, giving the island only a six-month reserve. One Hawaiian newspaper lamented that "without dynamite, Hawaii could not long survive as a modern, civilized community,"[36] and predicted a return to "grass shacks."[37] Toward that end, a group of island businessmen proposed building a dynamite factory on Oahu. Two years of zoning and other hearings were held and the proposal was approved. In the end, the plant was never built due to high costs and the recent introduction of new, nonnitroglycerin explosives.

In 1898 a group of Kansas businessmen attempted to acquire land and build a dynamite factory near the small town of Turner. Local opposition thwarted the plan. According to one account, the businessmen had repeatedly misrepresented their intentions, first proposing ranchland for the grazing of cattle and then a cheese factory, "because they knew if it was known that it was to be a dynamite factory they would find opposition."[38] In fact, many explosives plants were nixed or moved due to their proximity to populated areas.

Finally, here is a wild tale from Pittsburgh, Pennsylvania, in 1886. "Col. W. T. Hoblitzell, agent of the Atlantic Dynamite Company, was employed this morning to break up a lot of old iron stored on a vacant piece of ground near the Republic Iron Works, on Sidney Street south side. He commenced work about six o'clock on a ten-ton anvil and used an immense charge of dynamite for the purpose. A terrific explosion followed and huge pieces of the anvil were scattered about in every direction, one piece weighing 300 pounds, wrecked a house occupied by Mrs. McNamara, a block distant. Another piece, over 200 pounds, crashed through the dwelling of Emil Cristman, also a block away. It struck a bed in which three children were sleeping, but who escaped serious injury. Another house occupied by Mrs. Luny Filley, 500 feet away, was also wrecked. The only person injured was James Atman, who was walking on the sidewalk one block away."[39] Atman sustained a serious leg injury but was expected to recover.

GLOSSARY

"A" blasting powder: Potassium-nitrate-based blasting powder, essentially gunpowder. More powerful and expensive than "B" blasting powder, it was used only for very hard rock after the invention of "B" blasting powder in 1857 by Lammot DuPont. Its use for guns has been displaced by nitrocellulose-based smokeless powder.

absorbents: The media in which nitroglycerin is carried. Originally kieselguhr (dirt) and then combustible substances such as sawdust, and finally other explosives such as ammonium nitrate.

adobe charge: Synonym for mudcapping.

air shot: The technique of leaving a space between charges, or between charges and stemming material, with the goal of reducing the strength of the blast. Some of the explosive force is dissipated into the space; the method was used to blast out large blocks of coal, ideal for fueling smelting furnaces. The practice was also used to save on explosives, enabling boreholes to be loaded with fewer cartridges. Finally, air shots are used to reduce the overall amount of fragmentation.

ammonia dynamite: Synonym for ammonium-nitrate-based or extra dynamite.

"B" blasting powder: Sodium-nitrate-based blasting powder. By far the most widely used type of black powder.

black powder: Synonym for blasting powder.

blasting agent: Includes non-cap-sensitive mixtures such as slurries, emulsions, and granular ammonium nitrate mixes that do not contain nitroglycerin or any other sensitizer. Most blasting agents are insensitive to shock, heat, and mishandling in general and must be detonated by a cap-sensitive explosive such as a cartridge of dynamite or a TNT/PETN booster.

blasting paper: Thick kraft paper used to roll blasting-powder cartridges. Blasting paper was also used to make tamping bags and to sheath primed dynamite cartridges for protection.

blockholing: Placing a charge in a hole drilled in a boulder.

blowout: An improperly tamped charge that expends most of its energy out of the top of the borehole.

blistering: Another synonym for mudcapping.

booster: Any explosive used to detonate a less sensitive explosive, or to augment the explosive power of a charge.

brisance: Fancy French word for shattering power.

bug dust: Fine coal dust. Bug dust was dangerous because vibrations caused by blasting could cause the tiny particles to become suspended in the air. The mixture of fine combustible particles surrounded by oxygen could then be ignited by a candle, spark, or a subsequent blast.

bulldozing: Yet another synonym for mudcapping.

bus wires: The 14-gauge copper wires used as extensions of leading wire to connect blasting circuits. Bus wires can be uninsulated, making it easier to quickly connect other wires. However, insulated bus wires are used in damp environments to prevent grounding. Then a section of the wire is stripped before being connected to other wires. Another tack is to string uninsulated bus wires between wooden posts to keep the wires off the ground.

cap crimper, cap nipper: A pliers-like tool used to create an embossed ring around the circumference of a blasting cap with fuse inserted; the compression of the tube "captures" the fuse, preventing it from being pulled out of the cap. Large,

bench-mounted versions were used by large operations.

cap-sensitive: An explosive compound that can be detonated via blasting cap, as distinguished from a compound that needs the detonation of a larger explosion to detonate.

cartridge count: The number of sticks per 50 pounds. The denser the formula, the fewer the sticks.

chambering: Carving out a cavity at the bottom of a borehole with successively larger charges in preparation for the final, tightly packed charge. Also called springing.

chambering spoon: Synonym for dynamite spoon. Also called a miner's spoon.

charge: A primed explosive that is ready to blast; also the explosives that are detonated by the primer.

closed work: Underground mining.

collar: The top of a borehole.

contractor's powder: Nitroglycerin-coated black powder. Synonym for railroad powder.

cordeau: An early kind of detonating cord consisting of a long lead tube filled with an explosive compound.

coupler: A device for connecting two cartridges.

coyote tunneling (coyoteing): Tunneling and placing large quantities of explosives in the side of a formation.

deck shot: Synonym for air shot.

deflagration: Burning. Low explosives explode when confined via very quick deflagration.

detonating cord: Long, fuse-like rope filled with an explosive compound that instantaneously detonates the entire length of cord when initiated by a blasting cap or another section of detonating cord.

detonator: Technically a blasting cap; laymen use the term as a synonym for blasting machine.

dimension stone: Carefully blasted and cut slabs of rock such as granite used for construction and decorative purposes.

dope: Slang for absorbents.

double jacking: Manual drilling with one man holding a drill bit and another man striking it with a hammer.

dummy: A tamping bag filled with sand or clay; dummies were also made with blasting paper and especially with scrap paper. Also an inert cartridge used for training or as a movie prop.

dynamite punch: A nonsparking tool used for making holes in cartridges during priming. The punches were also called "powder punches."

dynamite spoon: Long-handled scoop used for removing debris at the bottoms and backs of boreholes.

exploder: Obsolete synonym for blasting cap, but also sometimes used to indicate a blasting machine.

extra dynamite: Dynamite with ammonium nitrate, first as an additive, then as the primary constituent.

fly rock: Any hazardous flying debris produced by a blast. Blasting mats were mandated in some jurisdictions to prevent fly rock.

free-running dynamite: Either nitroglycerin-coated black powder or nitroglycerin-sensitized ammonium nitrate. It was sold in bags to be poured into boreholes, and in cartridges for loading in horizontal and upward holes.

fuze: Pronounced "fuzee." Early synonym for electric blasting cap.

gopher tunnel: Synonym for coyote tunnel.

hand jacking: Boring holes using a hammer and drill steel.

hangfire: A charge that explodes after it is supposed to.

initiator: Synonym for blasting cap, but also for mechanical devices for firing shock tube.

jack bit: Synonym for a detachable drill bit. Not a synonym for drill steels for hand jacking.

kieselguhr: Crumbly soil that was used as the very first absorbent for nitroglycerin in dynamite.

magazine: Any storage container for explosives, from small boxes to large warehouses to underground chambers.

misfire: A charge that does not explode.

muck pile: Usable blasted material. Originally a term for shoveling manure, "mucking" came to be ironically applied to the scooping up of valuable rocks and minerals. Blasting is always intended to move material forward or downslope into mounds that were convenient for access by men, or by machinery such as loaders.

mudcapping: Placing a charge without boring a hole. The charge was typically covered with mud, which kept it in place and provided a small amount of confinement.

Nonel: Dyno Nobel trade name for its original signal tubing. Short for "nonelectric."

open work: Aboveground mining.

overbreakage: Fragmentation that occurs beyond the desired zone.

overburden: Usually worthless dirt and rock that lie on top of valuable coal and ore deposits and seams. Sometimes blasted overburden can be sold for fill applications.

permissible explosive: A compound and shell that have been formulated and designed to meet strict requirements by the US Bureau of Mines and its successors.

primary explosives: Explosive compounds that are sensitive to detonation and are used to initiate the explosion of relatively less sensitive, secondary explosives. Blasting caps are the main primary explosives. Dynamite is a secondary explosive.

prime: To prepare a stick of dynamite for detonation via the insertion of a blasting cap and arranging the fuse or wires for safe insertion into a borehole.

primer: A cartridge that has a blasting cap inserted. Also an obsolete synonym for blasting cap.

railroad powder: Nitroglycerin-coated black powder.

secondary blasting: Breaking up boulders of undesired size that have been produced by initial blasting.

shock tubing, signal tubing: A hollow, explosives-filled plastic tube that conveys a small explosion from initiator to blasting cap.

single jacking: Manual drilling with one man holding a drill bit and striking it with a hammer.

snake-holing: Placing a charge under a boulder or stump.

"spitting the fuse": Igniting a fuse. The term comes from the visible emission of flame from a correctly lit fuse.

springing a hole: Using successive charges of explosives placed in the bottom of a hole to widen the bottom for the ultimate charge.

squib: Bottle-rocket-like device for igniting charges of black powder. Squibs could not be used to detonate dynamite.

steels: Bits used for hand jacking. Also used to refer to shafts with attached bits for mechanical drilling.

stemming: Incorrectly used as a verb to indicate the procedure of confining a blast by filling the balance of a primed hole with appropriate material. Technically a gerund (i.e., a verb used as a noun), indicating the material that is tamped rather than the act of tamping. The Bureau of Mines explicitly advised against the use of "stemming" as a verb, preferring "tamping" instead.

subsoiling: For farming. Loosening deep hardpack by detonating dynamite in hard-to-plow fields.

tamping: The act of compacting cartridges and then adding and compacting stemming material.

triple jacking: Manual drilling with one man holding a drill bit and two men alternately striking it with hammers.

undercutting: Sawing or picking a deep groove in the base of a coal face next to the ground, parallel with the floor of the mine. Undercutting weakened the mass and made for much-easier blasting.

vertical farming: Synonym for subsoiling.

NOTES

Chapter 1

1. "Tests Show Economy in Using Large-Size Dynamite," *Engineering & Mining Journal-Press*, February 24, 1923, 378.

2. E. I. DuPont de Nemours, *Blasters' Handbook* (Wilmington, DE: E. I. DuPont de Nemours, 1958), 57.

3. E. I. DuPont de Nemours, *High Explosives Price List No. 8* (Wilmington, DE: E. I. DuPont de Nemours, November 17, 1936), 2.

4. E. I. DuPont de Nemours, *Blasters' Handbook* (Wilmington, DE: E. I. DuPont de Nemours, 1952), 21.

5. Giant Powder, *Explosives and Blasting Supplies* (San Francisco: Giant Powder, Consolidated, 1924), 11.

6. Ibid., 45.

7. B. C. Hay, "Explosives," *Monthly Bulletin of the Canadian Institute of Mining and Metallurgy*, January 1921, 220.

8. US Bureau of Mines, *An Analysis of Black Powder and Dynamite* (Washington, DC: USGPO, 1916), 13.

9. "Safety in Mining," *Mining and Scientific Press*, August 1915, 201.

10. E. I. DuPont de Nemours, *Blasters' Handbook*, 1958, 59.

11. *Journals of the 21st Legislative Assembly of the Territory of Arizona* (Phoenix: Press of Arizona Republic, 1901), 70.

12. E. I. DuPont de Nemours, *Blasters' Handbook* (Wilmington, DE: E. I. DuPont de Nemours, 1980), 80.

13. Eric Twitty, *Blown to Bits in the Mine* (Ouray, CO: Western Reflections, 2001), 87.

14. "A Dynamite Box Maker," *Barrel and Box*, September 1906, 89.

15. N. I. Steers, "Interesting History of the Blasting Cap," *Railways*, February 1925, 355.

16. Interstate Commerce Commission, *ICC Regulations for the Transportation of Explosives and Other Dangerous Articles by Freight and Express* (Washington, DC: USGPO, 1914), 240.

17. "A Dynamite Birdhouse," *DuPont Magazine*, October 1916, 5.

18. "Destroy Dynamite Boxes!" *DuPont Magazine*, November–December 1916, 7.

19. "Disposal of Empty Explosives Containers," *Engineering and Contracting*, July 21, 1920, 75.

20. Institute of Makers of Explosives, ime.org/content/1920-1929.

21. E. I. DuPont de Nemours, *Blasters' Handbook* (Wilmington, DE: E. I. DuPont de Nemours, 1949), 55.

22. "Atlas Powder," *American Eagle*, June 1949, 25.

23. E. I. DuPont de Nemours, *Blasters' Handbook*, 1958, 61.

24. Ibid.

25. Interstate Commerce Commission, *ICC Regulations for the Transportation of Explosives and Other Dangerous Articles by Freight and Express* (Washington, DC: USGPO, 1914), 240.

26. E. I. DuPont de Nemours, *Blasters' Handbook*, 1952, 148.

27. Ibid.

28. E. I. DuPont de Nemours, *Blasters' Handbook*, 1958, 151.

29. "Box Production," *Barrel and Box*, July 1913, 12.

30. Interstate Commerce Commission, *ICC Regulations for the Transportation of Explosives and Other Dangerous Articles by Freight and Express* (Washington, DC: USGPO, 1914), 238.

31. Institute of Makers of Explosives, *Rules for Handling, Storage, Delivery and Shipping Explosives* (New York: IME, 1920), 11.

32. Atlas Powder, *Atlas Explosives Products* (Wilmington, DE: Atlas Powder, 1957), 44.

33. US Department of the Interior, Bureau of Mines, *The Construction of the Hoover Dam* (Washington, DC: USGPO, 1933), 33.

34. G. M. Norman, "Lost Motion in Explosives Making Due to Needless Variety of Products," *Chemical and Metallurgical Engineering*, August 30, 1922, 458.

35. Ibid.

36. E. I. DuPont de Nemours, *Blasters' Handbook* (Wilmington, DE: E. I. DuPont de Nemours, 1966), 48.

37. Hercules Powder, advertisement, *Explosives Engineer*, January–February 1961, front inner cover.

38. Ibid.

39. "The Manufacture of Blasting-Gelatine," *Journal of the Society of Chemical Industry*, March 31, 1890, 267.

40. William Ascarza, "Kingman Area's Golconda Mine Was a Big Zinc Ore Producer," *Arizona Daily Star*, June 13, 2016, A5.

41. "Blast Kills Two," *Akron (OH) Beacon Journal*, February 3, 1954, 8.

42. Michael Kouris, *Dictionary of Paper* (Peachtree Corner, GA: TAPPI, 1996), 114.

43. "Explosives for 1909–1910," *Canal Record*, April 7, 1909, 251.

44. US Department of Commerce, *Packing for Domestic Shipment: Wood Boxes* (Washington, DC: USGPO, 1927), 11.

45. Earl Sackett, "The Ubiquitous Powder Box," *Engineering and Mining Journal*, August 1977, 100.

46. Ibid.

47. US General Services Administration, *Code of Federal Regulations, Title 49, Parts 71 to 90* (Washington, DC: USGPO, 1956), 537.

48. US Department of the Interior, Bureau of Mines, *Accident Statistics as an Aid to Prevention of Accidents in Metal Mines* (Washington, DC: USGPO, 1945), 23.

49. "Manhattan Indifference to Danger Shown by the Throngs Which Daily Pass Boxes on Sidewalks Filled with Enough Explosive to Wreck Five Skyscrapers," *New York Tribune*, August 29, 1915, 8.

50. Institute of Makers of Explosives, *Warnings and Instructions for Consumers in Transport, Storage, Handling, and Using Explosives* (Washington, DC: IME, 2000), 4.

51. Ibid.

52. US Bureau of Alcohol, Tobacco and Firearms, *ATF—Explosives Laws and Regulations* (Washington, DC: USGPO, 1990), 65.

53. William Ascarza, "Benson's Apache Powder Put the Boom in Arizona Mining," *Arizona Daily Star*, June 13, 2016, A2.

54. Ibid.

55. US Department of Transportation, Bureau of Mines, *Hazardous Materials Regulations* (Washington, DC: USGPO, 1976), 15,981.

56. "Think Safety," *Kenosha (WI) Evening News*, October 7, 1933, 6.

Chapter 2

1. Division of the Federal Register, *Code of Federal Regulations, Title 30: Mineral Resources, Parts 1–199* (Washington, DC: USGPO, 2013), 383.

2. Aetna Powder Company, advertisement, *Hardware Age*, July 29, 1920, 36.

3. US Department of the Interior, *Accidents from Hoisting and Haulage in Metal Mines* (Washington, DC: USGPO, 1945), 26.

4. "New Explosives Carrier," *Construction and Engineering Monthly*, October 1938, 32.

5. Ibid.

6. E. I. DuPont de Nemours, advertisement, *Coal Age*, February 1938, 84.

7. US Department of the Interior, Bureau of Mines, *Prevention of Accidents in Metal Mines* (Washington, DC: USGPO, 1945), 26.

8. A. E. Anderson, "Safe Practices in the Blasting of Coal," *Coal Industry*, August 1920, 368.

9. Division of the Federal Register, *Code of Federal Regulations, Title 49, Parts 1–99* (Washington, DC: USGPO, 1939), 1182.

10. Giant Powder, *Explosives and Blasting Supplies*, 107.

11. E. I. DuPont de Nemours, advertisement, *Tech Engineering News*, November 1923, 189.

12. N. S. Greensfelder, "Biography of the Hercules Powder Co.," *Colorado School of Mines Magazine*, August 1923, 17.

13. Institute of Makers of Explosives, "Mission Statement," ime.org/content/ime mission statement.

14. Eustace Weston, *Rock Drills: Design, Construction and Use* (New York: McGraw Hill, 1910), 3.

15. Thomas Foster, *Coal Miner's Pocketbook* (New York: McGraw Hill, 1916), 674.

16. International Library of Technology, *Rock Boring, Rock Drilling, Explosives and Blasting* (Scranton, PA: International Textbook, 1907), section 36, p. 19.

17. Arthur La Motte, *Blasters' Handbook* (Wilmington, DE: E. I. DuPont de Nemours, 1922), 36.

18. Ibid., 37.

19. Arthur La Motte, *Blasters' Handbook* (Wilmington, DE: E. I. DuPont de Nemours, 1925), 41.

20. "Successful Dynamiting at 35 Degrees below Zero," *Municipal and County Engineering*, February 1924, 53.

21. Atlas Powder, advertisement, *American City*, March 1922, 48.

22. Ibid.

23. E. I. DuPont de Nemours, *Blasters' Handbook*, 1952, 148.

24. Atlas Powder, advertisement, *Mining Congress Journal*, October 1954, 19.

25. "American Mining Congress Equipment Show Report," *Coal Age*, June 1951, 115.

26. Joseph T. Miller and Robert B. Brown, Blasting stemming plug, US Patent 5936187, filed September 19, 1997, and issued August 10, 1999.

27. "Advantages of Tamping Bags," *Dupont Magazine*, May 1916, 11.

28. Aetna Powder, *Aetna Dynamite* (Chicago: Aetna Powder, 1905), 8.

29. Ibid.

30. E. I. DuPont de Nemours, *High Explosives, First Section* (Wilmington, DE: E. I. DuPont de Nemours, 1920), 47.

31. US Department of the Interior, Bureau of Mines, *Safe Storage, Handling and Use of Commercial Explosives in Metal Mines, Nonmetallic Mines and Quarries* (Washington, DC: USGPO, 1946), 13.

32. Ensign-Bickford, *Primacord-Bickford Detonating Fuse* (Simsbury, CT: Ensign-Bickford, 1941), 5.

33. Ibid.

34. E. I. DuPont de Nemours, *DuPont Farmers' Handbook* (Wilmington, DE: E. I. DuPont de Nemours, 1920), 30.

35. E. I. DuPont de Nemours, *Blasters' Handbook*, 1958, 229.

36. La Motte, *Blasters' Handbook*, 1922, 56.

37. US Department of the Army, *Demolition Materials* (Washington, DC: USGPO, 1973), 2–8.

38. Ibid., 2–7.

39. Ideal Blasting Supply, idealblasting.com/adjustable-crimping-pliers.

40. William P. Frank, "Scrapping Ladies," *News Journal* (Wilmington, DE), September 29, 1942, 9.

41. David Schenkman, *Explosives Control Tokens* (Fayetteville, WV: National Scrip Collector's Association, 1989), 12.

42. Ibid.

43. Ibid.

44. Ibid.

45. US Department of the Interior, Bureau of Mines, *Progress Report on Investigations of Detachable Rock-Drill Bits* (Washington, DC: USGPO, 1937), 2.

46. Ibid.

47. Ibid.

48. "Inland Mine and Stone Co.," *Explosives Engineer*, December 1940, 379.

49. "Thawing Dynamite Went Off," *Kansas City Kansas Republic*, December 28, 1916, 1.

50. "Chicago's H20," *Chicago Tribune*, June 8, 1939, 10.

51. "Crimped!" *Daily Item* (Port Chester, NY), February 28, 1946, 16.

52. "Perils of Blasting Caps," *Tampa Tribune*, July 8, 1913, 37.

53. "Electric Blasting," *American Architect and Building News*, April 29, 1893, 76.

54. United States Senate Special Subcommittee of the Committee on Public Lands, *Investigation of Mine Explosion at Centralia, Illinois* (Washington, DC: USGPO, 1947), 236.

55. George Joseph Young, *Elements of Mining* (New York: McGraw Hill, 1946), 139.

56. "Escapes Death—Miner Is Pierced through Head with an Iron Rod," *Princeton (MN) Union*, October 26, 1905, 6.

57. E. I. DuPont de Nemours, *Agricultural Blasting: A Money-Making Profession* (Wilmington, DE: E. I. DuPont de Nemours, 1915), 6.

58. Ibid.

59. US National Park Service, *Handbook for the Storage, Transportation, and Use of Explosives* (Washington, DC: USGPO, 1999), 86.

60. Ibid.

Chapter 3

1. E. I. DuPont de Nemours, *Blasters' Handbook*, 1949, 271.

2. Ibid.

3. E. I. DuPont de Nemours, *Blasters' Handbook*, 1958, 347.

4. W. R. Ellis, "Dynamite: The New Aladdin's Lamp," *Ohio State Engineer*, January 1935, 4.

5. William S. Dutton, *DuPont: One Hundred and Forty Years* (New York, Scribner, 1951), 115.

6. Bennett Grotta, "Development and Application of Initiating Explosives," *Industrial and Engineering Chemistry*, February 1925, 134.

7. Institute of Makers of Explosives, ime.org/content/industrialexplosives applications.

8. US Department of Agriculture, *Clearing New Land* (Washington, DC: USGPO, 1902), 9.

9. E. L. D. Seymour, *Farm Knowledge, Volume III: Farm Implements and Construction* (New York: Doubleday Page,

1919), 299.

10. Atlas Powder, *Better Farming with Atlas Powder* (Wilmington, DE: Atlas Powder, 1923), 41.

11. E. I. DuPont de Nemours, *Blasters' Handbook*, 1952, 28.

12. E. I. DuPont de Nemours, *Red Cross Dynamite* (Wilmington, DE: E. I. DuPont de Nemours, 1912), 102.

13. "The Truth about Subsoiling with Red Cross Dynamite," *DuPont Magazine and Agricultural Blaster*, October 1913, 2.

14. Ibid.

15. US National Park Service, *Handbook for the Storage, Transportation, and Use of Explosives* (Washington, DC: USGPO, 1999), 86.

16. US Department of the Interior, Bureau of Mines, *Prevention of Accidents in Metal Mines* (Washington, DC: USGPO, 1945), 59.

17. Atlas Powder, *Better Farming with Atlas Powder*, 4.

18. Ibid., 5.

19. Ibid., 24.

20. US Department of the Interior, Bureau of Mines, *The Explosibility of Coal Dust* (Washington, DC: USGPO, 1911), 2.

21. Advertisement, *Coal Industry Yearly Index*, 1922, 208-A.

22. Institute of Makers of Explosives, "Permissible Explosives Approval," ime.org/uploads/public/issue%20Briefs-2018/Permissible%20ApprovalsMarch%202018.pdf.

23. Atlas Powder, *Explosives and Blasting Supplies* (San Francisco: Atlas Powder, 1924), 60.

24. US Department of the Interior, Bureau of Mines, *Active List of Permissible Explosives and Blasting Devices Approved before December 31, 1975* (Washington, DC: USGPO, 1976), 9.

25. "Dynamite—the Cutting Edge of Civilization," *America at Work*, September 20, 1923, 13.

26. E. I. DuPont de Nemours, *DuPont Farmers' Handbook* (Wilmington, DE: E. I. DuPont de Nemours, 1915), 144.

27. Atlas Powder, *Better Farming with Atlas Powder*, 7.

28. E. I. DuPont de Nemours, *DuPont Farmers' Handbook* (Wilmington, DE: E. I. DuPont de Nemours, 1915), 149.

29. S. R. Russell, "Use of Explosives in Production of Crushed Stone," *Engineering and Contracting*, April 19, 1922, 75.

30. E. I. DuPont de Nemours, *Blasters' Handbook*, 1958, 377.

31. Atlas Powder, *Atlas Explosives Products*, 1957, 4.

32. Ibid.

33. US Department of the Interior, Bureau of Mines, *Drilling and Blasting in Open-Cut Copper Mines* (Washington, DC: USGPO, 1927), 18.

34. S. R. Russell, "Use of Explosives in Production of Crushed Stone," *Engineering and Contracting*, April 19, 1922, 77.

35. La Motte, *Blasters' Handbook*, 1925, 72.

36. "The Use of Dynamite in Road Building," *Good Roads*, October 7, 1911, 179.

37. William S. Dutton, "Wonders That Are Done with Dynamite," *American Magazine*, June 1924, 80.

38. Arthur Pine Van Gelder and Hugo Schlatter, *The History of the Explosives Industry in America* (New York: Columbia University Press, 1927), 1042.

39. La Motte, *Blasters' Handbook*, 1922, 143.

40. Ibid.

41. US Department of Labor, *Wrecking and Demolition Operations* (Washington, DC: USGPO, 1961), 4.

42. H. E. Davis, "Dynamite in Fire Fighting," *Quarterly of the National Fire Protective Association*, July 1925, 29.

43. James Russel Wilson, *Complete Story of San Francisco's Terrible Calamity of Earthquake and Fire* (Sacramento, CA: Continental, 1906), 39.

44. E. I. DuPont de Nemours, *Blasters' Handbook*, 1958, 426.

45. Giant Powder, *Explosives and Blasting Supplies*, 29.

46. E. I. DuPont de Nemours, *High Explosives* (Wilmington, DE: E. I. DuPont de Nemours, 1911), 127.

47. La Motte, *Blasters' Handbook*, 1922, 89.

48. Ibid.

49. Atlas Powder, *Explosives and Blasting Supplies*, 26.

50. Giant Powder, *Explosives and Blasting Supplies*, 28.

51. Ibid.

52. La Motte, *Blasters' Handbook*, 1925, 98.

53. US Department of the Interior, *Avalanche Handbook* (Washington, DC: USGPO, July 1976), 128.

54. Ibid.

55. "Dynamite Blows Out Oil-Well Fires When Steam and Water Fail," *Popular Mechanics*, January 1925, 68.

56. Atlas Powder, *Atlas Explosives Products*, 1957, 18.

57. Atlas Powder, *Twenty-Five Years* (Wilmington, DE: Atlas Powder, 1937), 15.

58. Ibid.

59. Harold A Mathiak, *Pothole Blasting for Wildlife* (Madison: Wisconsin Conservation Department, 1965), 12.

60. Ibid.

61. "Blasting for Ornamental Landscape Work," *Park and Cemetery and Landscape Gardening*, March 1917, 75.

62. "Editorial," *Park and Cemetery and Landscape Gardening*, February 1920, 1.

63. "Methods of Digging Graves with Explosives," *Park and Cemetery and Landscape Gardening*, March 1920, 12.

64. Mac Bentley, "Sequoyah County to Stop Using Dynamite on Graves," *The Oklahoman*, February 19, 2004, 1.

65. "Digging a Ditch in a Flash," *Boone County (KY) Recorder*, October 28, 1915, 3.

66. "Uncovered Ball of Snakes," *Mount Carmel Item* (Pennsylvania), February 10, 1912, 3.

67. Jefferson Powder, advertisement, *Lumber Trade Journal*, December 15, 1913, 56.

68. US Department of Agriculture, *Clearing Land* (Washington, DC: USGPO, 1918), 16.

69. Ibid.

70. Atlas Powder, *Better Farming with Atlas Powder*, 94.

71. Ibid., 95.

72. N. S. Greensfelder, "The Manufacture of Naval Stores," *Compressed Air Magazine*, July 1922, 199.

73. E. I. DuPont de Nemours, *Blasters' Handbook* (Wilmington, DE: E. I. DuPont de Nemours, 1977), 246.

74. Atlas Powder, *Explosives and Rock Blasting* (Dallas: Atlas Powder, 1987), 447.

75. Ibid., 449.

76. E. I. DuPont de Nemours, *Blasters' Handbook* (Wilmington, DE: E. I. DuPont de Nemours, 1967), 351.

77. "Controlled Demolition, Inc.," en-wikipedia.org/wiki/Controlled Demolition Inc.

78. Adrian Dinnane, *DuPont: From the Banks of the Brandywine to Miracles of Science* (Wilmington, DE: E. I. DuPont de Nemours, 2002), 37.

79. Nicholas Azzara, "Can Local Man Save Us from Storms?," *Braden (FL) Herald*, June 1, 2007, 11-A.

80. Ibid.

81. "Dynamite Is Fickle," *The Enterprise* (San Mateo, CA), August 26, 1899, 4.

82. Ibid.

Chapter 4

1. Van Gelder and Schlatter, *The History of the Explosives Industry in America*, 548.

2. Ibid.

3. Aetna Powder, Handling Explosives

(Chicago, Aetna Powder, 1913), 10.

4. Weston Goodspeed and Charles Blanchard, *Counties of Porter and Lake, Indiana* (Chicago: F. F. Battey, 1882), 541.

5. "General Summary—Aetna Explosives," *United States Investor*, June 12, 1915, 18.

6. Van Gelder and Schlatter, *The History of the Explosives Industry in America*, 752.

7. Ibid., 548.

8. "Powder Plant to Open," *Arizona Daily Star*, February 8, 1922, 8.

9. "Apache Powder Company Is Essential to Arizona Industry," *Arizona Daily Star*, February 21, 1941, 68.

10. "Apache Co. Employees Handle Explosives in Safety," *Arizona Daily Star*, February 22, 1929, 47.

11. Ibid.

12. US Department of Labor, Bureau of Labor Statistics, *Monthly Review of the United States Bureau of Labor Statistics* (Washington, DC: USGPO, January 1918), 171.

13. Ibid.

14. Atlas Powder, *Atlas Explosives Products*, 1957, 21.

15. "McArthur's Austin Plant One of U.S. Most Modern," *Chillicothe (OH) Gazette*, December 17, 1963, 10.

16. Max Dorian, *The du Ponts, from Gunpowder to Nylon* (Boston: Little, Brown, 1962), 131.

17. Van Gelder and Schlatter, *The History of the Explosives Industry in America*, 563.

18. William S. Stevens, "The Powder Trust," *Quarterly Journal of Economics*, May 1912, 469.

19. Van Gelder and Schlatter, *The History of the Explosives Industry in America*, 1094.

20. Dinnane, *DuPont: From the Banks of the Brandywine to Miracles of Science*, 87.

21. E. I. DuPont de Nemours, *Blasters' Handbook*, 1952, 21.

22. "Unimax Technical Data Sheet," dynonobel.com/ResourceHub/Technical%20Information/North%20America/Dynamite/Unimax.pdf.

23. "About Dyno Nobel," dynonobel.com/aboutdynonobel.

24. Daniel Gleeson, "Dyno Nobel Helps BMA Caval Ridge Become Electronic Blasting Leader," *International Mining*, im-mining.com/2020/02/13/dyno-nobel-helps-bma-caval-ridge-become-electronic-blasting-leader.

25. Van Gelder and Schlatter, *The History of the Explosives Industry in America*, 408.

26. Ibid., 451.

27. Cleveland Chamber of Commerce, *Industries of Cleveland* (Cleveland, OH: Richard Edwards, 1879), 147.

28. Barry A. Brown, "Trustees Meeting August 11, 1870," *Minute Books of the California Powder Works* (Self-published, 2020, Open Source Collection, internetarchive.com).

29. US Farm Services Administration, "Farmable Wetlands Program," fsa.usda.gov/programs-and-services/conservation-programs/farmable-wetlands.

30. Hercules Powder, advertisement, *Engineering and Mining Journal*, September 1960, 213.

31. Hercules Powder, "Unigel Semigelatin Dynamite," *Bulletin TD 2131* (Wilmington, DE: Hercules Powder, February 1973), 1.

32. E. I. DuPont de Nemours, *Blasters' Handbook*, 1949, 43.

33. Van Gelder and Schlatter, *The History of the Explosives Industry in America*, 632.

34. "Cyanamid Buys Illinois Powder Subject to Vote," *Latrobe (PA) Bulletin*, August 5, 1957, 17.

35. American Cyanamid, advertisement, *Contractors and Engineers*, November 1957, 13.

36. Van Gelder and Schlatter, *The History of the Explosives Industry in America*, 632.

37. Independent Explosives, "About Us—Past and Present," expsco.com.

38. Van Gelder and Schlatter, *The History of the Explosives Industry in America*, 163.

39. George M. Cruikshank, *A History of Birmingham and Its Environs* (Chicago: Lewis, 1910), 310.

40. Van Gelder and Schlatter, *The History of the Explosives Industry in America*, 557.

41. Ira Elbert Bennett and John Hays Hammond, *History of the Panama Canal* (Washington, DC: Historical Publishing, 1915), 346.

42. Keystone National Powder, *Farming with Dynamite* (Emporium, PA: Keystone National Powder, 1912), 3.

43. Pennsylvania Auditor General, *Report of the Auditor General on the Finances of the Commonwealth of Pennsylvania for the Year Ending November 30, 1905* (Harrisburg, PA: Harrisburg Publishing, 1906), 460.

44. Aetna Powder, advertisement, *Hoard's Dairyman*, May 31, 1912, 694.

45. Ibid.

46. Ibid.

47. Aetna Powder, advertisement, *Breeder's Gazette*, March 6, 1912, 618.

48. "To Business Farmers," Aetna Powder, advertisement, *Kansas Farmer*, May 19, 1912, 9.

49. "Demonstration at Main Street," *Houston Post*, January 10, 1913, 3.

50. "General Summary—Aetna Explosives," *United States Investor*, June 12, 1915, 18.

51. "The Powder Stocks," *United States Investor*, June 10, 1915, 29.

52. "Industrials: Aetna Powder Co.," *Moody's Manual of Railroad and Corporate Securities, 1915, Volume II* (New York: Moody's Manual, 1915), 3644.

53. "Apache Powder Plant Is Inspected by Mining Men," *Copper Era and Morenci Leader* (Clifton, AZ), April 7, 1922, 3.

54. Ibid.

55. "Arizona Powder Plant," *Mohave County (AZ) Miner and Our Mineral Wealth*, December 30, 1921, 2.

56. "Apache Powder Plant Is Inspected by Mining Men," *Copper Era and Morenci Leader* (Clifton, AZ), April 7, 1922, 3.

57. Ibid.

58. "McArthur's Austin Plant One of U.S. Most Modern," *Chillicothe (OH) Gazette*, December 17, 1963, 10.

59. Ibid.

60. "Activities Increasing," *Wall Street Journal*, November 17, 1924, 1.

61. US Department of Commerce, National Technical Information Service, *Encyclopedia of Explosives and Related Items, Volume 5* (Washington, DC: USGPO, 1962), D1571.

62. Dyno Nobel, advertisement, *Daily Herald* (Provo, UT), June 26, 2005, F3.

63. "Divorce and Probate," *Deseret News* (Salt Lake City, UT), December 10, 1900, 7.

64. Giant Powder, "Blast Holes for Trees and Give the Roots More Pasture," advertisement, *Sacramento (CA) Daily Union*, May 20, 1916, 2.

65. "Opportunity for All Men of New Castle," *New Castle (PA) Herald*, May 28, 1918, 2.

66. Grasselli Powder, "Wanted," advertisement, *Pittsburgh Gazette-Times*, September 26, 1922, 14.

67. "Seen from the Hilltop," *Kane (PA) Republican*, July 28, 1930, 4.

68. "Hercules Powder Co., a Birmingham Institution, Makes Essential Article," *Birmingham (AL) News*, February 1, 1922, 11.

69. "Grafton," *Alton (IL) Evening Telegraph*, March 6, 1915, 8.

70. "Grafton," *Alton (IL) Evening*

Telegraph, June 14, 1915, 6.

71. "Man Who Was Bitten by Copperhead Snake Much Better," *Alton (IL) Evening Telegraph*, June 23, 1915, 3.

72. "Tornado at Alton," *The Times* (Streator, IL), May 21, 1915, 1.

73. "Man Hurt at Work," *Wilkes-Barre (PA) Record*, September 7, 1961, 20.

74. "Injured at Work," *The Tribune* (Scranton, PA), March 29, 1962, 1.

75. "'Nitro Man' Gets a Kick out of Dynamite," *Chicago Tribune*, July 4, 1987, 27.

76. Independent Explosives, "Immediate Openings," advertisement, *Wilkes-Barre (PA) Record*, July 22, 1973, 48.

77. Bernie Kohn, "No Layoffs Expected after Hercules Sale," *Poughkeepsie (NY) Journal*, February 21, 1985, 13.

78. "What Dynamite Will Do for the Farmer," *Hickory (NC) Democrat*, April 4, 1912, 4.

79. Jefferson Powder, "The Power Question Is Solved," advertisement, *Birmingham (AL) News*, March 27, 1913, 5.

80. "Another Response to 'Taxpayer,'" *Cameron County Press* (Emporium, PA), December 20, 1900, 5.

81. "Keystone Powder Mills," *Franklin Repository* (Chambersburg, PA), December 20, 1871, 4.

Chapter 5

1. Van Gelder and Schlatter, *The History of the Explosives Industry in America*, 548.

2. E. I. DuPont de Nemours, *Blasters' Handbook*, 1980, 79.

3. "War Salvaged Explosives," *Virginia Department of Agriculture and Immigration Yearbook 1922* (Richmond, VA: Davis Bottoms, 1922), 125.

4. E. I. DuPont de Nemours, "Agritol Replaces Pyrotol for Land Clearing," advertisement, *Cornell Countryman*, April 1928, 218.

5. E. I. DuPont de Nemours, *Agritol—the New Explosive for Land Clearing* (Wilmington, DE: E. I. DuPont de Nemours, 1929), 1.

6. E. I. DuPont de Nemours, advertisement, *Mining Congress Journal*, May 1931, 21.

7. Ibid.

8. "New Blasting Explosive," *Coal Trade Journal*, January 7, 1880, 87.

9. US Chamber of Commerce, Bureau of the Census, *Statistical Abstract of the United States—1950* (Washington, DC: USGPO, 1950) 801.

10. "New Blasting Agent," *Coal Age*, February 1933, 73.

11. E. I. DuPont de Nemours, advertisement, *Proceedings of the Illinois Mining*

Institute, 1933, advertising 39.

12. E. I. DuPont de Nemours, *Blasters' Handbook* (Wilmington, DE: E. I. DuPont de Nemours, 1942), 5.

13. Gordon S. Queensbury Jr., "Combat Engineering—Is the Focus Too Narrow?" *Army Engineer*, March 1990.

14. US Defense Threat Reduction Agency, *Draft Environmental Impact Statement for DTRA Activities on White Sands Missile Range, New Mexico* (Washington, DC: USGPO, January 2006), F-78.

15. Bennett and Hammond, *History of the Panama Canal*, 344.

16. "Guth's Station," *Allentown (PA) Democrat*, April 18, 1911, 2.

17. "Property Acquired—Large Order," *Wall Street Journal*, October 2, 1924, 5.

18. "New Explosive for Farm and Road Work," *Michigan Roads and Pavements*, August 14, 1924, 5.

19. "Campaign Urges Use of Pyrotol," *USDA Official Report* (Washington, DC: USGPO, October 28, 1925), 2.

20. Charles Sitz and J. B. Cole, *The Use of Explosives on the Farm* (Blacksburg: Virginia Polytechnic Institute, May 1926), 19.

21. E. I. DuPont de Nemours, "Announcing Agritol," advertisement, *Mount Pleasant (MI) Daily Times*, March 16, 1928, 6.

22. E. I. DuPont de Nemours, *Blasters' Handbook*, 1942, 7.

23. Ralph Hinz, "Explosives in Mines," *Birmingham (AL) News*, January 18, 1952, 18.

24. "Mystery Death of Woman Laid to Coal Blasting Device," *Times Dispatch* (Richmond, VA), January 2, 1935, 4.

25. "Plenty of Dynamite Used Even During Peaceful Days," *East Hampton (NY) Star*, January 7, 1943, 3.

26. US Department of Commerce, National Technical Information Service, *Encyclopedia of Explosives and Related Items, Volume 2* (Washington, DC: USGPO, 1962), C78.

27. US Department of the Army, *Technical Manual 9-1300-214—Military Explosives* (Washington, DC, USGPO, November 1967), 7–85.

28. Ibid.

29. "DuPont to Phase Out Dynamite for 'Tovex.'" *Moberly (MO) Monitor-Index & Evening Democrat*, February 6, 1974, 21.

30. Ibid.

31. Ibid.

Chapter 6

1. Van Gelder and Schlatter, *The History of the Explosives Industry in America*, 693.

2. "Notes on the Literature of Explosives," *Proceedings of the US Naval Institute, Volume XI* (Annapolis, MD: US Naval Institute, 1885), 279.

3. Van Gelder and Schlatter, *The History of the Explosives Industry in America*, 692.

4. Ibid.

5. Ibid., 687.

6. Ibid., 552.

7. Ibid., 698.

8. Ibid., 663.

9. Ibid., 697.

10. Ibid., 671.

11. Ibid., 699.

12. Ibid.

13. Ibid., 697.

14. Ibid., 695.

15. Ibid., 344.

16. Ibid., 634.

17. Ibid., 633.

18. Ibid., 473.

19. Ibid., 354.

20. Ibid., 446.

21. Ibid., 447.

22. US Patent Office, *Official Gazette of the United States Patent Office* (Washington, DC: USGPO, 1911), 481.

23. Van Gelder and Schlatter, *The History of the Explosives Industry in America*, 675.

24. Ibid., 676.

25. Ibid.

26. Ibid., 700.

27. Ibid., 662.

28. Ibid., 646.

29. Ibid., 407.

30. "Giant Powder—Facts about the Manufacture of Dynamite," *Boston Globe*, April 13, 1884, 9.

31. "The Tribune Wanted in New York," *Fort Scott (KS) Daily Tribune*, January 29, 1887, 4.

32. Ibid.

33. "XLVIIIth Congress—2nd Session," *Portland (ME) Daily Press*, January 26, 1885, 1.

34. Ibid.

35. Alfred Chandler Jr. and Stephen Salsbury, *Pierre DuPont and the Making of the Modern Corporation* (Washington, DC: Beard Books, 2000), 261.

36. "Back to Grass Shacks?," *Honolulu Advertiser*, May 3, 1952, 4.

37. Ibid.

38. L. P. Hewitt, "The Dynamite Plant at Turner," *The Press* (Kansas City, KS), January 21, 1898, 1.

39. "Dreadful Dynamite," *Knoxville (TN) Daily Chronicle*, May 15, 1886, 1.

BIBLIOGRAPHY

Adams, William, Virginia Wrenn, and L. S. Horton. *Production of Explosives in the United States during the Calendar Year 1940*. US Department of the Interior, Bureau of Mines. Washington, DC: USGPO, 1942.

"Advantages of Electric Blasting in Health, Safety, and Economy." *Mining Science*, February 1914, 47.

Aetna Explosives. *Aetna Explosives and Their Characteristics*. New York: Aetna Explosives, 1920.

Aetna Powder. *Blasting Stumps and Boulders*. Chicago: Aetna Powder, 1912.

Akhaven, Jacqueline. *The Chemistry of Explosives*. London: Royal Society of Chemistry, 2022.

Allison, Douglas. "Commercial Explosives—Their Selection and Use." *Engineering and Mining Journal Press*, February 2, 1924, 197.

Anderson, A. E. "Eliminating Missed Shots in Coal Mining." *Coal Industry*, March 1921, 157.

Anderson, A. E. "The Galvanometer and Its Advantages in Electric Blasting." *Colorado School of Mines Magazine*, October 1915, 1.

Anderson, A. E. "Packing, Transportation and Storage of Explosives," *Mining Science*, June 20, 1912, 531.

Andrej, George. *Rock Blasting*. New York: E. & F. N. Spon, 1878.

Atlas Powder. *Atlas Explosives*. Wilmington, DE: Atlas Powder, 1940.

Atlas Powder. *Atlas Explosives Products*. Wilmington, DE: Atlas Powder, 1949.

Atlas Powder. *Atlas Explosives Products*. Wilmington, DE: Atlas Powder, 1957.

Atlas Powder. *Explosives and Rock Blasting*. Dallas: Atlas Powder, 1987.

Austin Powder. *Austin Powder—150 Years of Explosives*. Cleveland, OH: Austin Powder, 1983.

Bacchus, Thomas. *Dynamite—the New Aladdin's Lamp*. Wilmington, DE: Hercules Powder, 1922.

Barab, Jacob. "Fumes Encountered in Mining Operations and Handling of Explosives." *Coal Review*, September 13, 1922, 22.

Barab, Jacob. "Modern Explosives in Industry." *Mining Congress Journal*, May 1936, 54.

Barbour, Percy. "Explosives Used in War and Metal Mining." *Engineering and Mining Journal*, September 25, 1915, 5.

Barnett, E. deBarry. *Explosives*. New York: D. Van Nostrand, 1919.

"Blasting by Electricity." *Brick and Clay Record*, August 3, 1915, 16.

"Blasting Costs in American Quarries." *Engineering and Contracting*, January 15, 1919, 7.

Bowne, Stephen. *Dynamite: A Most Damnable Invention*. New York: Thomas Dunne Books, 2005.

Brunton, David, and John Davis. *Modern Tunneling with Special Reference to Mine and Water-Supply Tunnels*. New York: John Wiley, 1914.

Brunton, David, and John Davis. *Safety and Efficiency in Mine Tunneling*. US Department of the Interior, Bureau of Mines. Washington, DC: USGPO, 1914.

Bumstead, Dale. "The New Hired Hand—Dynamite." *DuPont Magazine*, July 1913, 8.

Business America. *The History of the E. I. du Pont de Nemours Powder Company*. New York: Banker and Investor Magazine, 1912.

California Cap. *Detonators for High Explosives*. Oakland: California Cap, 1932.

Chellson, H. C. "From Gunpowder to Modern Dynamite." *Engineering and Mining Journal*, May 1926, 231.

Cook, Melvin. "Modern Blasting Agents." *Science*, October 21, 1960, 1105.

Cook, Melvin. *The Science of High Explosives*. New York: Reinhold, 1988.

Cooper-Key, Aston. *A Primer of Explosives for the Use of Local Inspectors and Dealers*. New York: Macmillan, 1905.

"Correct Methods of Priming Dynamite." *Mining & Engineering World*, July 25, 1914, 151.

Cosgrove, John. *Rock Excavation and Blasting*. Pittsburgh, PA: National Fireproofing, 1913.

Cullen, A. C. "Explosives, Part 5: Methods of Firing Explosives." *Coal Mine Management*, December 1922, 44.

Daw, Albert, and Zacharias Daw. *The Blasting of Rock in Mines, Quarries, Tunnels, Etc*. New York: Spon and Chamberlain, 1909.

De Blois, L. A. "About Explosives." *DuPont Magazine*, September 1923, 4.

De Kalb, Courtney. *Manual of Explosives*. Toronto: Ontario Bureau of Mines, 1900.

Dick, Richard, Larry Fletcher, and Dennis V. D'Andrea. *Explosives and Blasting Procedures Manual*. Washington, DC: USGPO, 1983.

"Disadvantages of Small Diameter Dynamite Cartridges." *Explosives Engineer*, May 1925, 65.

Drinker, Henry. *A Treatise on Explosive Compounds, Machine Rock Drills and Blasting*. New York: John Wiley & Sons, 1883.

"Dynamite." *Mining Science*, April 7, 1919, 318.

"Dynamite—a Constructor of Civilization." *Explosives Engineer*, April 1923, 32.

E. I. DuPont de Nemours. *Agricultural Blasting: A Money-Making Profession*. Wilmington, DE: E. I. DuPont de Nemours, 1915.

E. I. DuPont de Nemours. *Blasters' Handbook*. Wilmington DE: E. I. DuPont de Nemours, 1942–80.

E. I. DuPont de Nemours. *Blasting Pole and Post Holes*. Wilmington, DE: E. I. DuPont de Nemours, 1915.

E. I. DuPont de Nemours. *Ditching and Field Clearing with Dynamite*. Wilmington, DE: E. I. DuPont de Nemours, 1950.

E. I. DuPont de Nemours. *DuPont Explosives for Shale and Clay Blasting*. Wilmington, DE: E. I. DuPont de Nemours, 1916.

E. I. DuPont de Nemours. *DuPont Products*. Wilmington, DE: E. I. DuPont de Nemours, April 1916.

E. I. DuPont de Nemours. *Explosives for Quarrying*. Wilmington, DE: E. I. DuPont de Nemours, 1920.

E. I. DuPont de Nemours. *Handbook of Explosives for Farmers, Planters, and Ranchers*. Wilmington, DE: E. I. DuPont de Nemours, 1910.

E. I. DuPont de Nemours. *Increasing Orchard Profits with DuPont Red Cross Dynamite*. Wilmington, DE: E. I. DuPont de Nemours, 1915.

E. I. DuPont de Nemours. *Subsoiling with Dynamite*. Wilmington, DE: E. I. DuPont de Nemours, 1911.

E. I. DuPont de Nemours. *This Is DuPont*. Wilmington, DE: E. I. DuPont de Nemours, 1949.

Ellsworth, H. E., and J. K. Brandon. "Manufacture of Safety Fuse." *Cement, Mill and Quarry*, July 5, 1922, 45.

Ellsworth, John. *100 Years: Being the Story of Safety Fuse in America since 1836*. Simsbury, CT: Ensign-Bickford, 1936.

Engineering and Mining Journal Staff. *Handbook of Mining Details*. New York: McGraw Hill, 1912.

Ensign-Bickford. *Ignitacord*. Simsbury, CT: Ensign-Bickford, 1957.

Ensign-Bickford. *Primacord Detonating Fuse: What It Is . . . How to Use It*. Simsbury, CT: Ensign-Bickford, 1960.

Ensign-Bickford. *Quarrycord*. Simsbury, CT: Ensign-Bickford, n. d.

"Explosives for Use in Quarrying." *Stone*, January 1912, 138.

Farm Bureau. *Improving Crop Yields by the Use of Dynamite*. New York: New York Central & Hudson River Railroad, 1911.

Fay, Willard. "Dynamite Makers." *Colliers*, March 25, 1916, 21.

Field, Simon. *Boom! The Chemistry and History of Explosives*. Chicago: Chicago Review Press, 2017.

Fisher, C. H. "Explosives in Safe Keeping." *Crushed Stone Journal*, April–May 1940, 31.

Foster, C. Le Neve. *The Elements of Mining and Quarrying*. London: Charles Griffin, 1903.

Friend, Robert. *Explosives Training Manual*. Wilmington, DE: ABA, 1975.

Fry, C. L. *Prices of Explosives*. US War Industries Board. Washington, DC: USGPO, 1919.

Giant Powder. *Better Farming with Giant Farm Powders*. Giant, CA: Giant Powder, 1920.

Gleasner, Diana. *Dynamite*. New York: Walker, 1982.

Gokhale, Bhalchandra. *Rotary Drilling and Blasting in Large Surface Mines*. Boca Raton, LA: CRC, 2011.

Greensfelder, N. G. *Eliminating Waste in Blasting*. Wilmington, DE: Hercules Powder, November 1922.

Greensfelder, N. G. "Shot-Firing by Electricity." *Contractor's & Engineer's Monthly*, January 1923, 61.

Hall, Clarence, and Spencer Howard. *Magazines and Thaw Houses for Explosives*. US Department of the Interior, Bureau of Mines. Washington, DC: USGPO, 1912.

"Hercules Acquires Aetna Co." *Hardware World*, August 1921, 152.

Hercules Powder. *Hercules Blasting Agents & Blasting Supplies*. Wilmington, DE: Hercules Powder, November 1959.

Hercules Powder. *Hercules Detonators*. Wilmington, DE: Hercules Powder, 1927.

Hercules Powder. *Hercules Dynamite on the Farm—Ditching Dynamite*. Wilmington, DE: Hercules Powder, November 1933.

Hercules Powder. *Hercules Explosives & Blasting Supplies*. Wilmington, DE: Hercules Powder, November 1935.

Hercules Powder. *High Explosives Price List No. 5-M*. Wilmington, DE: Hercules Powder, November 17, 1936.

Hercules Powder. *Progressive Cultivation Facts for Farmers*. Wilmington, DE: Hercules Powder, 1913.

Herr, Fred. "Using a Dynamite Explosion to Blow Out an Oil-Well Fire." *Popular Mechanics*, December 1922, 8.

Higgins, Edwin. *The Prevention of Accidents from Explosives in Metal Mining*. US Department of Commerce, Bureau of Mines. Washington, DC: USGPO, 1926.

Hunter, Charles. "What Happens When Dynamite Explodes?" *Salt Lake Mining Review*, September 15, 1924, 25.

Hunter, Herbert. *History of Explosives*. US Department of the Army, Ordinance Department. Washington, DC: USGPO, 1919, 41.

Ilsley, L. C., and A. B. Hooker. *Electric Shot Firing in Mines, Quarries, and Tunnels*. US Department of Commerce, Bureau of Mines. Washington, DC: USGPO, 1926.

Institute of Makers of Explosives. *Explosives for Agriculture*. New York: IME, 1931.

Institute of Makers of Explosives. *The Use of Explosives for Agriculture and Other Purposes*. New York: IME, 1917.

Jackson, Charles, and E. D. Gardner. *Stoping Methods and Costs*. US Department of the Interior, Bureau of Mines. Washington, DC: USGPO, 1936.

Jaeger, Charles. *Rock Mechanics and Engineering*. London: Cambridge University Press, 1972.

Jones, E. B. "Proper Storage of Blasting Supplies." *American Contractor*, December 16, 1922, 27.

Keystone Consolidated Publishing. *Mining Catalog (Metal and Quarry Edition), 1921*. Pittsburgh, PA: Keystone Consolidated Publishing, 1921.

Keystone Consolidated Publishing. *The Keystone Catalog (Metal and Quarry Edition), 1926*. Pittsburgh, PA: Keystone Consolidated Publishing, 1926.

Keystone National Powder. *Farming with Dynamite*. Emporium, PA: Keystone National Powder, 1912.

Kneeland, Eric. *Preliminaries of Coal Mining*. New York: McGraw Hill, 1926.

La Motte, Arthur. *Blasters' Handbook*. Wilmington, DE: E. I. DuPont de Nemours, 1918–39.

La Motte, Arthur. "The Care and Operation of Blasting Machines." *DuPont Magazine*, August 1914, 19.

Leland, E. "Checking Supplies Underground at Pilares Mine." *Engineering and Mining Journal-Press*, August 1, 1925, 165.

Levy, S. I. *Modern Explosives*. London: Sir Isaac Pitman & Sons, 1920.

MacFarland, David, and Guy Rolland. *Handbook of Electric Blasting*. Wilmington, DE: Atlas Chemical Industries, 1965.

Marshall, Arthur. *Explosives, Volumes I–III*. London: J. & A. Churchill, 1917.

Maurice, William. *The Shot-Firers Guide*. London: George Tucker, 1910

Maxim, Hudson. *Dynamite Stories and Some Interesting Facts about Explosives*.

New York: Hearst's International Library, 1916.

McConahy, J. Nelson. *Dynamite! A Blaster's History*. Charleston, SC: CreateSpace, 2015.

McDaniel, Allen. *Excavation: Machine Methods and Costs*. New York: McGraw Hill, 1919.

Moil, A. *Mining Interests*. London, Methuen, 1904.

Mordecai, Thomas. *Professional Standards for Preparing, Handling, and Using Explosives*. Boulder, CO: Paladin, 1995.

Munroe, Charles. *Permissible Explosives Defined*. US Department of Commerce, Bureau of Mines. Washington, DC: USGPO, 1927.

Munroe, Charles. "Storage and Handling of Explosives in Mines." *Engineering & Mining Journal*, February 19, 1916, 349.

National Fuse and Powder. *Igniter Cord*. Denver, CO: National Fuse and Powder, n. d.

Nishman, C. E., and C. E. Kiessling. *Rock Drilling*. US Department of the Interior, Bureau of Mines. Washington, DC: USGPO, 1940.

Nuckolls, A. H. *Chemistry of Combustions (Explosives)*. Chicago: American School of Correspondence, 1917.

Pamely, Caleb. *The Colliery Manager's Handbook*. London: Crosby, Lockwood & Son, 1898.

Pearson, David, and Ron Bommarito. *Antique Mining Equipment and Collectibles*. Atglen, PA: Schiffer, 2016.

Peele, Robert. *Mining Engineers Handbook*. New York: John Wiley & Sons, 1918.

Pennsylvania Lines. *Intensive Farming and Use of Dynamite*. Baltimore: Lord Baltimore, June 1911.

Posedel, Jennifer, and Stephen Lawton. *Hercules*. Charleston, SC: Arcadia, 2011.

"Practical Hints on Handling Explosives." *Pacific Miner*, June 1910, 212.

"The Principles of Blasting." *The Contractor*, July 1, 1912, 42.

Quinan, William. *High Explosives*. London: Sir Isaac Pitman & Sons, 1912.

Ramsey, Albert, and H. Westin. *A Manual on Explosives*. New York: E. P. Dutton, 1917.

Rice, George. *Coal Dust Explosions*. US Department of the Interior, Bureau of Mines. Washington, DC: USGPO, 1911.

"Rock Drilling by Electric Power." *Mining Science*, December 26, 1907, 624.

Rutledge, John. *The Use and Misuse of Explosives in Coal Mines*. US Department of the Interior, Bureau of Mines. Washington, DC: USGPO, 1916.

"Safety Explosives." *Mines and Minerals*, October 1908, 113.

Scanlon, William. "Blasting in Tunnels and Shaft Sinking." *Coal Industry*, April 1921, 203.

Schellen, Heinrich. *Magneto-electric and Dynamo-electric Machines*. New York: D. Van Nostrand, 1884.

"Scientific Principles in Use of Dynamite." *Coal Age*, March 22, 1913, 2.

Snelling, Walter, and Christian Storm. *The Analysis of Black Powder and Dynamite*. US Department of the Interior, Bureau of Mines. Washington, DC: USGPO, 1916.

Stoneking, J. B. "Advantages of Electrical Blasting." *Concrete*, September 1919, 41.

Stoneking, J. B. "Detonators for Blasting." *DuPont Magazine*, April 1910, 10.

Storm, Christian. *The Analysis of Permissible Explosives*. US Department of the Interior, Bureau of Mines. Washington, DC: USGPO, 1916.

Stovall, D. H. "Carelessness with Dynamite Is Inexcusable." *Ores and Metals*, January 20, 1907, 39.

"Tamping and Stemming." *DuPont Magazine*, April–May 1915, 11.

Twitty, Eric. *Blown to Bits in the Mine*. Ouray, CO: Western Reflections, 2001.

Twitty, Eric. *Riches to Rags*. Montrose, CO: Western Reflections, 2005.

"The Unreasoning Fear of Explosives." *DuPont Magazine*, November–December 1922, 1.

US Department of Agriculture, Forest Service. *Missoula Blasts & Explosives*. Washington, DC: USGPO, December, 1997.

US Department of Commerce, Bureau of Public Roads. *Presplitting—a Controlled Blasting Technique for Rock Cuts*. Washington, DC: USGPO, 1966.

US Department of the Interior, Bureau of Mines. *Destruction of Damaged, Deteriorating, or Unwanted Commercial Explosives*. Washington, DC: USGPO, 1945.

US Department of the Interior, Bureau of Mines. *Permissible Explosives, Mining Equipment and Apparatus Approved Prior to January 1, 1923*. Washington, DC: USGPO, 1923.

US Department of the Interior, Bureau of Mines. *Stemming in Metal Mines Progress Report 7*. Washington, DC: USGPO, 1943.

US Department of Labor, Mine Safety Health Administration. *Blasting Incidents in Mining*. Washington, DC: USGPO, August 1988.

US Department of Labor, Mine Safety Health Administration. *Blasting Requirements—Surface Coal*. Washington, DC: USGPO, 1994.

US Department of the Navy, Ordinance Department. *Use of Explosives in Underwater Salvage*. Washington, DC: USGPO, February 28, 1956.

US Department of the Treasury, Bureau of Alcohol, Tobacco, and Firearms. *Your Guide to Explosives Regulations*. Washington, DC: USGPO, 1976.

"Use of Paper Tamping Bags Saves Explosives in Blasting." *Engineer and Contractor*, March 19, 1919, 282.

Van Gelder, Arthur, and Hugo Schlatter. *The History of the Explosives Industry in America*. New York: Columbia University Press, 1927.

Van Winkle, R. N. "Blasting Practice in Quarries and Open Pit Mines." *Pit and Quarry*, July 15, 1925, 75.

Williams, William. *History of the Manufacture of Explosives for the Great War: 1917–1918*. Washington, DC: USGPO, 1919.

Voynick, Stephen. *The Making of a Hardrock Miner: An Account of a Worker in Copper, Molybdenum, and Uranium Mines in the West*. Berkeley, CA: Howell-North Books, 1978.

Woodbury, C. A. "Research in the Dynamite Industry." *DuPont Magazine*, April–May 1942, 1.

Young, Otis, and Robert Lenon. *Black Powder and Hand Steel—Miners and Machines on the Old West Frontier*. Norman: Oklahoma University Press, 1973.

Young, William. *Elements of Mining*. New York: McGraw Hill, 1923.

INDEX